Greek and Roman Networks in the Mediterranean

How useful is the concept of "network" for historical studies and the ancient world in particular? Using theoretical models of social network analysis, this book illuminates aspects of the economic, social, religious, and political history of the ancient Greek and Roman worlds.

Bringing together some of the most active and prominent researchers in ancient history, this book moves beyond political institutions, ethnic, and geographical boundaries in order to observe the ancient Mediterranean through a perspective of network interaction. It employs a wide range of approaches, and to examine relationships and interactions among various social entities in the Mediterranean. Chronologically, the book extends from the early Iron Age to the late Antique world, covering the Mediterranean between Antioch in the east to Massalia (Marseilles) in the west.

This book was published as two special issues in the *Mediterranean Historical Review* (vol. 22, 2007).

Irad Malkin is Professor of Greek History and Chair of the History department at Tel Aviv University, Israel. He holds the Cummings Chair for Mediterranean History, and is co-Editor of the *Mediterranean Historical Review.*

Christy Constantakopoulou is lecturer in Ancient History at the School of History, Classics and Archaeology of Birkbeck College, London.

Katerina Panagopoulou is lecturer in Ancient History at the Department of History and Archaeology, University of Crete, Rethymno, Greece.

Greek and Roman Networks in the Mediterranean

Edited by Irad Malkin,
Christy Constantakopoulou and
Katerina Panagopoulou

Routledge
Taylor & Francis Group
LONDON AND NEW YORK

First published 2009 by Routledge
2 Park Square, Milton Park, Abingdon, Oxon, OX14 4RN

Simultaneously published in the USA and Canada
by Routledge
270 Madison Avenue, New York, NY 10016

Routledge is an imprint of the Taylor & Francis Group, an informa business

Typeset in Times by Value Chain, India
Printed and bound in Great Britain by TJI Digital, Padstow, Cornwall

British Library Cataloguing in Publication Data
A catalogue record for this book is available from the British Library

ISBN 10: 0-415-45989-3
ISBN 13: 978-0-415-45989-1

CONTENTS

To Frank Cass, a man with a Mediterranean vision

Notes on Contributors

Hugh Bowden is a senior lecturer in Ancient History at King's College London. He is the author of *Classical Athens and the Delphic Oracle* (Cambridge 2005). He has worked widely on religion in the ancient world, and his current research is on mystery cults.

Anna Collar is a Ph.D. student at the University of Exeter. She is interested in Pagan monotheism, transmission of religious ideas, ancient history, archaeology, and archaeological theory.

Christy Constantakopoulou is Lecturer in Ancient History in the School of History, Classics and Archaeology of Birkbeck College. She is interested in geographical readings of history and in the Aegean Sea and its islands in the Archaic, Classical and Hellenistic periods. She has recently published a monograph entitled *The Dance of the Islands: Insularity, Networks, the Athenian Empire, and the Aegean World* (Oxford: Oxford University Press, 2007) and is now working on a project on Aegean networks in the Hellenistic period.

John K. Davies, FBA, was Rathbone Professor of Ancient History and Classical Archaeology at the University of Liverpool from 1977 until 2003. His main interests lie in Greek history of all periods, especially its administrative, cultic, legal, economic and financial aspects. Recent publications include numerous chapters in collective volumes and the joint editorship (with Z. H. Archibald and V. Gabrielsen) of *Making, Moving, and Managing: The New World of Ancient Economies, 323–31 B.C.* (Oxford: Oxbow Books, 2005).

Panagiotis Doukellis is Associate Professor of Roman History at the University of Aegean (Mytilini). His main fields of investigation are Roman Greece, landscape history, and the history of political ideas. He is the editor of the collective volumes *The Greek Landscape* (Athens: HESTIA Kollaros, 2005); *Perception and Evaluation of Cultural Landscapes*, in collaboration with L. G. Mendoni (Athens and Paris: Research Centre for Greek and Roman Antiquity, 2005); *Föderalismus in der Antike*, in collaboration with P. Siewert, L. Aigner-Foresti, and G. Zecchini (Munich, 2005); and *Esclavage antique et discriminations socio-culturelles*, in collaboration with V. I. Anastasiadis (Bern: Peter Lang, 2006).

Vincent Gabrielsen is Professor of Ancient History at the University of Copenhagen. His research is centred on the social and economic history of the Classical Greek and Hellenistic worlds. His publications include *Financing the Athenian Fleet: Public Taxation and Social Relations* (Baltimore and London: Johns Hopkins University Press, 1994); *The Naval Aristocracy of Hellenistic Rhodes* (Aarhus: Aarhus University Press, 1997); *Making, Moving and Managing: The New World of Hellenistic Economies 323–31 B.C.*, ed. with Z. Archibald and J. K. Davies (Oxford: Oxbow, 2005); and *The Black Sea in Antiquity: Regional and Interregional Economic Exchanges Between the Black Sea and the Mediterranean*, ed. with J. Lund (Aarhus: Aarhus University Press, 2007).

Simon Hornblower is a Grote Professor of Ancient History at UCL. His most recent books are *Thucydides and Pindar* (2004) and (as co-editor) *Pindar's Poetry, Patrons and Festivals* (2007). He is now completing the final volume of a three-volume commentary on Thucydides.

Yannis Lolos is Assistant Professor of Ancient History, Department of History, Archaeology and Social Anthropology, University of Thessaly, Volos. Main interests include landscape archaeology; Greek countryside; Sikyon. Major publications include *Land of Sikyon; Archaeology and History of a Greek City-state* [Hesperia suppl. 39] (2008); 'The Sanctuary of Titane and the City of Sikyon', *Annual of the British School at Athens* 100 (2005): 275–98; 'Greek Roads: A Commentary on the Ancient Terms', *Glotta* 79 (2003): 137–74; 'The Hadrianic Aqueduct of Corinth with an appendix on the Roman aqueducts in Greece', *Hesperia* 66 (1997): 271–314.

Irad Malkin is Professor of Greek History at Tel Aviv University where he is currently the Chair of the Department of History. He holds the Cummings Chair for Mediterranean History and Cultures and is co-Founder (1986) and co-Editor of the *Mediterranean Historical Review*. His major publications include *Religion and Colonization in Ancient Greece* (Brill: Leiden, 1987); *Myth and Territory in the Spartan Mediterranean* (Cambridge: Cambridge University Press, 1994), paperback edition, Cambridge University Press, 2003; French translation, 1999); *The Returns of Odysseus: Colonization and Ethnicity* (Berkeley and Los Angeles: University of California Press, 1998; Italian translation, 2004; Hebrew translation, 2004); *Ethnicity and Identity in Ancient Greece* (in Hebrew, Tel Aviv, 2003); *Networks in the Archaic Mediterranean* (Forthcoming). He has also edited *Ancient Perceptions of Greek Ethnicity* (Washington, DC: Center for Hellenic Studies and Harvard University Press, 2001) and *Mediterranean Paradigms and Classical Antiquity* (London: Routledge, 2005. Special Issue of the *Mediterranean Historical Review* 18, 2003).

Robin Osborne is a professor of Ancient History at the University of Cambridge. His work ranges widely over Greek history, archaeology and art.

His recent publications include *Greek History* (2004), *Greek Historical Inscriptions 404–323 B.C.* (ed. with P. J. Rhodes, 2003), *Classical Archaeology* (ed. with S. E. Alcock, 2007), and *Art's Agency and Art History* (ed. with J. J. Tanner, 2007).

Dimitris Paleothodoros is currently Assistant Professor of Classical Archaeology at the University of Thessaly. He is a specialist of Athenian pottery and iconography. His more recent work includes the monograph *Epictetos* (Leuven-Namur-Duddley-Boston, 2004) and articles on Dionysiac iconography (Etruscan Studies 9–10, 2004–7) and the diffusion of Attic pottery in the Mediterranean (most recently in Eulimene 2005–6).

Katerina Panagopoulou is Lecturer of Ancient History at the University of Crete/Greece and Temporary Lecturer in Greek Culture at the Greek Open University. Her main research interests lie in Classical, Hellenistic and Roman Economic and Social History, in Greek and Roman coinage (especially of northern Greece and Asia Minor), in Greek Art and in Hellenistic kingship. Publications include: 'Between Necessity and Extravagance: Silver as a Commodity in the Hellenistic Period', *Annual of the British School at Athens* 102 (2007): 1–29, pl. 1–3; 'Cross-Reading Images: Iconographic "Debates" between Antigonids and Ptolemies during the Third and Second Centuries BC', *Eulimene* 6–7 (2005–6): 157–73 (summary at www.shef.ac.uk); 'The Antigonids: Patterns of a Royal Economy', in *Hellenistic Economies*, edited by Z. Archibald, J. K. Davies, V. Gabrielsen, and G. J. Oliver, 313–62 (London: Routledge, 2001); *The Early Antigonids: Coinage, Money and the Economy*, Royal Numismatic Society Special Monograph Series (forthcoming).

Selene Psoma is senior researcher at the National Hellenic Research Foundation (Institute for Greek and Roman Antiquity) and was elected Senior Lecturer in Ancient History at the University of Athens. Her main fields of investigation are Numismatics and Epigraphy of Macedonia and Thrace. She published a monography on *Olynthus and the Chalcidic League* (Stuttgart: Steiner, 2001) and collaborated in the publication of the Inscriptions of Aegean Thrace (Athens: Research Centre for Greek and Roman Antiquity, 2005) and of a volume of *Sylloge Nummorum Graecorum* (Athens: The Academy of Athens, 2005). Her recent work is mainly on coinages of the so-called Thasian Peraia (*Revue Numismatique* 2003; Agoranomia. Essays in Honour of J. H. Kroll [New York: American Numismatic Society, 2006]); numismatic terminology (*Revue Numismatique* 2006; Zeitschrift für *Papyrologie und Epigraphik* 160, 2007); federal coinages (*Klio* 89, no. 1, 2007); history of Macedonia (*Villes et cours en Macidoine* [Athens: Research Centre for Greek and Roman Antiquity, 2006]), etc. She is preparing the publication of the excavation coins of Maroneia and of the Classical city at the peninsula of Molyvoti.

Dominic Rathbone is Professor of Ancient History at King's College London. His main current research interests are Roman Egypt, the agrarian history of Roman Italy, and private and public finances in the Roman empire. His most recent papers are: 'Roman Egypt', in *The Cambridge Economic History of the Greco-Roman World*, edited by W. Scheidel, I. M. Morris, and R. P. Saller, 698–719 (Cambridge: Cambridge University Press, 2007); 'Military Finance and Supply', in *The Cambridge History of Greek and Roman Warfare*, edited by P. Sabin, H. van Wees, and M. Whitby, vol. II, 158–76 (Cambridge: Cambridge University Press, 2007); with P. Temin, 'Financial Intermediation in First-century A.D. Rome and Eighteenth-century England', in *Pistoi dia tèn technèn. Bankers, Loans and Archives in the Ancient World: Studies in Honour of Raymond Bogaert*, edited by K. Verboven, K. Vandorpe, and V. Chankowski, 371–419 (Leuven: Peters, 2008).

Gary Reger is Professor of History at Trinity College in Hartford, Connecticut, USA. He is the editor, with Hugh Elton, of *Regionalism in Hellenistic and Roman Asia Minor* (Bordeaux: Ausonius, 2007), and the author of articles on the Hellenistic economy and Mylasa in Karia, most recently 'Hellenistic Greece and Western Asia Minor', Chapter 17 of the *Cambridge Economic History of the Greco-Roman World*, edited by Walter Scheidel, Ian Morris, and Richard Saller, 460–83 (Cambridge: Cambridge University Press, 2007), and 'The Pseudo-Rhodian Drachms of Mylasa Revisited', with Richard Ashton, in *Agoranomia: Studies in Money and Exchange Presented to John H. Kroll*, edited by Peter Van Alfen, 125–50 (New York: American Numismatic Society, 2006).

Ian Rutherford is a professor of Greek at the University of Reading. He is interested in Ancient Greek poetry, Ancient Greek religion, the Bronze Age Aegean and Anatolia, cultural interaction, particularly with relation to Greece and Egypt, pilgrimage, multilingualism, and canonization. He has recently co-edited with J. Elsner *Pilgrimage in Greco-Roman and Early Christian Antiquity: Seeing the Gods* (Oxford, 2005).

Bella Sandwell is a lecturer in Ancient History at the University of Bristol. She works on religious interaction in Late Antiquity, particularly in the writings of Libanius and John Chrysostom. She has written a number of articles on these authors, and her book. *Religious Identity in Late Antiquity*, was published in August 2007.

Michael Sommer is a lecturer in Ancient History at the University of Liverpool. He is interested in Roman Empire, culture contact, history of the Levant, and economic history. His recent publications include the following monographs: *Der römische Orient: Zwischen Mittelmeer und Tigris* (Darmstadt: Wissenschaftliche Buchgesellschaft, 2006); *Roms orientalische*

Steppengrenza: Palmyra – Edessa – Dura-Europos – Hatra: Eine Kulturgeschichte von Pompeius bis Diocletian (Stuttgart: Franz Steiner, 2005); and *Die Phönizier: Handelsherren zwischen Orient und Okzident* (Stuttgart: Kröner, 2005).

Maria Stamatopoulou is University Lecturer in Classical Art and Archaeology at the University of Oxford and Tutorial Fellow in Archaeology at Lincoln College. Her research to date has focused on Thessaly and its funerary archaeology from the Archaic to the Hellenistic periods. She is working on archaeological material from Macedonia and Thessaly and is currently preparing for publication the Archaeological Society of Athens's excavations in the cemeteries of Demetrias and Pharsalos. Recent publications include: *Excavating Classical Culture: Recent Archaeological Discoveries in Greece*, co-edited with M. Yeroulanou (Oxford: Beazley Archive and Archaeopress, 2002); *Architecture and Archaeology in the Cyclades: Papers in Honour of J. J. Coulton*, co-edited with M. Yeroulanou (Oxford: Archaeopress/British Archaeological Reports, 2005); 'Thessalian Aristocracy and Society in the Age of Epinician', in *Pindar's Poetry, Patrons and Festivals: From Archaic Greece to the Roman Empire*, edited by S. Hornblower and C. Morgan, 309–41 (Oxford: Oxford University Press, 2007); 'A New Funerary Gold Leaf from Pherai' (together with R. Parker), in *Archaiologikē Ephēmeris* 2004 (2007): 1–32; ''Totenmahl' Reliefs of the 4th–2nd Centuries BC and the Archaeology of Feasting in a Funerary Context', in *Cultural Messages in Ancient Greece and Rome*, edited by O. Hekster and St. Mols, BABesch supplement (Leiden: Peeters, in press).

Riet van Bremen is a senior lecturer in ancient history at ULC. She works on Hellenistic Asia Minor, and has a special interest in the political, social and religious history of Karia. With J.-M. Carbon she is currently editing Hellenistic Karia. Proceedings of the first International Hellenistic Karia Conference, Oxford 29 June–2 July 2006. Among her recent publications are: 'Plotina Augusta to all her Friends: the Empress Plotina and the Epicurean Philosophers at Athens', *Chiron* 35 (2005), and 'The Entire House is Full of Crowns: Greek Agones and the Celebration of Victory', in S. Hornblower and C. Morgan (eds.), *Pindar's Poetry, Patrons and Festivals, From Archaic Greece to the Roman Empire* (2007).

Kostas Vlassopulos is a lecturer in Greek History at the University of Nottingham. He is the author of *Unthinking the Polis: Ancient Greek History beyond Eurocentrism* (CUP, 2007). He is interested in Greek social and Greek social and political history, historiography, and comparative history.

Acknowledgements

The conference on Networks would not have been possible without the generous contributions of Costamare Shipping Company, the Alexander S. Onassis Foundation, and the Department of History and Archaeology at the University of Crete/Greece. We are deeply indebted to all of the above institutions for this. We should also like to thank Aegean Airlines, the Institute of Mediterranean Studies, and Birkbeck College for financial, administrative and overall support. Particular thanks are also due to the 25th and 23rd Ephoreia of Prehistoric and Classical Antiquities, not least to Dr M. Vlazaki and to Ms Eftuchia Protopapadaki (25th Ephoreia) and to Dr A. Vassilakis (23rd Ephoreia), who substantially enriched the activities and contributed to the conference debates; finally, to the Municipality and to the Prefecture of Rethymno/Crete and, not least, to the postdoctoral and doctoral students at the Department of History and Archaeology/University of Crete, who generously contributed to the success of the conference in various ways. Additional support for this publication comes from the Israel Science Foundation, the Maxwell Cummings Family Chair for Mediterranean History and Cultures at Tel Aviv University, the S. Daniel Abraham Centre for International and Regional Studies, Tel Aviv University, and the Morris E. Curiel Institute for European Studies, Tel Aviv University.

Introduction

Irad Malkin, Christy Constantakopoulou
and Katerina Panagopoulou

Picture Herakleia and Schinoussa, two small islands in the Aegean: they face each other, each little port confronting its twin. Schinoussa's port is protected towards the northwest, that of Herakleia to the southeast. Once a threatening wind rises, fishermen and yachtsmen rush to move their boats from one island to another. Over the years, a quiet symbiosis has developed between the islanders, with the sea serving as a shared space. The mental horizons of the islanders seem to comprehend their 'place' in tripartite terms: the respective villages of Herakleia and Schinoussa; their ports and the immediate lines of exchange between them; and the Aegean circle and the Mediterranean at large.

The situation can be said to be a 'fractal' of Mediterranean networks. While the scale at Herakleia and Schinoussa is small and local, the dynamics of connectivity can be extended to the entire Mediterranean. The content (or 'flow') of what passes between the 'nodes' of Mediterranean networks can vary greatly, yet patterns of practice tend to recur. Fernand Braudel[1] has repeatedly claimed that the Mediterranean is an 'exchange', and that no port can be studied without the implication of at least another port 'speaking' to it from somewhere else. The terms of the 'conversation' are variable: they may include the movement of goods and people, of ideas and armies, of cults and languages. Braudel speaks of the lines crisscrossing the Mediterranean in terms of *réseaux*, or 'networks'.

In tandem with Braudel's research on his magisterial *The Mediterranean*, Shlomo Dov Goitein was painstakingly writing a 'sociographic' portrait, *A Mediterranean Society*, as revealed from the documents of the Fustat (Cairo) Geniza.[2] Without the two of them making contact, these two Mediterranean historians have provided contemporary historiography with the basic guidelines for a network approach to the Mediterranean. Goitein's networks of commerce, marriage, and religious prescriptions (to name but a few) reveal a practice of connectivity in the Mediterranean at the time when, according to Henri Pirenne's famous thesis, the Mediterranean was hopelessly split between the two hostile civilizations of Islam and Christendom.[3] What arises from Goitein's work is an important implication that concerns the difference between practice and expressed values: while the two religions were openly inimical, Mediterranean networks kept functioning.

Neither scholar, has taken the concept of networks significantly beyond its descriptive value. However, by insisting on the *longue durée* of Mediterranean patterns (networks implied), Braudel eventually arrived at far-reaching conclusions with regard to the nature of civilizations.[4] He confronted the question of whether civilizations can be defined primarily in terms of 'networks of practice' (albeit not his terms), even though expressed values might seem contradictory. For example, Braudel points to direct continuity from the Phoenician presence in north Africa to the success of Islam in that region (and its failure in the Latin Mediterranean), or the continuity from the Greek substrate in the eastern Mediterranean to the Byzantine, and then to the Christian Orthodox civilization. With regard to the ancient Greeks, it may be useful to ask whether ancient 'Greek civilization' arose as an outcome of network-dynamics (with no 'Hellas' as a nation state, no marked 'centre' and no defined 'margins').[5] For such questions to be applied meaningfully to historical issues the very notion of a 'network' and its implications need a more sophisticated discussion, observing networks both descriptively (revealing their existence) and as a tool of analysis.

Networks can be discussed simply as connections between 'nodes', like joining dots on the map. However, 'connections' are easy to find and connecting lines might seem impressionistic or arbitrary. It is no news that the Greeks, for example, lived around the shores of the Mediterranean and the Black Sea, and that they had contacts with each other. 'The shores of Greece are like hems stitched on to the lands of barbarian peoples', says Cicero,[6] and Plato famously compares Greeks to frogs living around a pond.[7] Such descriptive networks are still in need of a question for which they may provide an answer. Our purpose in this volume is to qualify networks, to understand their duration, function, scope, overlapping, and historical implications.

Aside from historiographical, 'Mediterraneanist', and 'civilizational' angles networks may also be studied by examining characteristic of networks in general.[8] Such characteristics may be gleaned from current studies of 'network theory', with a strong emphasis on *Social Networks Analysis* (see below). It is interesting to note how today, with the rise of general interest in the concept of networks in both the exact and the social sciences, a convergence with the 'networks' of 'Mediterranean Studies' (as set out especially by Braudel and Goitein), can become fruitful and meaningful.

In the past fifty years interest in the 'Mediterranean' as a heuristic concept has grown considerably, with a major new work *The Corrupting Sea*, by Peregrine Horden and Nicholas Purcell, and new re-assessments and critiques.[9]

Mediterranean *réseaux* are no longer seen as lines drawn on a Euclidean plane, but are interpreted along more sophisticated models of connectivity. Such models seem apt for a postmodern mindset; they are conditioned by the extensive use of the World Wide Web; and suffused with philosophical notions that argue against 'centre–margin' thinking, such as that of the '*rhizome* network' in Gilles Deleuze and Félix Guatari.[10] These days the 'network' concept is filtering everywhere, changing sets of values and breaking up assumed hierarchies of centres and peripheries. It is characteristic of both postmodern and postcolonial thinking; it presents a new vision of geography and human space. Most significantly, network thinking has transformed the botanical metaphor, so influential in the nineteenth and twentieth centuries, of the tree: 'roots', 'stem', and 'branches' were used to denote genealogies of kinship, races, languages, and entire civilizations.

The present volume applies the *problématique* of the Social Networks Analysis theory to discussions of ancient history. In order to construct a paradigm for ancient history based on this theory, in the first place we need to examine what Social Networks Analysis has to offer, and in what ways the key elements of this theory can be applied effectively to some major issues of ancient history.

The concept of networks and the Social Network Analysis approach have been used in sociology and the sciences (mostly physics) for some years now, but their impact in history has been almost negligible. Social network analysis emphasizes the importance of relations between social entities, and examines the patterns and implications of these relations.[11] For example, the term 'Six degrees of separation', has become a commonplace, pointing to connectivity among people that can be reduced to a trail of no more than six persons (the six degrees). This became famous after three students created a board game (and an internet version of this) called Six Degrees of Kevin Bacon, proving that any actor in Hollywood can be connected to Kevin Bacon by a trail of a maximum of six other people.[12] Another example of the pervasiveness of network approaches, similarly essential to Social Network Analysis, is that of a 'Small World'. Duncan Watts and Steven Strogatz showed how few added (random) links could alter the connectivity of a network thus transforming it into a 'Small World'.[13] Their paper has been one of the most influential papers in the last decade, yet some of the conclusions they reached have not yet made an impact in historical studies.

Social networks analysis has been recently used in archaeology with some very interesting results: for example, Sean Graham's work on the Antonine Itineraries showed how social network analysis, applied through agent-based computer stimulation, resulted in an understanding of the fundamental differences of the regions of Iberia, Gaul, Italy, and Britain in terms of connectivity, clustering, and overall structural organization.[14] 'Networks' were the subject of Irad Malkin's Nellie Wallace Lectures in Oxford (2006), with a special emphasis on the Archaic Mediterranean and questions of Greek identity. In ancient history, although the idea of Mediterranean connectivity and interaction has been at the forefront of recent developments,[15] no theoretical approach using social network analysis has been attempted, except for concepts such as 'peer polity interaction' (originally borrowed from archaeology), that have been fruitfully used to examine networking of social entities.[16]

The chapters that follow illustrate the first response of distinguished ancient historians to this challenge, following a conference held in Rethymno, Crete, in May 2006. The meeting offered the opportunity to address these questions, in order to assess the methodological tools offered to us by Mediterranean-historiographical approaches and the Social Networks Analysis theory. It ultimately convinced some of us of the need to shift from the study of individuals and historical structures into that of social relationships, with a special emphasis on networks, in order not only to elucidate social structures *per se*, but also to better understand the dominant motors, agents, and processes that generated new structures.

The broad outlines of the paradigm we propose to adopt have been laid down by structural analysts. Focusing on the underlying *relationships* among the elements of a social system that constrain their interaction and shape patterns of behaviour, they argue that examining relations between social actors offers the best analysis of social categories (based on gender, age, social status, ethnicity, etc.) and of bounded groups. Social structuralists have relatively recently represented these structures in terms of networks; that is, as sets of *nodes*, i.e., members of a social system, and as sets of *ties* depicting the interconnections among nodes and which hence regulate the flows of resources, symmetrical friendships, transfers, and relationships.[17] It is within this general framework that we propose to examine the pattern of the creation and development of networks in Antiquity.

The Social Networks Analysis theory capitalizes on the existence of relationships, i.e., on the establishment of links among interacting units. In other words, small-scale social interaction is regarded as key to the creation and development of networks. The lasting patterns of relations among social actors, i.e., individuals, form *structures*. It is within this structural environment, through the *paths* created by the members of groups, that information and ideas, resources and services can be transmitted through groups.[18] If most nodes within a network are *linked* (through lines/paths) and very few nodes are isolated, then this network may be defined as *dense*. In fact, the more inclusive a network is, the denser it will be.[19] Such patterns can be replicated among entire groups on a broader scale, and complex, hierarchical relationships of power and interdependency developed among them, illustrating an important aspect of Mediterranean networks.

It is important to consider how central a node is in a network, by comparison to the various nodes in it, in other words, to determine how well connected a node is within its local environment. A node may be *locally* central; compared with points within the same network, it may be *relatively* central, i.e., 'between' other points in the same graph. If it is at short distances from many other points in the graph, it is *globally* central. Points with the lowest centrality may be regarded as *peripheral points* in a network.[20]

The study of networks is enhanced through their deconstruction into their components. In fact, the independent study of cliques, clusters, components, cores, and circles is likely to reveal the nature and function of a network. The aforementioned groupings form respective cyclic components, that is, maximal sub graphs connected with a line that returns to its own starting point.[21] *Hangers*, i.e., nodes hanging to a cyclic component; *bridgers*, i.e., nodes hanging to two or more cyclic components; *isolated trees*, i.e., chains of points not connected to any cyclic components; and *cut-points*, whose removal would divide the sub-graph into two or more subsets without connections, enabled contacts within and among networks. The 'structural equivalence' of social positions, i.e., the fact that structurally equivalent agents played the same role in their respective networks, made possible overlaps between cliques within a network and beyond.[22] Thus networks were developed within a single structure, or they expanded between (and among) different structures; most importantly, certain of these initiated transitions into new structures and patterns.

The dynamics of group relations within a network are often illustrated in graphs. Far from being sets of lines connecting points, graphs describe the properties of the patterns formed by lines. *Sociograms*, in particular, may be described as maps of social structures; they retain the mathematical properties of a graph and allow for new features to be added to the graph eventually.[23]

Some of these concepts are compatible with ancient historical patterns. For one thing, the social and political structures that were developed within specific geographical boundaries form a key area of study also in ancient history. Since some of these structures were shaped by the kind of political institutions, kinship obligations, friendship, and work relations about which we have some evidence allows ancient historians to investigate how social relationships interlocked in Antiquity. In other words, in what way the interactions latent in one relationship affected those occurring in others. We thus attempt to employ the methodological tools developed by structural analysts, in order to explore issues such as: how can ancient networks be identified? What forms did they take and on what scale did they exist? How were they structured and sustained? Were there any common patterns in their structure (economic, social, etc.)? In what ways did proximity/distance affect their development and growth? May any elements of network change or of network continuity be traced? How did networks affect individuals, communities and societies? Can one trace channels of interchange among different networks? In what ways were networks encouraged and promoted? Which factors and/or practices maintained them?

The present volume offers a spectrum of the themes that might gain in perspective from a network approach. In some cases the confrontation with network terminologies is explicit, in others just implied. All chapters, however, try to adopt a network approach to the issues they discuss. With regard, specifically, to the potential of the Social Networks Analysis, one may select different units of analysis and explore the levels of their interaction, as we do here. Ties of kinship, for one thing, formed dense networks;[24] economic exchange created traffics of communication based on transactions; important patterns of information were diffused along with the flow of resources. This book also illustrates how culture – specifically philosophical and religious ideas – spread along similar patterns of communication. 'Physical' networks, maritime or land networks and communication, also had an impact upon the societies that were somehow connected or divided through them; while travellers needed to manipulate the dominant networks they were likely to encounter on their way towards their final destination. The expansions and shrinkages of networks are also worthy of particular investigation: it is important to define whether the links involved in the transmission of information were strong or weak, and to investigate the mechanisms through which these were sustained through time.[25]

It is to be emphatically underlined that focusing on networks does not entail simply adding on a new term to existing debates. Our ultimate aim is not to show that things were connected (they were), but to explore issues that the limits of current paradigms fail to address effectively. In other words, our choice of the networks paradigm is deliberate. It is driven from the need to understand both 'Mediterranean' and social structures, to analyse the factors that determined their shape and the flow of communication within them, and ultimately to understand the motors that led to transitions towards new structures. In fact, shifting from the concept of 'structures' to that of 'networks' (social and others) not only allows us greater temporal flexibility but also makes it easier for us to understand and to interpret the transmission of information, the transfer of recourses, and, not least, historical transitions.

We should also be aware of the limitations of applying tools borrowed from the Social Networks Analysis approach. Inevitably certain issues remain outside its scope. Insofar as the definition of the number of samples and of the type of categories we choose to elaborate on remain largely arbitrary, depending on the scope and goals of our research, we must always make the latter explicit when explaining the methodology of our selection of samples; not least, we should highlight any grey areas, i.e., the areas in which the Social Networks Theory is not applicable. We also need to understand how social networks might shape the behaviour of their members.

Emphasis in the present volume is attached to the qualitative rather than to the quantitative aspects of the Social Networks theory, for two reasons: in the first place, for some of us, identifying and applying the implications of networks and network-theories is a novel experiment. Furthermore, the fragmented numeric information about Antiquity that is available to us does not allow for mathematical calculations of the 'structural variables', that is, of the quantities that measure the development of group relations within a paradigm.[26] It is desirable, however, to pursue with quantifications, where possible, in the future.[27] We will also need to supplement this type of research with graphs, such that do not oversimplify, which is an art in itself. These graphs may clarify significantly the structure and function of networks.[28]

Prima facie the engagement in the study of networks appears to be a response to current trends and to theories borrowed from sciences and sociology. The present volume shows, we hope, that drawing webs of connections and illustrating their actual scale, and multiple dimensions may help us to overcome impasses to which other paradigms have led us: by bringing to the fore the history of interactions and interconnections through the social networks theory, we may end up with deeper perceptions of the Mediterranean world in Antiquity and what bound its civilizations together. It is on these grounds that the shift of our attention from focal points to the effects and patterns of the interactions among them opens up for ancient historians a new set of research agenda, worth pursuing in the years to come.

- - - - - - - - - - -

'Networks are present everywhere. All we need is an eye for them.'[29]

The idea of a conference on networks as a new direction in the field of ancient history emerged in 2003. Christy Constantakopoulou and Katerina Panagopoulou set out to examine the usefulness of a network approach to history in the extended Greek world; indeed, the conference itself, which took place in Rethymno, Crete, on 26–28 May 2006, was entitled 'Networks in the Greek World'. Soon, however, it became apparent that we were looking at an even wider picture: the Phoenicians, the Romans, the Near East, and Late Antiquity became subjects of papers and brought the 'Mediterranean', rather than the extended Greek world, into the centre of the debates. We were initially thinking along the lines of certain recent developments in ancient history, which examined the usefulness of interaction and connectivity in a pan-Mediterranean scale, and we were wondering whether the concept of network and the relatively new approach developed in the Social Sciences of Social Network Analysis had anything to offer in the ongoing discussions in our field. We intended to bring together scholars working on different fields in ancient history in order to discuss, in the most hospitable environment of Crete, this new kind of approach.

In the opening of the conference we stressed that we did not aim to provide a prescriptive definition of a network, let alone create a uniform approach to what appeared to us a relatively unexplored and fascinating new paradigm. Similarly, we believe that there is an attractive diversity in the degrees of scholarly engagement with the concept of networks. We can talk about nodes, systems, and the role of agents in the systems, but we are also essentially interested in the transmission of ideas, a subject which, to a great extent, has attracted interest in terms of network analysis.

The *Mediterranean Historical Review* (two of its special issues form the basis for this book), has taken centre stage in Mediterranean studies since its first appearance in 1986.[30] From its inception the periodical adopted both a comparative and comparable approach, happily allowing for Braudelian *Longue durée* approach without Braudel's distaste for *histoire événementielle.* With its special issue on *Mediterranean Paradigms*, the subject of networks became an explicit interest of its editors. Now, with the publication of *Networks in the ancient Mediterranean*, we present for the first time a wide spectrum of network-related historical questions, applied to the ancient Mediterranean. It is hoped that this volume will generate discussion not only among Classical scholars but also among historians of other periods for whom the network approach may prove significant, perhaps creating new networks of scholarly engagement and debate.

Acknowledgements:

The conference on Networks would not have been possible without the generous contributions of Costamare Shipping Company, the Alexander S. Onassis Foundation, and the Department of History and Archaeology at the University of Crete/Greece. For this, we are deeply indebted. We should also like to thank Aegean Airlines, the Institute of Mediterranean Studies, and Birkbeck College for financial, administrative and overall support. Particular thanks are also due to the 25th and 23rd Ephoreia of Prehistoric and Classical Antiquities, not least to Dr M. Vlazaki and to Ms Eftuchia Protopapadaki (25th Ephoreia) and to Dr A. Vassilakis (23rd Ephoreia), who substantially enriched the activities and contributed to the conference debates; finally, to the Municipality and to the Prefecture of Rethymno/Crete and, not least, to the postdoctoral and doctoral students at the Department of History and Archaeology/University of Crete, who generously contributed to the success of the conference in various ways. Additional support for this publication comes from the Maxwell Cummings Chair for Mediterranean History and Cultures at Tel Aviv University and the Israel Science Foundation.

References

Alcock, Susan E. 'Alphabet Soup in the Mediterranean Basin: the Emergence of the Mediterranean Serial'. In *Rethinking the Mediterranean*, edited by W. V. Harris, 314–36. Oxford: Oxford University Press, 2005.

Braudel, Fernand. *A History of Civilizations*. Trans. Richard Mayne. New York: Lane/Penguin Press, 1994.

———. *The Mediterranean and the Mediterranean World in the Age of Philip II* (*La Méditerrannée et le Monde Méditerranéen à l'époque de Philippe II*, 1st ed. 1949, 2nd revised ed. 1966), 2 vols. (vol. 2 published 1973). London and New York: Harper and Row.

Barabási, A.-L. *Linked: The New Science of Networks*. Cambridge, MA: Perseus, 2002.

Berkowitz, S. D., and B. Wellman, eds. *Social Structures: A Network approach*. Cambridge: University Press, 1988.

Carrington, Peter J., John Scott, and Stanley Wasserman, eds. *Models and Methods in Social Network Analysis*, Structural Analysis in the Social Sciences 27. New York: Cambridge University Press, 2006 (repr.).

Cicero *De Republica*, 2.9.

Deleuze, Gilles, and Felix Guatari. *A Thousand Plateaus: Capitalism and Schizophrenia*, translated by B. Massumi. Minnesota: University of Minnesota Press, 1987.

Ginnelli, M., G. Fass, and B. Turtle. *Six Degrees of Kevin Bacon*. New York: Plume, 1996.

Goitein, S. D. *A Mediterranean Society: The Jewish Communities of the Arab World as Portrayed in the Documents of the Cairo Geniza*, 5 vols. Berkeley, Los Angeles and London: Berkeley University Press, 1999.

Graham, S. 'Networks, Agent-Based Models and the Antonine Itineraries: Implications for Roman archaeology'. *Journal of Mediterranean Archaeology* 19 (2006): 45–64.

Granovetter, M. S., 'The strengh of weak ties', *American Journal of Sociology* 78 (5): 1360–80.

Harris, W. V., ed. *Rethinking the Mediterranean*. Oxford: Oxford University Press, 2005.

Horden, P., and N. Purcell. *The Corrupting Sea. A Study of Mediterranean History*. Oxford: Blackwell, 2000.

Ma, J. 'Peer Polity Interaction in the Hellenistic Age'. *Past and Present* 180 (2003): 9–39.

Malkin, Irad. 'Networks and the rise of Greek civilization in the Archaic period'. Nellie Wallace Lectures in Oxford (2006, forthcoming).

———, ed. *Mediterranean Paradigms and Classical Antiquity*. Special issue of *Mediterranean Historical Review*. London: Routledge, 2005.

———. 'Networks and the emergence of Greek identity'. *Mediterranean Paradigms Classical Antiquity*, Special Issue of the *Mediterranean Historical Review* 18 (2003): 56–74.

Plato, *Phaedo* 109b

Scott, John. *Social Network Analysis. A Handbook*. 2nd ed. London–Thousand Oaks–New Delhi: SAGE Publications, 2000.

Wasserman, S., and K. Faust. *Social Network Analysis: Methods and Applications*. Cambridge: Cambridge University Press, 1994.

Watts, Duncan. J. *Small Worlds. The Dynamics of network between Order and Randomness*. Princeton: Princeton University Press, 1999.

———. *Six Degrees: The Science of a Connected Age*. London: Vintage, 2003.

Watts, D. J., and S. Strogatz. 'Collective dynamics of "small-world" Networks'. *Nature* 393 (1998): 440–42.

SELECT BIBLIOGRAPHY

Barth, Fredrick. *Ethnic Groups and Boundaries.* London: Little Brown & Co., 1969.

Berkowitz, S. D., and B. Wellman, eds. *Social Structures: A Network approach.* Cambridge: University Press, 1988.

Carrington, Peter J., John Scott, and Stanley Wasserman, eds. *Models and Methods in Social Network Analysis*, Structural Analysis in the Social Sciences 27. New York: Cambridge University Press, 2006 (repr.).

Crozier, M., and E. Friedberg, *L'acteur et le système.* Paris 1977. Eng. ed., *Actors and Systems: the Politics of Collective Action.* Chicago: University of Chicago Press, 1980.

Forsé, M., ed. 'Les réseaux sociaux'. *L'année sociologique,* vol. 47.1 (1997) :???.

Granovetter, Mark S. 'The strengh of weak ties'. *American Journal of Sociology* 78 (5): 1360–80.

Harris, W. V., ed. *Rethinking the Mediterranean.* Oxford: Oxford University Press, 2005.

Horden, P., and N. Purcell. *The Corrupting Sea. A Study of Mediterranean History.* Oxford: Blackwell, 2000.

Ma, John. 'Peer Polity Interaction in the Hellenistic Age'. *Past and Present* 180 (2003): 9–39.

Malkin, I., ed. *Mediterranean Paradigms and Classical Antiquity.* London: Routledge, 2005.

Scott, John. *Social Network Analysis. A Handbook.* 2nd ed. London–Thousand Oaks–New Delhi: SAGE Publications, 2000.

Wasserman, Stanley, and Katherine Faust, eds. *Social Network Analysis. Methods and Applications.* Cambridge–New York: Cambridge University Press, 1994.

Notes

1 Braudel, *The Mediterranean and the Mediterranean World in the Age of Philip II.*

2 Goitein, *A Mediterranean Society.*

3 Pirenne, *Mohammed and Charlemagne.*

4 Braudel, *A History of Civilizations,* The title of another magisterial work.

5 Malkin, 'Networks and the emergence of Greek identity'.

6 Cicero, *De Republica* 2.9.

7 Plato, *Phaedo* 109b.

8 Malkin, 'Networks and the emergence of Greek identity'.

9 Notably Harris, *Rethinking the Mediterranean*, and Malkin, ed. *Mediterranean Paradigms and Classical Antiquity.*

10 Deleuze and Guatari, *A Thousand Plateaus: Capitalism and Schizophrenia.*

11 Wasserman and Faust, *Social Network Analysis*, 3–10.

12 Mike Ginelli, Craig Fass, and Brian Turtle created Six Degrees of Kevin Bacon. The story of the game and the logic behind it is now included in Ginelli, Fass and Turtle, *Six Degrees*. The Oracle of Bacon can be found in http://www.fast-rewind.com/bacon.htm.

13 Watts and Strogatz, 'Collective dynamics of "small-world" networks'. See also Watts, *Small Worlds*, and *Six Degrees*, esp. 69–100.

14 Graham, 'Networks, Agent-Based Models and the Antonine Itineraries'.

[15] See for example Horden and Purcell, *Corrupting Sea*; Harris, ed. *Rethinking the Mediterranean*; and Malkin, ed. *Mediterranean Paradigms.*
[16] See in particular Ma, 'Peer Polity Interaction in the Hellenistic Age'.
[17] Berkowitz and Wellman, *Social Structures*, 4, 480; Wasserman and Faust, *Social Network Analysis*, 18–20.
[18] Scott, *Social Network Analysis*, 9–26.
[19] Ibid, 70–74.
[20] Ibid, 81–88.
[21] Ibid, 106.
[22] Ibid, 123–74; Wasserman and Faust, *Social Network Analysis*, 109–14, 347.
[23] Scott, *Social Network Analysis*, 63–67, 148–53.
[24] On Wellman's view that kinship networks tended to be the densest networks, see Scott, *Social Network Analysis*, 78–79.
[25] Granovetter, 'The Strength of Weak Ties'.
[26] Wasserman and Faust, *Social Network Analysis*, 1.
[27] Cf., for instance, Carrington, Scott and Wasserman, *Models and Methods.*
[28] On the importance of network measurements, see Wasserman and Faust, *Social Network Analysis*, 8.
[29] Barabasi, *Linked*, 7.
[30] See Susan E. Alcock, 'Alphabet Soup in the Mediterranean Basin: the Emergence of the Mediterranean Serial" in Harris 2005: 314–36.

Beyond and Below the Polis: Networks, Associations, and the Writing of Greek History

Kostas Vlassopoulos

The study of Greek history in the last few decades has been dominated by the concept of the polis. This trend has taken two separate, but closely connected, forms. On the micro-level, the view of the polis as an adult male citizen club tends to emphasize a number of insuperable polarities as fundamental features of Greek social, economic, political, and cultural history: citizen vs. foreigner, male vs. female, free vs. slave, Greek vs. barbarian.[1] On the macro-level, the concept itself of the Greek polis came ultimately to play the role of the subject of Greek history. The polis was and is still seen as *the* Greek form of state, *the* Greek form of society, *the* Greek form of economy; and as such was what separated the Greeks from the other peoples of the Mediterranean and the Near East, thereby allowing modern scholars to focus on specifically Greek economic, social, cultural, and political history. The many different Greek communities, scattered all over the Mediterranean and the Black Sea, could be seen as so many different replicas of the common denominator. And instead of the bewildering variety between the developments of different Greek communities, modern scholars could substitute a unified history based on the rise, acme, and decline of the polis.

Two main problems arise with this standard approach. The first is that, in an age of disillusionment with structuralism, the static polarities seem particularly problematic.[2] But the more important problem is the inability of the polis concept to explain both the wide diversity and the fascinating unity of the Greek world.[3] We have no analytical standpoint apart from the level of the polis to understand the Greek world as a whole.[4] Instead, we accept the wider Greek world as a matter of fact; in this framework, everybody who spoke Greek is considered somehow part of Greek culture and civilization. The assumption behind this approach is that culture and society are closed, homogeneous, bounded entities. Greek culture is understood as a closed, bounded entity juxtaposed to other closed, bounded entities. This has also created problematic reifications: Greek history is written as a national history, and is separated from other national histories in the Mediterranean (e.g., Egyptian or Persian). Even when scholars take into account the interactions between Greeks and other peoples, they tend to understand this as an interaction between two closed, bounded entities.[5]

What I wish to propose in this paper is that the concept of *network* can allow us to move beyond this orthodox model of writing Greek history.[6] The concept of network enables us to explore two levels apart from the polis; the level of social, economic, political, and cultural interaction *below* the polis, and the level *beyond*. I will propose that in order to study the networks below the polis, we need to employ the concept of *koinônia* (association); and that in order to study networks beyond the polis, we need to employ the concept of the world-system.

Below the Polis: The *Koinôniai*

Networks are processes that connect people and places in order to move goods, people and ideas/technologies. But how do they actually work at a micro-level? Networks function by bringing people into association with each other, in order to buy and sell, to work, to drink and eat, to participate in worship, to discuss, to fight, to travel, etc. This concept of association plays a fundamental role in the work of Aristotle, who sees the polis as a *koinônia* encompassing other *koinôniai*:[7]

> But all *koinôniai* are parts, as it were, of the *koinônia* of the polis. Travellers for instance associate together for some advantage, namely to procure some of their necessary supplies. But the *politikê koinônia* too, it is believed, was originally formed, and continues to be maintained, for the advantage of its members. … Thus the other *koinôniai* aim at some particular advantage; for example, sailors combine to seek the profits of seafaring in the way of trade or the like; comrades in arms the gains of warfare, their aim being either plunder or victory over the enemy, or the capture of a city. … All these *koinôniai* then appear to be parts of the *koinônia* of the polis.[8]

Of course an association can take an institutional form, like a *deme* (township), or a tribe; but there are many other *koinôniai*, ranging from the temporary and the informal, to the highly institutionalized. It is regrettable that modern historians have not paid enough attention to this Aristotelian concept, as it offers a number of great advantages.[9] The first advantage of the concept of *koinônia* is its ability to overcome static polarities and divisions, such as those between masters and slaves, citizens and metics (alien residents), men and women, Greeks and barbarians. Of course, these distinctions did exist and they did play an important role; but the problem is that we have made of them unsurpassable states of being, when in reality these concepts are images and identities that are defined by constant challenge and renegotiation, depending on the context.[10] The advantage of Aristotle's concept, then, is that it allows us to focus on the concrete experience of people in the various forms of *koinôniai* in which they associate. In the case of the Greek and barbarian identities, what is the actual form of such a polarity when they both participate in a *koinônia* aboard a ship? And what happens when a free man and a slave both participate in a *koinônia* for work (in the shipyards, for instance), under the same conditions? What occurs when a citizen and a metic drink together and converse in a tavern or a barber's shop? And last but not least, what takes place when men and women participate together in a *koinônia* of religious practice?

A further advantage of Aristotle's *koinônia* is its conceptual flexibility. In contrast with the modern conception of isomorphism between society, economy, and the state, the concept of *koinôniai* recognizes that the boundaries of different *koinôniai* are different, and not necessarily overlapping. The boundaries identifying a household, for instance, are very different from those of guest-friendship, or for a trading agreement, a religious or scholarly community, or a group of *systratiôtai* (military comrades). All these relationships, which are described as *koinôniai* by Aristotle, are part of the *politikê koinônia*, but each one has its own boundaries, directions, and rhythms of change.

Let us look at a number of examples that show the value of the concept of *koinônia*. I will restrict my examples to cases of *koinôni*ai between Greeks and foreigners. Much has been written about the Greeks' perceptions of the Other, their racism, and their contempt for Barbarians.[11] While of course there is a lot of truth in these accounts, it seems to me that these issues are rarely discussed with real people in mind, nor are they subjected to the sort of analysis that historians with better evidence are accustomed to applying.

Let us imagine a Greek and a Phoenician drinking a cup of wine after work in Piraeus, and that the former is Athenian: how would he formulate his discussion on the superiority of Athenian democracy over Oriental despotism, and what would the Phoenician reply? Would the Greek speak to the Phoenician in the same way as he would with other Greeks? How would his personal, daily contact with the Phoenician influence his own perceptions? Given that in Piraeus, Greek citizens and Phoenician aliens lived in the same neighbourhoods, worked in the same streets, buried their dead in the same graveyards, and their children played together—how did this day-to-day experience influence and formulate people's perceptions in general?[12]

Although we do not have the direct evidence to answer these questions, nor do we have records of actual encounters and discussions, certain bits and pieces of evidence have survived that offer some clues. For example, an official Athenian inscription honouring the king of Phoenician Sidon also stipulates the following: 'Those Sidonians that while residing in Sidon and having political rights there (*politeuomenoi*) come to Athens for trade should be exempt from the *metoikion* tax, from acting as *choregoi* and from being registered for an *eisphora*.'[13] The Athenians in this inscription recognize that, while a Near Eastern community might be governed by a king, its citizens were not without some political rights.[14] So our Phoenician might argue that the Athenians themselves recognized that the commonly held idea of Oriental despotism was far from the actual reality. How did the Athenians know that some Sidonians had political rights in Sidon? Surely, the Greek discourses on Oriental despotism that we find in Greek texts did not provide them with the material to raise such a possibility. Is not this inscription a result of these actual encounters that I am postulating?

Further evidence is found in Xenophon's *Oeconomicus*, in which Ischomachus describes the following encounter:

> Once I had an opportunity of looking over the great Phoenician merchantman, Socrates, and I thought I had never seen tackle so excellently and accurately arranged. For I never saw so many bits of stuff packed away separately in so small a receptacle. ... I found that the steersman's servant, who is called the mate, knows each particular section so exactly that he can tell, even when away, where everything is kept and how much there is of it, just as well as a man who knows how to spell can tell how many letters there are in Socrates and in what order they come. Now, I saw this man in his spare time inspecting all the stores that are wanted, as a matter of course, in the ship. I was surprised to see him looking over them, and asked what he was doing. 'Sir,' he answered, 'I am looking to see how the ship's tackle is stored, in case of accident, or whether anything is missing, or mixed up with other stuff. For when God sends a storm at sea, there's no time to search about for what you want, or to serve it out if it's in a muddle. For God threatens and punishes careless fellows, and you're lucky if he merely refrains from destroying the innocent; and if he saves you when you do your work well, you have much cause to thank heaven for.'[15]

Ischomachus (or Xenophon) was clearly impressed by the huge Phoenician vessel; from the detailed way it is described one can easily assume that the event was well remembered. So intrigued was Ischomachus that he went deliberately to make inquiries with the mate of the Phoenician ship, and one wonders in which language Xenophon believed them to have conversed: did Ischomachus speak Phoenician, or did the Phoenician sailor speak Greek? Were there also Greek sailors aboard the Phoenician ship? And what else can the two men have talked about? Clearly, after such a detailed conversation, no Greek would continue to sustain the view of Barbarian inferiority or incapacity.

Another fascinating example occurs in the account of Athenogenes, an Egyptian metic involved in selling perfumes in late fourth-century Athens.[16] The cunning

Athenogenes conspires with Antigone, a prostitute, to sell to a wealthy young Athenian citizen two male slaves along with their heavily indebted perfume workshop. In this case, the presumed superiority of the dominant Greek citizen is completely subverted. Although the details of the story are not of direct concern here, what happens later on is quite revealing (§29–31):

> During the war against Philip, he left the city just before the battle and did not serve with you at Chaeroneia. Instead, he moved to Troezen, disregarding the law, which says that a man who moves in wartime shall be indicted and summarily arrested if he returns. The reason for the move, it seems, was this: he thought that the city of Troezen would survive, whereas he had passed a sentence of death on ours. ... He is so degraded and so true to type wherever he is, that even after his arrival at Troezen, when they had made him a citizen, he became the tool of Mnesias the Argive and, after being made a magistrate by him, expelled the citizens from the city.[17]

Here we have an Egyptian perfume-seller with the obligation to fight for Athens, along with thousands of other foreigners living in Athens, and who instead escapes to a tiny, obscure place like Troezen, where he is enrolled as a citizen and even becomes a magistrate.[18] How common was such an event? To judge from the tone of the passage, it does not seem very extraordinary—indeed, the outrage is for his having disenfranchised other citizens, not for his having managed to become one himself. What did Athenogenes bring with him from his Egyptian cultural baggage when he became a citizen and a magistrate? We need a larger horizon.[19]

Elsewhere we find an Athenian grave inscription dating from the third quarter of the fifth century, bearing an epigram in Homeric hexameters, which reads thus: 'This is the beautiful tomb of Manes, the son of Orymaios, the best of Phrygians there ever were in wide Athens. And by Zeus, I never saw any woodcutter better than me. He died in war.'[20] We have here the epigram of a foreigner, a Phrygian, who is a manual worker and clearly proud of his craft and his manual skills. Furthermore, it is reasonable to suppose that he died fighting for Athens during the Peloponnesian War, or equally possible in some other previous military engagement. Thousands of metics fought alongside Athenians every time Athens was at war.[21] How would our Phrygian converse with his Athenian comrades-in-arms while serving in the Athenian regiment? How would he converse with his Athenian fellow-workers? What would a low-class Athenian think when reading this epigram on his way to his workshop?

To conclude, the concept of the network takes the form of *koinônia* (association) in the micro-level below the polis. This concept can allow us to move beyond the static polarities that we employ when writing Greek history, and move towards more nuanced accounts.

Beyond the Polis: Networks and the World-System

How can the concept of network help us with levels beyond the polis? Many modern historians believe that the polis is the key to the underlying unity of the Greek world and of all Greek culture, and that it is somehow possible thereby to make a clear distinction between the Greek world and all the other cultures of the Mediterranean and the Near East. These same historians have also been reluctant to create proper methodological and analytical tools for studying the variety of Greek cultures, their links, and forms of interaction. We still lack the analytical tools for studying all those intermediate levels, the complex links between them, and the ways in which the variety of communities and levels scattered all over the Mediterranean came to form a single, overarching system with its own rules.

Athenian monumental funerary art of the Classical period provides a good example of the problem and my proposed solution.[22] One of the most impressive monuments of fourth-century Athenian art is the funerary monument of Nikeratos—a metic from the city of ancient Istria on the Black Sea; the monument is clearly inspired and imitates the famous Mausoleum of Halicarnassus. But the creation of this new form of funerary art, which fuses Greek temple architecture and sculpture with Near Eastern decorative themes and funerary monuments, is neither simply an imitation, nor another illustration of Athenian creativity; it was based on the long experimentation between Greek and non-Greek artistic practices that took place in the wider Mediterranean Greek world.

Working for both Greek and non-Greek customers, Greek artists in Sinope (the Black Sea), Cyrene (North Africa), and the Greek cities of Asia Minor experimented for a long period, fusing the tradition of Greek public art with the various non-Greek traditions of monumental tombs, which were to be seen in their adjacent areas. In the fourth century this experimentation found its way to Athens, giving rise to one of the most impressive artistic achievements of the Greek world, clearly demonstrating the interaction between various components of Greek and other Mediterranean cultures, the role of the wider Greek world as a laboratory of experimentation and interaction, the introduction of the new practices to Athens, and not least the role of metics in Athenian culture.

The above example should lead us to a number of conclusions. First, to talk of the art of an individual polis (Athens) is misleading, as the artistic monuments created in Athens are the product of complex processes taking place over time within the wider Mediterranean world, without belonging to or emanating from a particular centre. Second, the unity of Greek culture cannot be attributed solely to the emergence and development of the polis. Composed of a multitude of different, fragmented polities scattered all over the Mediterranean and the Black Sea, the Greek 'world' was created and fostered by the vast numbers of people in movement, namely, sailors, traders, soldiers, artists, physicians, poets, and intellectuals. As Filippo Cassola has put it: 'The activities [of these thousands of mobile people] were enough to create a connective network which embraced the whole Greek world and caused an exchange of experiences that guaranteed not the homogeneity of the culture, but the reciprocal comprehension and the reciprocal interest among the inhabited centres.'[23] But if we accept that this is the case, we need to reach a further conclusion: the networks of mobilizing and moving these people, the networks which create the unity of the Greek world, were not controlled only by Greeks, nor did they involve only Greeks.

These three observations can best be answered by adopting the concept of the world-system, which allows us to study how the various networks function beyond the level of the individual polity.

What is a world-system? It is important to clear up some significant misconceptions that gather round this notion.[24] The first is that a world-system implies by necessity a distinction between a dominant core and an exploited periphery (and semi-periphery). Therefore, if it is impossible to find a clear distinction between a centre and an exploited periphery, then there exists no world-system. A dominant centre and an exploited periphery is only one possible form of a world-system. While the modern world-system is undoubtedly structured on a centre/periphery basis, this does not mean that all world-systems in the past have had the same characteristics, or that there were no world-systems before the emergence of the modern one. Indeed, a variety of different forms are feasible: Janet Abu-Lughod has convincingly argued that the medieval world-system of the thirteenth and fourteenth centuries CE had a form of concentric circles, instead of a single centre and periphery.[25]

The second misconception is that direct links are required for a world-system to qualify as such. A sceptic might ask: what exactly do we gain in comprehension if we say that Paestum (Italy), Thebes (Greece), and Olbia (Black Sea) belong to the same world-system.[26] But although direct contacts might as well develop within a world-system, they are by no means mandatory. What makes Croton, Aegina, Athens, Samos, and Persia all components of the same world-system is not their direct contacts; it is rather the networks that already existed in the sixth century—and which prompted the famous doctor Democedes to traverse the whole Mediterranean from his native Croton to Persia—that link these places together into a 'system'.[27]

If we accept Wallerstein's minimum definition that '[a system] is a world-system, not because it encompasses the whole world, but because it is larger than any juridically defined political unit'[28] then the crucial issue is that a single community or polity (a Greek polis) cannot be a self-sufficient unit of analysis.[29] The term 'world-system' is used here to define and analyse a larger frame of historical reference. There are various forms of interactions and processes that one could call world-systems; they range from low- to high-intensity systems; and from anarchic to centrally organised. There need then be two primary qualifications: a world-system does not necessarily encompass the whole world; there can be several coexisting world-systems; and the extent of each of them can change from period to period and so can only be historically reconstructed. Moreover, a world-system is a system, but not necessarily a highly structured and coherent one; again, its intensity can be described only in concrete historical analysis, and not in a priori theory. So, a world-system can indeed be a (highly structured) system of the (whole) world, as it is nowadays; but it can also take historically contingent forms varying in extent, structure, and intensity.

The components of such a world-system comprise three main aspects: *processes* (the movement of goods, people, and ideas and/or technologies); the various *centres* that organize and control these processes; and the *forms of change* of this system. The first two aspects are dealt with in detail in a forthcoming article,[30] so my analysis here will focus on the forms of change.

A world-system can be said to exist when there is a manifest presence of processes, exchanges, and interactions linking many individual groups, communities, and polities; and when these processes, exchanges and interactions, moving people, goods, and ideas, range beyond the boundaries of a single group, community, or polity. Within this framework, three different world processes can be identified: processes moving people; processes moving goods; and processes moving ideas/technologies.[31] Any study of these processes would have to take several issues into account, such as the scale and extent of these networks through time and space. Another issue is the form of the networks that underlie these processes (e.g., is the network based on diaspora communities? How much power do the network agents have? Are they supported by state agencies or not?). A final issue would be how these different processes are interlinked: does the movement of people follow the same path as the movement of ideas, or are the different networks separate and with different routes and directions?

A world-system has a number of different centres, one being its religious centres, which bring together communities, forge links of common identity, disseminate types of worship and ideologies; in this respect, the importance of Delphi and Olympia is well documented.[32] Second come the *emporia*: those commercial nodes that organize, attract, and channel the movement of goods, people, ideas, and technical know-how.[33] Then there are the centres of cultural, scientific, and academic practices, such as the courts of the Sicilian tyrants[34] or of some Anatolian dynast,[35] or the philosophical schools of fourth-century Athens,[36] or the Cnidian and Coan centres of medicine.[37] Added to these of course were political centres, like Athens or Syracuse.

Any analysis of the various centres of such a world-system would have to take into account three main issues. The first is the *relationship* between different types of centre, that is, to discern how religious centres relate to political centres (e.g., Delphi and Athens); or how a commercial centre managed to transform itself into a political and cultural centre (e.g., Rhodes in the Hellenistic period).[38] The second issue concerns *scale*; the size, catchment area, and extent of influence of a given centre never remained static. For example, Naucratis and Alexandria are good illustrations of the increase in population, power, and extent of network links that took place from the Archaic to the Hellenistic period.[39] A third issue concerns influence and competition. A centre can attempt not only to attract the activities and processes on which it is based, but to control them too; moreover, a centre might try to transform its influence in one field into covering other fields; or, alternatively, its role as a centre in one particular field might necessitate the creation of centres in other fields too.

Finally, we need to take into account the forms of change that take place within a world-system. Failure to differentiate between them has often resulted in misleading conclusions. Following Robert Gilpin,[40] we can define three forms of change. The first concerns 'the nature of the actors or diverse entities that compose an international system', and could be called *actor change* (e.g., the change from the archaic political system composed of Greek polities, to the classical one composed of Greek polities and a large empire (Persia), to the Hellenistic one with the emergence and domination of the large Hellenistic kingdoms); the second type of change is 'a change in the form of control or governance of an international system; this type of change will be called *control change*' (e.g., the change from the tri-polar balance between Athens, Sparta and Persia in the fifth century, to the anarchic balance of the evolving fourth century); and, third, 'a change may take place in the form of regular interactions or processes among the entities in an ongoing international system; this type of change will be labelled simply *interaction change*' (e.g., the change from mainly cultural and scattered political and military exchanges between Greek polities in the Archaic period, to a new form of relationships, where a border conflict in one region might lead to a general war among all the Greek polities).[41]

The failure to differentiate between the different forms of change in the political world-system creates acute problems. Instead of positing all-encompassing (and misleading) forms of change, like the emergence or the decline of the polis, it is better to look at the different underlying forms of change that might be taking place at the same time. The so-called fourth-century crisis and the decline of the polis is a good example.[42] After the defeat of the Greek cities by Philip in 338 BCE, we see a control change for the large Greek *poleis*, like Athens and Sparta, who lose their ability to control the system; but there is no control change for the vast majority of Greek *poleis*, who were already under the control of various powers by the early fifth century. If anything, the actor change, with the emergence of the new Hellenistic monarchies, meant for many medium and small Hellenistic *poleis* a control change in the opposite direction: many *poleis* in the Hellenistic world did benefit from being controlled by a monarchy, instead of being controlled by other *poleis*, as in the Classical period.[43]

In conclusion: the concept of network can help us to write a new kind of Greek history, both below and beyond the polis. We live in a globalized world, with fragmented identities, extensive networks, high mobility, and a world-system that seems out of any single power's control. The ancient Greek world provides a fascinating arena for studying many of the most troubling issues of our own times. Whether we will be able to benefit from this reciprocal *problematique* remains to be seen.

Notes

[1] For the role of these polarities in Greek historians, see Cartledge, *The Greeks*.

[2] See the critical attitudes to structuralism in Hunter, *Policing Athens*; Loraux, *Experiences of Tiresias*; Cohen, *Athenian Nation*.

[3] See Whitley, *Style and Society*, 194.

[4] The concept of 'city-state culture', adopted by Hansen in the more recent publications of the Copenhagen Polis Centre, is though a positive step forward; see Hansen, 'Introduction'.

[5] Among recent protests for such attitudes, see Dougherty and Kurke, *Cultures within Greek Culture*.

[6] For a similar plea, see Malkin, 'Networks'.

[7] See Sakellariou, *Polis-State*, 214–82.

[8] *Nicomachean Ethics* 1160a.4–6.

[9] For exceptions, see Murray, 'Polis and Politeia'; Ober, 'Polis as Society'.

[10] Aristotle's great discovery was indeed diversity and multiplicity, where other Greek thinkers were trying to see the underlying unity behind the apparent diversity; at the same time he used the concept of hierarchy to subordinate diversity to his normative ideas; see Saxonhouse, *Fear of Diversity*, 189–95. We can keep his explorations of diversity, without therefore accepting his concept of hierarchy.

[11] See for example Hall, *Inventing the Barbarian*; Hartog, *The Mirror of Herodotus*. Cartledge, *The Greeks*, 51–77, offers a well-articulated presentation of such arguments.

[12] For the fundamental importance of actual experience for any historical account and supposition, see the classic Thompson, *Poverty of Theory*.

[13] *IG*, II^2, 141.2.30–6.

[14] See the interpretation of Culasso Gastaldi, *Le prossenie ateniesi*, 113–14.

[15] *Oeconomicus* 8.11–6.

[16] Hypereides, *Against Athenogenes*.

[17] For the law mentioned by Hypereides, see Adak, *Metöken als Wohltäter Athens*, 67–72, whose interpretation I accept.

[18] See Whitehead, *Hypereides*, 287–88, 339–41.

[19] Our colleagues studying the Bronze Age Aegean have been more open-minded in this respect: see Knapp, 'Thalassocracies'; Cline, 'Tinker, Tailor, Soldier, Sailor'.

[20] *IG*, I^3 1361. Of course, predictably the interest of the few ancient historians who have dealt with this inscription has focused on dating, topography, and the language and oral features of the epigram. No wonder the political and social implications of this document have never been seriously addressed. But see now the commentary in Bäbler, *Fleissige Thrakerinnen*, 159–63.

[21] Adak, *Metöken als Wohltäter Athens*, 67–72.

[22] For the following, see Hagemajer Allen, 'Becoming the Other'

[23] Cassola, 'Chi erano i Greci', 10.

[24] Cf. Shipley, 'Distance, Development, Decline'.

[25] Abu-Lughod, *Before European Hegemony*.

[26] An example of this attitude can be seen in Finley, *Ancient Economy*, 177–78.

[27] Herodotus 3.125–38.

[28] Wallerstein, *Modern World-system*, 15.

[29] Wallerstein, *Unthinking Social Science*.

[30] See Vlassopoulos, 'Between East and West'.

[31] See Charpin and Joannès, *La circulation des biens*, for a similar perspective on the Near East.

[32] On their emergence, see Morgan, *Athletes and Oracles*; see also Rougemont, 'Delphes et les cités grecques'; Sanchez, *L'Amphictionie des Pyles*.

[33] Bresson and Rouillard, *L'emporion*.

[34] Dunbabin, *Western Greeks*, 298–99.

[35] Hornblower, *Mausolus*.

[36] Ostwald and Lynch, 'Growth of Schools'.

[37] Sherwin-White, *Ancient Cos*, 256–89.

[38] Gabrielsen, *Naval Aristocracy*.

[39] See the articles in Nicolet, *Mégapoles méditerranéennes*.

[40] Gilpin, *War and Change*, 39–40; in my essay I have changed his terms to make them more comprehensible.

[41] See the case of Lokrians and Phokians in the 390s and the eruption of the Corinthian War; *Hellenica Oxyrhynchia* 13.3–4.

[42] A classic treatment is Mossé, *La fin*. See the critique of Davies, 'The Fourth Century Crisis'.

[43] See Gauthier, 'Grandes et petites cités'.

References

Abu-Lughod, Janet L. *Before European Hegemony. The World-system AD 1250–1350*. New York and Oxford: Oxford University Press, 1989.

Adak, Mustafa. *Metöken als Wohltäter Athens: Untersuchungen zum sozialen Austausch zwischen ortsansässigen Fremden und der Burgergemeinde in klassischer und hellenistischer Zeit (ca. 500–150 v. Chr.)*. Munich: Tuduv, 2003.

Bäbler, Balbina. *Fleißige Thrakerinnen und wehrhafte Skythen: Nichtgriechen im klassischen Athen und ihre archäologische Hinterlassenschaft*. Stuttgart: B.G. Teubner, 1998.

Bresson, Alain and Pierre Rouillard, eds. *L'Emporion*. Paris: Publications du Centre Pierre Paris, 1993.

Cartledge, Paul. *The Greeks. A Portrait of Self and Others*. Oxford: Oxford University Press, 2002.

Cassola, Filippo. 'Chi erano i Greci?'. In *I Greci. Storia, cultura, arte, societa. Vol. II: Una storia greca. I. Formazione*, edited by S. Settis. Turin: Einaudi, 1996.

Charpin, D., and F. Joannès, eds. *La circulation des biens, des personnes et des idées dans le proche orient ancien: actes de la XXXVIIIe rencontre Assyriologique internationale*. Paris: Editions Recherche sur les Civilisations, 1992.

Cline, Eric H. 'Tinker, Tailor, Soldier, Sailor: Minoans and Mycenaeans Abroad'. In *Politeia: Society and State in the Aegean Bronze Age*, edited by R. Laffineur and W. D. Niemeier. Liege and Austin, TX: Program in Aegean Scripts and Prehistory, 1995.

Cohen, Edward. *The Athenian Nation*. Princeton: Princeton University Press, 2000.

Culasso Gastaldi, Enrica. *Le prossenie ateniesi del IV secolo a.C.* Alessandria, 2004.

Davies, John K. 'The Fourth Century Crisis: What Crisis?'. In *Die athenische Demokratie im 4. Jahrhundert v. Chr. Vollendung oder Verfall einer Verfassungsform?*, edited by W. Eder. Stuttgart: Franz Steiner, 1995.

Dougherty, Carol, and Leslie Kurke, eds. *The Cultures within Ancient Greek Culture: Contact, Conflict, Collaboration*. Cambridge: Cambridge University Press, 2003.

Dunbabin, Thomas J. *The Western Greeks*. Oxford: Oxford University Press, 1948.

Finley, Moses I. *The Ancient Economy*. Berkeley and Los Angeles: University of California Press, 1985.

Gabrielsen, Vincent. *The Naval Aristocracy of Hellenistic Rhodes*. Aarhus: Aarhus University Press, 1997.

Gauthier, Phillipe. 'Grandes et petites cités: hégémonie et autarcie'. *Opus* 6–8 (1987/89): 187–202.

Gilpin, Robert. *War and Change in World Politics*. Cambridge: Cambridge University Press, 1981.

Hagemajer, Allen, K. 'Becoming the "Other": Attitudes and Practices at the Attic Cemeteries'. In *The Cultures within Ancient Greek Culture: Contact, Conflict, Collaboration*, edited by C. Dougherty and L. Kurke. Cambridge: Cambridge University Press, 2003.

Hall, Edith. *Inventing the Barbarian: Greek Self-definition through Tragedy*. Oxford: Oxford University Press, 1989.

Hansen, Mogens H. 'Introduction. The Concepts of City-state and City-state Culture'. In *A Comparative Study of Thirty City-state Cultures*, edited by M. H. Hansen. Copenhagen: Kongelige Danske Videnskabernes Selskab, 2000.

Hartog, François. *The Mirror of Herodotus: The Representation of the Other in the Writing of History*. Translated by Janet Lloyd. Berkeley: University of California Press, 1988.

Hornblower, Simon. *Mausolus*. Oxford: Oxford University Press, 1982.

Hunter, Virginia J. *Policing Athens: Social Control in the Attic Lawsuits, 420–320 B.C.* Princeton: Princeton University Press, 1994.

Knapp, Bernard A. 'Thalassocracies in Bronze Age Eastern Mediterranean Trade: Making and Breaking a Myth'. *World Archaeology* 24 (1993): 332–47.

Loraux, Nicole. *The Experiences of Tiresias: The Feminine and the Greek Man*. Translated by Paula Wissing. Princeton: Princeton University Press, 1995.

Malkin, Irad. 'Networks and the Emergence of Greek Identity'. *Mediterranean Historical Review* 18. (2003): 56–74.

Morgan, Catherine. *Athletes and Oracles: The Transformation of Olympia and Delphi in the Eighth Century BC*. Cambridge: Cambridge University Press, 1990.
Mossé, Claude. *La fin de la démocratie athénienne: Aspects sociaux et politiques du déclin de la cité grecque au IVe siècle avant J.C.* Paris: Presses Universitaires Francais, 1962.
Murray, Oswyn. 'Polis and *Politeia* in Aristotle'. In *The Ancient Greek City-state*, edited by M. H. Hansen. Copenhagen: Munksgaard, 1993.
Nicolet, Claude, ed. *Mégapoles méditerranéennes. Géographie urbaine retrospective*. Paris: Maisonneuve, 2000.
Ober, Josiah. 'The Polis as a Society: Aristotle, John Rawls, and the Athenian Social Contract'. In *The Ancient Greek City-state*, edited by M. H. Hansen. Copenhagen: Munksgaard, 1993.
Ostwald, M., and J. P. Lynch. 'The Growth of Schools and the Advance of Knowledge'. In *Cambridge Ancient History, VI: The Fourth Century BC*. Cambridge: Cambridge University Press, 1994.
Rougemont, Georges. 'Delphes et les cités grecques d'Italie du sud et de Sicile'. In *La Magna Grecia e i grandi santuari della madrepatria: atti del trentunesimo Convegno di studi sulla Magna Grecia*. Taranto: Istituto per la storia e l'archeologia della Magna Grecia, 1992.
Sakellariou, Michaïl B. *The Polis-State: Definition and Origin*. Athens: De Boccard, 1989.
Sanchez, Pierre. *L'Amphictionie des Pyles et de Delphes: recherches sur son rôle historique, des origines au IIe siècle de notre ère*. Stuttgart: Franz Steiner, 2001.
Saxonhouse, Arlene. *Fear of Diversity. The Birth of Political Science in Ancient Greek Thought*. Chicago: University of Chicago Press, 1992.
Sherwin-White, Susan M. *Ancient Cos. An Historical Study from the Dorian Settlement to the Imperial Period*. Göttingen: Vandenhoeck und Ruprecht, 1978.
Shipley, Graham. 'Distance, Development, Decline? World-systems Analysis and the "Hellenistic" Word'. In *Centre and Periphery in the Hellenistic World*, edited by P. Bilde, T. Engberg-Pedersen, L. Hannestad, J. Zahle, and K. Randsborg. Aarhus: Aarhus University Press, 1993.
Thompson, Edward Palmer. *The Poverty of Theory: Or an Orrery of Errors*. London: Merlin Press, 1978.
Vlassopoulos, Kostas. 'Between East and West: The Greek Poleis as Part of a World-System'. *Ancient West and East* 6 (2007): 91–111.
Wallerstein, Immanuel. *The Modern World-system. I: Capitalist Agriculture and the Origins of the European World Economy in the Sixteenth Century*. New York: Academic Press, 1974.
———. *Unthinking Social Science: The Limits of Nineteenth-century Paradigms*. Oxford: Blackwell, 1991.
Whitehead, David. *Hypereides: The Forensic Speeches*. Oxford: Oxford University Press, 2000.
Whitley, James. *Style and Society in Dark Age Greece*. Cambridge: Cambridge University Press, 1991.

Network Theory and Theoric Networks

Ian Rutherford

In ancient Greece, city-states sent out religious delegations, usually termed '*theoriai*', to take part in common extraterritorial festivals or for other reasons. The members of these delegations, called '*theoroi*', were proxies who acted on behalf of their communities.[1] In some cases the sanctuary too sent out *theoroi* to announce a festival, possibly to ensure synchronization of calendars, and these were hosted in the cities they visited by special *theoroi*-receivers, or *theorodokoi*.[2] So the typical structure is centripetal, with a sanctuary in the middle, and a large number of cities sending *theoriai* to it. There were many of these systems, differing greatly in size, some claiming to cover the whole of the Greek world, others more limited in scope.

The evidence for this practice is largely epigraphic, and it is not evenly preserved. The best documented period is the Hellenistic one, from when we have incomplete records of visits of *theoroi* to several sanctuaries, including Samothrace and Delos. In some cases, it is possible to infer visits by *theoroi* on the basis of other types of epigraphic document, such as proxeny lists.[3] Catalogues of *theorodokoi* from Delphi and other places help to fill out the picture. Two general points are worth making about the data. First, what is recorded is only a very small fraction of what must have gone on, and some well-visited sanctuaries (like that of Artemis at Ephesos) leave virtually no records at all. Secondly, the data that survives is probably heavily biased in favour of the larger religious networks—the ones to which the vocabulary of '*theoria*' is most commonly applied. Smaller, more local festival networks probably existed in large numbers all over the Greek world, but they tend not to be recorded as much. An exception is Hellenistic Crete, for which numerous surviving interstate treaties provide evidence for regional religious networks and bilateral systems.[4]

A 'theoric network' can be broadly defined as a network that underlies the operation of *theoria*. In my earlier work on this subject, I tended to conceive these organizations as 'catchment areas', a term derived from geography and used in studies of pilgrimage in modern India.[5] But I have come over to the view that 'network' is a better concept, partly because the evidence for ancient Greece documents relations between sanctuaries and particular cities, rather than areas; and partly because the concept of a 'catchment area' implies that the movement is from periphery to centre only, whereas in a theoric network it may move in both directions.

Theoria and theoric networks are closely related to at least two other types of interstate religious activity: first, amphictionies, which are generally smaller and more tightly organized, with participation more or less obligatory for members;[6] and secondly networks that link perceived colonies and perceived mother cities, such as the one that existed between Miletus and its 'colonies', and which obligated the latter to send tribute to the former.[7] Thus colony–metropolis networks can be characterized as hierarchical in so far as one central member dominates all others. By contrast, in at least some theoric networks, such as those centred round the major Panhellenic sanctuaries, the relations between cities and sanctuary seem to have been obligation-free, and in that respect we could perhaps characterize such theoric networks as 'non-hierarchical'.[8]

Earlier I defined a 'theoric network' as a network that underlies the operation of *theoria*, but this definition needs further refinement. One application of the term would be networks based around a single sanctuary or a single festival at a sanctuary. Thus, the Delphic theoric network comprises the city of Delphi itself, and the cities that send *theoriai* to it, as well as the individuals who have an institutional role in the process, such as the *theorodokoi*. The size of the Delphic network is vividly illustrated by an extant list of *theorodokoi* from the late third century BCE, which has room for about 700 cities, arranged in terms of the routes (a sort of network in themselves) taken by the *theoroi* who announced the festival.[9] I would imagine that far fewer than 700 cities actually sent *theoriai* to Delphi in any one year,[10] but that this factor makes little difference to the status of the network: the institutional infrastructure of *theorodokoi* in all the cities itself constitutes a sort of network.

Some theoric networks have as their scope the whole Greek world, but others are more limited. For the sanctuary of Samothrace, we have comparatively good records, registering *theoroi* who were made *proxenoi*, dating probably from the third to first centuries BCE. Most of them came from the cities in Asia Minor and Thrace, an interesting contrast to the distribution of the cities that send *mustai* to Samothrace for initiation, more of whom came from mainland Greece.[11] Another example of a smaller theoric network is that of Delos during the period of Delian independence in the third and second centuries BCE. The Delian temple inventories from this period record *theoroi* who funded the performances of the Delian Maidens. The *theoroi* tend to emanate from Alexandria and a group of states, many of them islands, in the south-western Aegean, a surprise considering the strongly Ionian associations of Delos in the Classical Period. Since in this case we have dates, we can roughly trace how the activity in the theoric network, the 'theoric traffic', changes over time.[12]

Related to this is another theoric network, based round Alexandria and centred on the Ptolemaia which was set up by Ptolemy Philadelphos in honour of his father Ptolemy Soter around 278 BCE. Much of the evidence for this comes from funerary vases found at Hadra near Alexandria, some of which contained the ashes of *theoroi* who died while performing their missions.[13] In this case the *theoroi* come from the area of the southern Aegean dominated by the Ptolemaic Empire, including one from Phalasarna in Crete, a remarkable thing in view of the tendency of Cretan cities not to be involved in Panhellenic *theoria*.[14] This theoric network was a conspicuous expression of homage towards Ptolemaic control in the Aegean region in this period, and to that extent it should be characterized as 'hierarchical', unlike the theoric networks centred round the major Panhellenic sanctuaries mentioned earlier.

In some cases the religious cult that characterizes the sanctuary may come to be replicated in some of the *theoroi*-sending cities. So the cult of the 'Samothracian gods' is attested in many Greek cities, including some that are known to have sent *theoroi* to Samothrace,[15] and there was probably a correlation between a city sending a *theoria* to Samothrace and it having its own cult of those deities. Similarly, in the case of the Delian network in the period of Independence, we find the cult of Delian Apollo established in some of the cities that send the most *theoroi* to Delos, such as the islands of Kos and Rhodes. At Kos, it seems that *theoroi* to Delos may have departed on the occasion of a local festival of Apollo Dalios attended by delegates from neighbouring city-states, so that we have a sort of micro-network within the network.[16] In Sicily, the cult of Apollo Archegetes may have served a similar function.[17]

Alongside such 'sanctuary-specific' theoric networks, there is a broader use of the term theoric network, defined as covering all the major sanctuaries at once—a massive coordinated system, with major festivals staggered over a four-year period. We know, for example, that the same individuals sometimes served their cities in a theoric capacity with respect to different sanctuaries.[18] More importantly, we know that there was a perceived reciprocity among different festivals. For example, a fragmentary decree from third-century Kos seems to refer to *theoroi* sent by Kos to festivals at Samothrace and at Itonos in Thessaly (the latter arguably a site of special significance for Kos, which was in some accounts a colony of Thessaly). These *theoroi* are supposed to announce the major Koan festival of Asclepius *en route*, at Khios in the case of those going to Samothrace, and at Argos in the case of those going to Thessaly (which Kent Rigsby has recently argued is not Peloponnesian Argos but Phthiotic Argos in Thessaly).[19] So attending other festivals and announcing one's own are two inextricably linked parts of the operation of the same 'hypernetwork'. No doubt that tendency may get stronger in the Hellenistic period, when festival culture became more intense, but the hypernetwork is probably part of the picture from much earlier on.

If we want to apply 'network theory' to theoric networks, the first decision to make is what sort of 'network theory' to apply. The last five years have seen an explosion in interest in network theory, typified by the work of Stephan Strogatz, whose book *Sync* examines questions of how synchronized or simultaneous action within groups of individuals can be explained in terms of an underlying network, which reaches a certain 'tipping point' that catalyses a cultural change. Other researchers have applied network theory to the problems of human consciousness, neural nets and even quantum mechanics, leading to the hypothesis of the so-called 'quantum chorus'.[20] From another angle, we have the actor-network theory of Bruno Latour, which applies a network approach to institutions and scientific progress.[21]

For the purpose of this paper I will be talking primarily about Social Network Analysis (SNA), a form of network theory that goes back to the 1950s.[22] In SNA social networks are represented either in mathematical form, as a matrix, or in the form of a graph, with 'ties' or 'nodes' connected by 'links' or 'vertices'. (The nodes are usually people, or at least people of a certain type.) The 'ties' are instances of a given type of relationship, which is usually perceived as transitive, so that it is represented in graphs with a one-directional arrow. Sometimes the tie is reciprocal, in which case it is represented by a bidirectional arrow. The number of links connecting to a node is known as its 'degree'; the number of links feeding into a node makes up the 'indegree', while the number of arrows leaving is the 'outdegree'.[23] Network theory can also cover relations between people and events or things: the term for that is an 'affiliation network'. You can also take account of differing values of link: strong relations, weak relations, and so on (representing values by adding numbers to the arrows).

Once the graph or matrix is established, it can be analysed. Truths that were hidden in the maelstrom of empirical data are supposed to become clearer in the light of mathematical abstraction. So one can discern structures—modules, cliques, bridges, centres, and peripheries. One can determine how long the distance is, the 'reach', between any two given nodes. One may look for 'clusters', which means, in the technical language, a group of at least three nodes that are all interlinked,[24] or at the

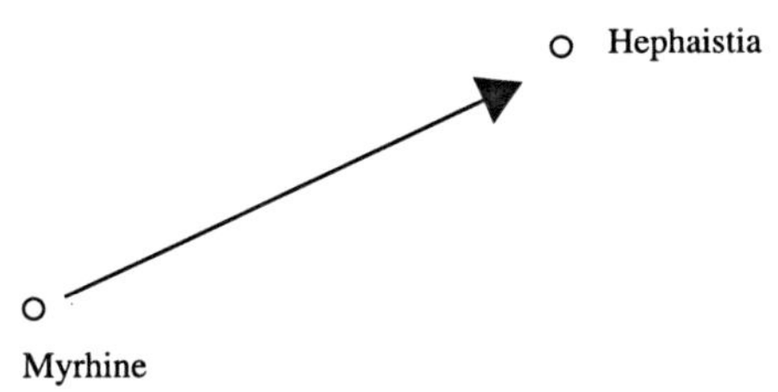

Figure 1 Graph of theoria on Lemnas.

role of different nodes in networks: ones have a lot of links can be said to have 'centrality' or 'betweenness' or 'prestige' or, using Bourdieu's term, 'social capital'.[25]

Two of the more important principles that have been established recently in SNA are the following. First, it was shown by the mathematician Albert-Laszló Barabási that a particularly common way for networks to be organized is for some of the nodes to have far more links than others; in mathematical terms, the probability distribution of links to nodes in a network seems to obey what is called a 'power law', so that the most popular nodes, the 'hubs' in the network, have an exponentially large number of links compared with the average. Such networks are called 'scale-free', because the pattern of the distribution of the links does not follow any particular scale, and they are contrasted with networks with a single scale in which every node has roughly the same number of links.[26] The second principle worth considering is that of the strength of weak ties, which concerns the problem of how a large and extended networks can have a high degree of internal connectivity, even though most of the links within it are between adjacent nodes; Strogatz and Watts proved by mathematical analysis that even a very small number of long-distance ties can be enough to allow the network to be well connected.[27]

We move now to the question whether SNA helps us understand theoric networks in any way. My answer to this will be that it does, at least to a limited degree, in two major respects. First, SNA gives us a convenient way of representing theoric networks, as well as some useful technical terms. Secondly, SNA may help us to identify some abstract features of the operation networks that might otherwise go unnoticed.

SNA usually takes as its primary object people, individuals, but extending it to cities seems reasonable, using the nodes to represent cities, and the tie to mean 'sends a *theoria* to'. For example, to take the simplest possible case, in the third century BC the Athenian cleruch city of Murrhine on the west coast of the island of Lemnos sent a *theoria* across the island to the festival of the Horaia at the sanctuary of the Kabeiroi at Hephaistia in the north.[28] This is shown graphically in Figure 1. And the corresponding matrix is:

	Myrrhine	Hephaistia
Myrrhine	x	1
Hephaistia	0	x

Note that I am presupposing in this that *theoria* is a relation between cities, i.e., the destination of a *theoria* is a city. This makes sense because in Ancient Greece, sanctuaries and festivals tend to be controlled by the local city.[29] In principle we might make the destination a sanctuary or a festival, assigning it to a different category from the other nodes, but that would make things harder when we come to analyse reciprocal relations between cities.[30]

Consider now the more complex cases of the theoric networks from the Aegean that we happen to have some data for, starting with the one based around Hellenistic

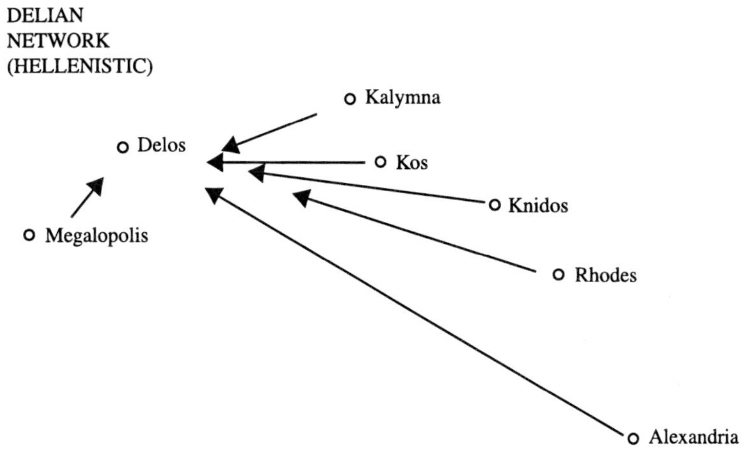

Figure 2

Delos — and note that with data of this limited quantity and quality, diagrams seem to be clearer than matrices (see Figure 2).

We could add values to the lines, representing the number of attested *theoriai* for different periods, as shown in Figures 3 and 4.

Next, consider the case of the Samothracian theoric network, this time a simplified diagram comprising only a few of the *theoria*-sending cities, and note that in this case the dates of most of the delegations are not known precisely (Figure 5).

And finally the theoric network attested for the Alexandrian Ptolemaia is shown in Figure 6. (This time, no numbers are included, because no city is attested as having sent a *theoria* more than once.)

These diagrams give the impression that such festival networks are exclusively two-dimensional phenomena, albeit the size and scope may vary over time (as in the case of

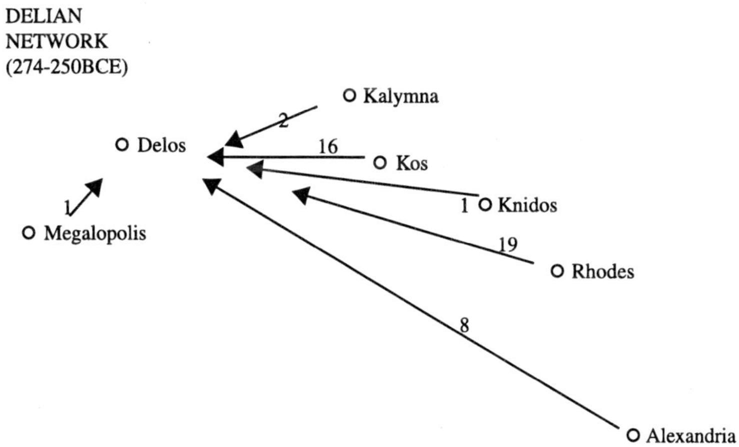

Figure 3

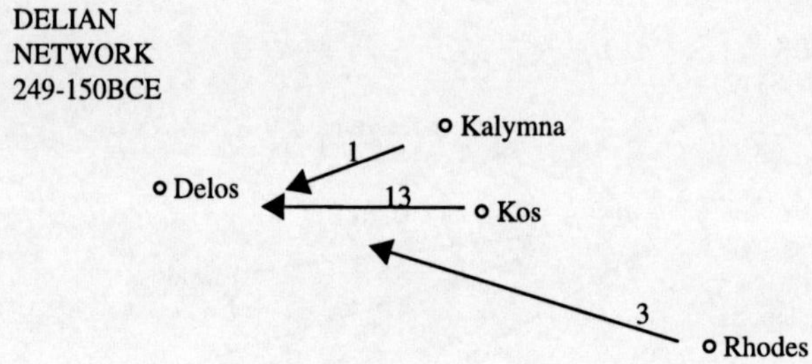

Figure 4

the Delian network). Note, however, that coordination involved in the operation of such networks has a diachronic dimension as well, since the different cities engage with the centre on a synchronized timetable. Recent work in network theory has pointed to synchronized activity as a key factor in the operation of networks.[31]

The festival networks described above have a simple structure consisting of a centre and numerous nodes linked to it by unidirectional ties. *Theoria* is most readily thought of as a unidirectional relation between a city and a sanctuary, but it is sometimes reciprocal. Partly this happens in a sort of virtual sense: for example, Delphi wants to make sure that each Pythia and Soteria festival is attended by numerous *theoriai* from all over Greece, so it sends its own *theoriai* around Greece to announce them. These announcing-*theoriai* fall short of the status of true *theoriai*, but conceptually there is an analogy between the two types. Then in some cases there is true reciprocity, where two cities send a *theoros* to attend each other's festivals; for example, in the Hellenistic networks Kos is known to have sent a *theoria* to Alexandria, and we also know that Alexandria sent one to Kos, and we know that Kos exchanged *theoriai* with Thessaly

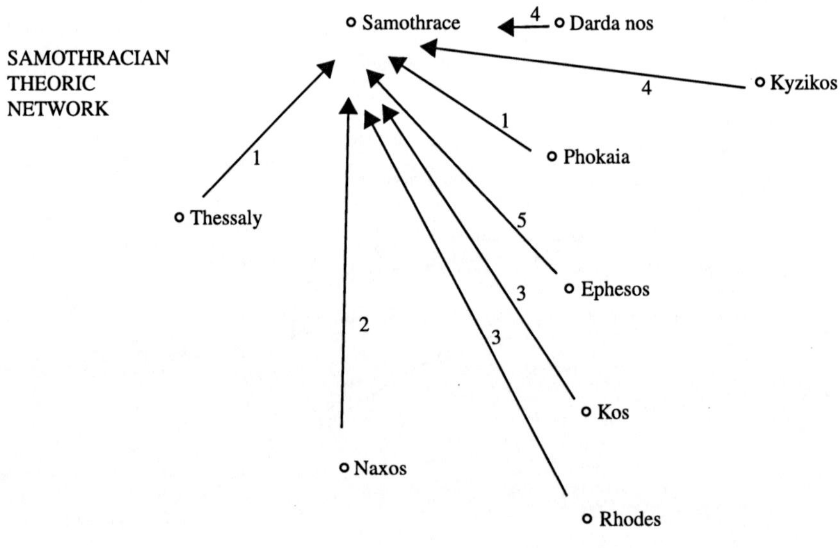

Figure 5

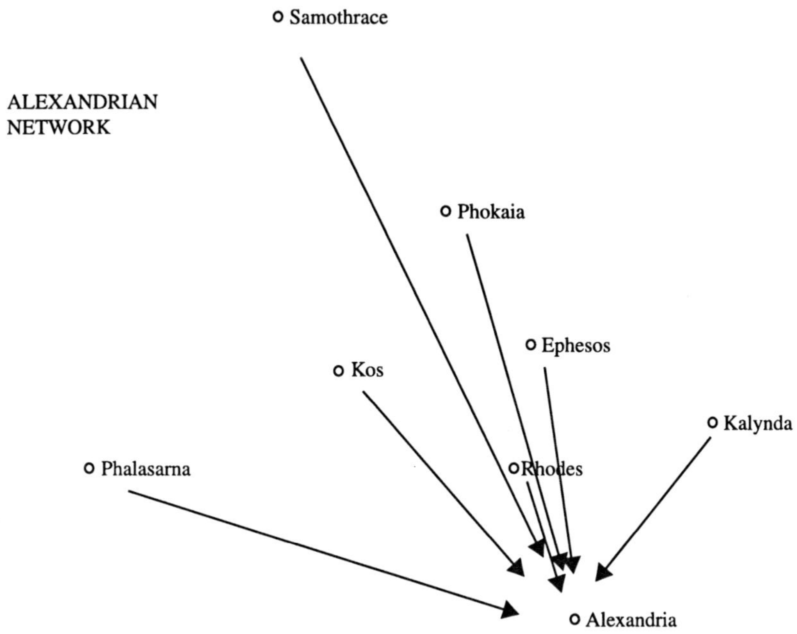

Figure 6

and Samothrace as well.[32] Thus, we may want to think about using double arrows in some cases to signify this fact, and perhaps two types of double arrow to signify the two types of reciprocity: thus, in the case of full reciprocity, we could use a true double arrow as shown in Figure 7.

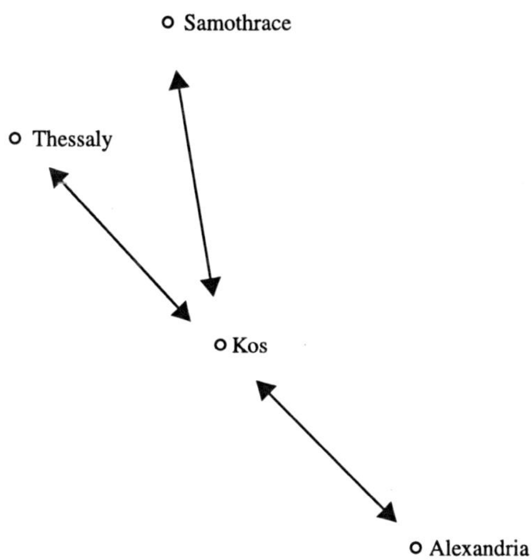

Figure 7 Fully reciprocal theoria.

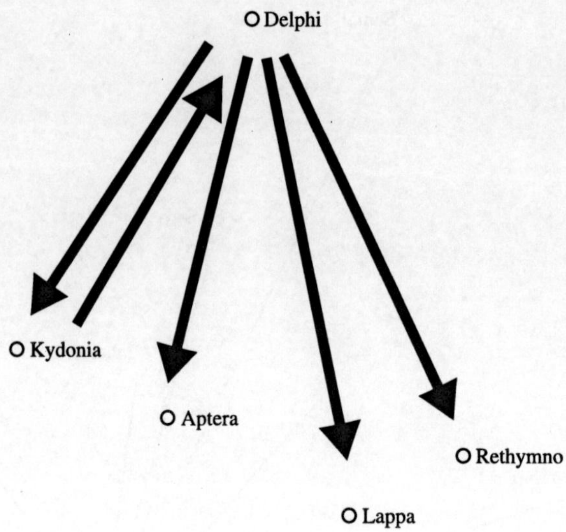

Figure 8 Reciprocity between festival announcers and festival delegates.

In the case of reciprocity between festival announcers and festival delegates, we could signify the two strands of the relationship with different types of arrow, perhaps a weaker line for the announcers. So for example in the case of Delphic festival announcers in the Kydonia-Rethymno region of northern Crete, we might get a map as shown in Figure 8.

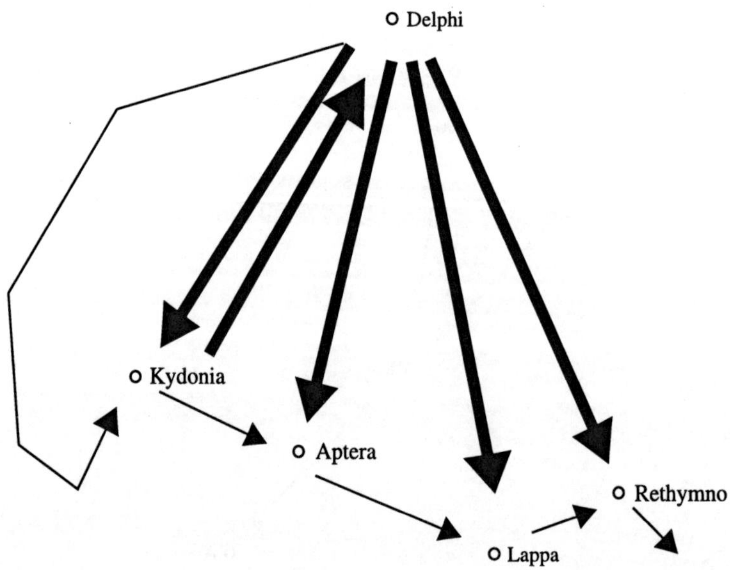

Figure 9 As figure 8, with addition of the route followed by the festival announcers.

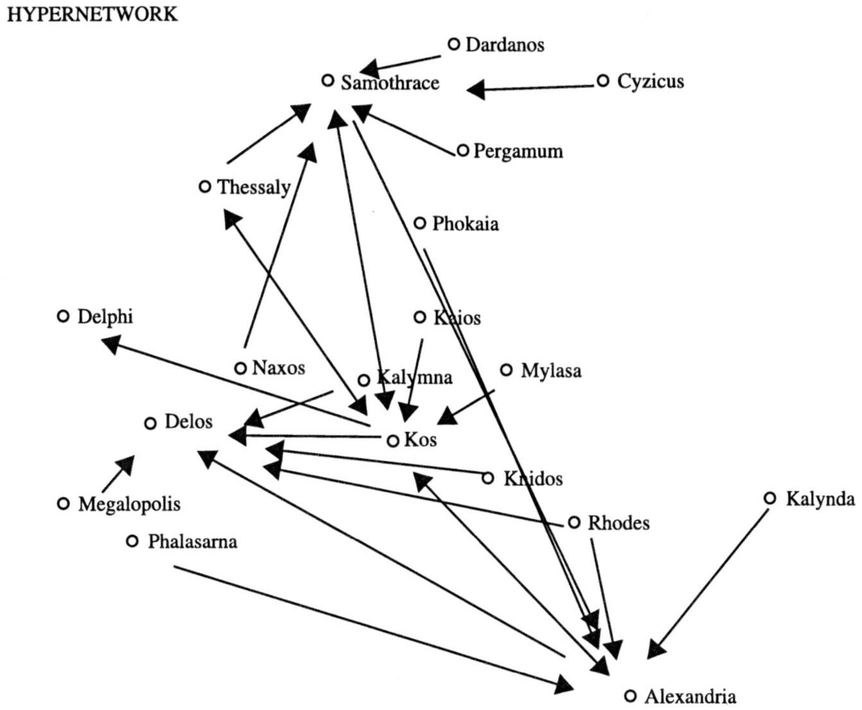

Figure 10

For the sake of completeness, we ought to take account of yet a third type of ties, i.e., the ties constituted by the routes that the festival delegates follow; we might represent this by a thinner line, as shown in Figure 9.

The next step is to apply the same techniques of representation to the 'hypernetwork'. For example, if we combine the attested members of the networks of Delos, Samothrace (some of them only in this case), Alexandria and Kos outlined above, we might end up with something like Figure 10.

This graph of the 'hypernetwork' shows clearly that some cities have a higher 'degree' than others, i.e., they have more links. Some receive more *theoriai*, i.e., they have a higher 'indegree', to use the terminology; those correspond to cities that we knew already to be associated with major super-regional sanctuaries. But some also send more *theoroi* than others, i.e., they have a higher 'outdegree', as, for example, Phokaia sending *theoroi* to both Samothrace and Alexandria, while Rhodes sends them to both Samothrace and Delos. Finally, some states both receive and send, for example Kos.[33] Even without the large volume of Koan epigraphy that survives, the Delian temple inventories would attest more sustained *theoria* from Kos than from anywhere else.[34] In fact, in respect of its 'hyperconnectivity', Kos seems to resemble several other prominent island-states.[35] The overall impression we get is somewhat reminiscent of Barabási's 'scale-free' network, in which the degree distribution between nodes is not even, and some nodes have an exponentially larger number of links than others.

Even in the case of the comparatively rich Koan dossier, our data is far from complete, and its overall patchiness to some extent precludes us from understanding of the finer points of these networks. For example, SNA is very interested in 'clusters', defined as sets of at least three nodes, each of which is linked to all of the others. The hypernetwork outlined above contains no instance of a cluster (although there are cases of bilateral reciprocity), and they are difficult to find anywhere in the surviving evidence. The only documented case of cluster in a theoric network is from an early Hellenistic decree from Euboea setting out regulations for the organization of festivals in the Euboean Confederation.[36] There was a yearly sequence of festivals over an eight-month period from autumn to spring: four instances of the Dionysia held in the four main cities of Karystos, Eretria, Chalcis, and Oreos; then four instances of the Demetreia in honour of Demetrios Poliorketes held in same four cities in reverse order. Each festival in every city is supposed to be attended by *theoroi* from the other three, a rotating pattern implying complete reciprocity between all four cities. This is not a typical network, though, but a special arrangement designed for the internal operation of an unusual regional federation. The 'hypernetwork' of Hellenistic festival culture must surely have contained within it many clusters, since there are likely to have been many *theoria*-inclined cities like Kos, and the reason that they are not explicitly attested is probably that the data is so incomplete.

In spite of the limitations of the ancient evidence, there are, I believe, cases where the insights of SNA may prove productive. For example, another recently developed principle of SNA—the so-called 'strength of weak ties'—can be illustrated by the relationship of Hellenistic Crete and the Panhellenic sanctuaries of the mainland. We have fairly good evidence for regional networks connecting different cities in central and western Crete, and we also know that festival announcers from Delphi visited Crete, announcing the Delphic festivals in dozens of Cretan cities.[37] But we have comparatively little evidence that Cretan *theoroi* visited either Delphi (despite the tradition that the Delphic sanctuary of Apollo was founded from Crete), or other Panhellenic sanctuaries.[38] Perhaps this is no more than a false impression created by the epigraphical evidence, which records links with Panhellenic sanctuaries, in so far as they are part of formal institutional arrangements, and not ordinary contacts. On the other hand, it may well be true that the cities of Crete played a smaller role in the broader Panhellenic festival network than most areas of Greece, something that Pindar already hints at in his epinician for Ergoteles of Himera (the *Twelfth Olympian*), who would never have achieved success on the international athletic circuit if he had stayed in his native Gortyn.[39]

We might then assume that the Cretans were well and truly isolated, but here the Strogatz–Watts principle of 'the strength of weak ties' may prove illuminating: if there was good connectivity *within* Crete at least, even the existence of a few or occasional links between Cretan cities and the Panhellenic sanctuaries would be enough to provide a high degree of connectivity between Crete as a whole and the broader Greek world. The implication of this for the Panhellenic theoric network as a whole could be put like this: the hypothesis that the theoric network served as a primary way of connecting all cities in the Greek world might seem to require that cities usually sent *theoriai* to the Panhellenic festivals held at the major sanctuaries. In fact, it is questionable whether they all sent them regularly, and where we do have reasonably good evidence (as in the case of the naopoeic inscriptions from fourth-century Delphi),[40] it seems to indicate that only a small fraction of cities sent them. However, the hypothesis that the theoric network served as a primary way of connecting Greek cities can still stand, as long as we assume that those cities that did send *theoriai* also had close links with other cities in their region which may not have done so.

Overall, then, SNA seems to provide a way of describing theoric networks, and a number of valuable insights into their structure and functioning. At the same time, however, there are two major limitations that should be mentioned. The first is the limitation of the ancient data, already mentioned: to use SNA the way it is supposed to be used, we are safer relying on primary sources, covering a broader range of sanctuaries, and documenting the movements of *theoroi* more fully. It would also be helpful to know more about patterns of attendance for smaller regional or local festivals.

The second limitation is that although SNA provides valuable insights, I wonder whether there are some fundamental questions about the operation of theoric networks that SNA does not seem able to address. For example, there is the issue of hierarchy in religious networks, mentioned already. Some religious networks seem to be hierarchical in the sense that active participation in them is mandated by central religious or political authority; others, such as the networks organized around the traditional Panhellenic sanctuaries, are not hierarchical in this sense. SNA is able to establish a crude numerical hierarchy in terms of the 'degree' of any specific node, i.e., the number of links going into or leaving the node. But it does not provide any guidance, as far as I can see, about how the level of obligation involved in participation might be taken account of, or modelled. Possibly this might be done by distinguishing different types of tie between cities, one for obligatory relations and one for non-obligatory ones, but I leave the details of how that is to be worked out for another paper.

Acknowledgements

I would like to thank the other participants for many useful comments. I would also like to thank the two anonymous readers. All errors of fact or judgement that remain are, of course, my own.

Notes

[1] On *theoria*, Rutherford, 'Theoria'; Dillon, *Pilgrims*; Ziehen, 'Theoria'; Ziehen, 'Theoros'.

[2] On *theoroi*-announcers, Perlman, *City and Sanctuary*; Boesch, *THEOROS*; Daux, 'Théore et théorodoque'.

[3] For one attempt, see Rutherford, 'The Keian Theoria'; also *SIG*^{3}.585 with Dittenberger's comments *ad loc.*

[4] Chaniotis, *Die Verträge*; see below.

[5] Catchment area: see Bhardwaj, *Hindu Places of Pilgrimage.*

[6] Amphictionies: see Tausend, *Amphiktyonie und Symmachie.*

[7] Colony and metropolis, in general: Graham, *Colony and Mother City*; for Miletus: Günther, *Das Orakel von Didyma*; Debord, *Aspects Sociaux*, 18–20; Ehrhardt, *Milet und seine Kolonien.*
[8] Sommer, this volume.
[9] Plassart, 'La liste des théorodoques'.
[10] Some guidance is provided by the accounts of the *naopoioi* from fourth-century Delphi, which record visitors who made token donations towards the rebuilding of the temple; the geographical range of visitors is surprisingly small: see Rutherford, 'The Keian Theoria'.
[11] Cole, *Theoi Megaloi.*
[12] See Bruneau, *Recherches*, 93–113.
[13] Ptolemaia: see RE s.v.; on Hadra vases: Cook, *Inscribed Hadra Vases*, nos. 1, 3, 4, 7, 8, 10.
[14] Androcles of Phalasarna: see below.
[15] See the excellent survey in Cole, *Theoi Megaloi.*
[16] See Rutherford, 'The Koan-Delian Ritual Complex'; Sokolowski, *Lois sacrées*, no. 156b.
[17] Cf. Malkin, 'Networks', 62; Malkin, 'Apollo'; for a recent parallel from Cyrene, see Dobias-Lalou, 'Voyageurs Cyrénéens'.
[18] Androkles of Phalasarna; Perlman, *City and Sanctuary*, 88 and 91.
[19] Text in Boesch, *THEOROS*, 17; Rigsby, 'Theoroi for the Koan Asklepieia'.
[20] Strogatz, *Sync*, 127ff.
[21] See Latour, *Reassembling the Social.*
[22] See other papers this volume.
[23] See Wasserman and Faust, *Social Network Analysis*, 125ff.
[24] Ibid., 291ff.
[25] For Bourdieu, 'The Forms of Capital'.
[26] Barabási, *Linked*, 65ff.; the principle was first formulated in Albert and Barabási, 'Emergence of Scaling in Random Networks'.
[27] See Strogatz and Watts, 'Collective Dynamics of "Small-World" Networks'.
[28] See Parker, 'Athenian Religion Abroad'.
[29] Sourvinou-Inwood, 'What Is Polis Religion', 297–98. In the case of the Samothracian network we may want to go for the second option, since the Samothracian *polis* seems to have sent its own *theoroi* to the festival: see Cole, *Theoi Megaloi.*
[30] There may, however, be cases where it makes more sense to assign the destination of the *theoroi* to a different category, and that could be accommodated within the framework of network theory also. The term for a network which does not plot the relation between a group of agents, but represents how they are affiliated to an entity of a different category is 'affiliation network'.
[31] For synchronization in networks, see Strogatz, *Sync.*
[32] See below for Kos.
[33] The Koan Asklepieia: *SEG* 33.671 is Koan decree in honour of Kaphisophon, a Koan doctor practising in Alexandria, who led an Alexandrian *theoria* to Kos (220–210 BC); see Herzog, 'Ein koische Artz'; also *SEG* 27.510, a decree for a doctor who heals *theoroi* visiting the *Asklepieion.* The celebrity of the Koan Asklepieia is also implied in the decrees recording recognition of Koan inviolability, for which see Herzog and Klaffenbach, *Asylieurkunden* and Rigsby, *Asylia*, 106–153.
[34] See Bruneau, *Recherches*; Rutherford, 'The Koan-Delian Ritual Complex'.
[35] See Horden and Purcell, *The Corrupting Sea*, 224–30; Constantakopoulou, 'Proud to be an Islander'.
[36] IG12.9.207, especially 18ff. On the Euboean confederation: Le Guen, *Les associations* 1.48, n. 169.
[37] Plassart, 'La liste des théorodoques', cols. 3 (bottom)–4 (top).
[38] For Delphi it comes down to three pieces of evidence: a Delphic *theoros* is attested from Kydonia (IG2.2.844); the city of Elyros in the south-west is known to have had its own

thearodokos at Delphi (IC2.13.1; interestingly, nearby Tarrha also had an old link with Delphi: Pausanias, *Perieg.* 10.16); and a decree from Praisos in the east allows for the possibility of delegations going from there to Olympia and Delphi: IC3.6.7b).

[39] Hornblower, *Thucydides and Pindar*, 191–96.

[40] See Rutherford, 'The Keian Theoria'.

References

Barabási, A.-L. *Linked. How Everything is Connected to Everything Else and What it Means for Business, Science and Everyday Life*. London: Plume, 2003.

Barabási, A.-L., and R. Albert. 'Emergence of Scaling in Random Networks'. *Science* 286 (1999): 509–12.

Bhardwaj, Surinder M. *Hindu Places of Pilgrimage in India (A Study in Cultural Geography)*. Berkeley: University of California Press, 1973.

Boesch, Paul. *THEOROS. Untersuchung zur Epangelie griechischer Feste*. Berlin: Mayer and Muller, 1908.

Bourdieu, Pierre. 'The Forms of Capital'. In *Handbook of Theory and Research for the Sociology of Education*, edited by J. C. Richard. New York: Greenwood Press, 1986.

Bruneau, Phillip. *Recherches sur les cultes de Delos à l' époque hellenistique et à l'époque impériale*. [Bibliothèque de l'École française d'Athènes et de Rome, 217]. Paris: Boccard, 1970.

Chaniotis, Angelos. *Die Verträge zwischen kretischen Poleis in der hellenistischen Zeit*. Stuttgart: Steiner, 1996.

Cole, Susan G. *Theoi Megaloi: The Cult of the Great Gods at Samothrace*. Leiden: Brill, 1984.

Constantakopoulou, Christy. 'The Dance of the Islands: Perceptions of Insularity in Classical Greece'. Ph.D. thesis, University of Oxford, 2002.

———.'Proud to Be an Islander: Island Identity in Multi-Polis Islands in the Classical and Hellenistic Aegean'. *Mediterranean Historical Review* 20 (2005): 1–34.

Cook, Brian F. *Inscribed Hadra Vases in the Metropolitan Museum of Art*. New York: Metropolitan Museum of Art, 1966.

Daux, George. 'Théore et théorodoque'. *Revue d'études Grecques* 80 (1967): 293ff.

Debord, Pierre. *Aspects Sociaux et Économiques de la Vie Religieuse dans l' Anatolie Gréco-Romaine* (EPRO 88). Leiden: Brill, 1982.

De Nooy, Wouter, Andrej Mrvar, and Vladimir Batagelj. *Exploratory Social Network Analysis with Pajek*. Cambridge: Cambridge University Press, 2005.

Dillon, Matthew. *Pilgrims and Pilgrimage in Ancient Greece*. London: Routledge, 1997.

Dobias-Lalou, Catherine. 'Voyageurs Cyrénéens'. In *Voyageurs et Antiquité Classique*, edited by H. Duchêne. Dijon: Editions universitaires de Dijon, 2003.

Ehrhardt, Norbert. *Milet und seine Kolonien*. Frankfurt am Main: Peter Lang, 1988.

Elsner, J., and I. Rutherford. *Pilgrimage in Greco-Roman and Early Christian Antiquity. Seeing the Gods*. Oxford: Oxford University Press, 2005.

Graham, Alexander J. *Colony and Mother City in Ancient Greece*. Manchester: Manchester University Press, 1971.

Günther, Wolfgang. *Das Orakel von Didyma in hellenistischer Zeit [Ist. Mitt. Beiheft 4]*. Tübingen: E. Wasmuth, 1971.

Hall, Jonathan M. *Hellenicity. Between Ethnicity and Culture*. Chicago: Chicago University Press, 2002.

Herzog, R. 'Ein koischer Artz am Ptolemäerhof. Kaphisophon Sohn des Leibarztes Philippos', *Parola di Passata* 38 (1983): 61–4.

Herzog, Rudolf, and G. Klaffenbach. *Asylieurkunden aus Kos*. Berlin: Akademie-Verlag, 1952.

Horden, Peregrine, and Nicholas Purcell. *The Corrupting Sea. A Study of Mediterranean History.* Oxford: Blackwell, 2000.
Hornblower, Simon. *Thucydides and Pindar. Historical Narrative and Power in Epinikian Poetry.* Oxford: Oxford University Press, 2004.
Latour, Bruno. *Reassembling the Social: An Introduction to Actor-Network Theory.* Oxford: Oxford University Press, 2005.
Le Guen-Pollet, Brigitte. *Les associations de Technites dionysiaques à l' époque hellénistique.* 2 vols. Nancy: Association pour la diffusion de la recherche sur l'antiquité, 2001.
Ma, John. 'Peer Polity Interaction in the Hellenistic Age'. *Past and Present* 180 (2003): 9–39.
Malkin, Irad. 'Apollo Archegetes and Sicily'. *Annali della Scuola Normale Superiore di Pisa* 3.16 (1986): 959–72.
———.'Networks and the Emergence of Greek Identity'. *Mediterranean Historical Review* 18 (2003): 56–74.
Parker, Robert C. T. 'Athenian Religion Abroad'. In *Athenian Democratic Accounts Presented to David Lewis,* edited by R. Osborne, and S. Hornblower. Oxford: Oxford University Press, 1994.
Perlman, Paula. *City and Sanctuary in Ancient Greece. The Theorodokia in the Peloponnese* (*Hypomnemata* 121). Göttingen: Vandenhoeck und Ruprecht, 2000.
Plassart, Andre. 'La liste des théorodoques'. *Bulletin de correspondance helleniue* 45 (1921): 1ff.
Rigsby, Kent. *Asylia. Territorial Inviolability in the Hellenistic World.* Berkeley: University of California Press, 1996.
———.'Theoroi for the Koan Asklepieia.' In *The Hellenistic Polis of Kos. State, Economy and Culture,* edited by K. Höghammer. Uppsala: Uppsala Universitet, 2004.
Rutherford Ian. 'Theoria'. *Der Neue Pauly,* 12/1 (Stuttgart: J.B. Metzler, 2002): 398–400.
———.'The Keian Theoria to Delphi: Neglected Data from the Accounts of the Delphic *Naopoioi* (CID2.1–28)'. *Zeitschrift für Papyrologie und Epigraphik* 147 (2004): 107–14.
———.'The Koan-Delian Ritual Complex: Apollo and Theoria in a Sacred Law from Kos'. In *Apolline Politics and Poetics* (forthcoming).
Scott, John. *Social Network Analysis. A Handbook.* London: Sage, 2000.
Sokolowski, Francizek. *Lois sacrées de cités grecques.* Paris: Boccard, 1969.
Sourvinou-Inwood, C. 'What Is Polis Religion'. In *The Greek City from Homer to Alexander,* edited by P. Murray, and S. Price. Oxford: Oxford University Press, 1990.
Strogatz, Steven. *Sync, The Emergence of Spontaneous Order.* Harmondsworth: Penguin, 2003.
Strogatz, S., and D. Watts. 'Collective Dynamics of "Small-World" Networks'. *Nature* 393 (1998): 440–2.
Tausend, Klaus. *Amphiktyonie und Symmachie. Formen zwischenstaatlicher Beziehungen im archaischen Griechenland* (Historia, Einzelschrift 73). Stuttgart: Franz Steiner, 1992.
Wassermann, Stanley, and Katherine Faust. *Social Network Analysis: Methods and Applications.* Cambridge: Cambridge University Press, 1994.
Watts, Duncan. *Six Degrees. The Science of a Connected Age.* London: Vintage, 2003.
Ziehen, Ludwig. *REA* 10. 2228-s. *Theoria,* 2239-s. *Theoros,* Stuttgart: J. B. Metzler, 1934.

Did the Delphic Amphiktiony Play a Political Role in the Classical Period?

Simon Hornblower

There can be few more attractive candidates for the modern label 'network' than the ancient Delphic amphiktiony.[1] It was an organization of geographically distinct communities, some of them distant; a network with two hubs, Delphi and Anthela. It met twice a year, spring and autumn, to transact essentially religious business. I shall not set out uncontested facts or events. I shall instead concentrate on the two words 'essentially religious'. Was the amphiktiony in the two centuries before Alexander the Great ever more than religious? Could it be, was it ever, manipulated for political ends? I shall argue that it could be, and that it was.

A few years ago John Davies, reviewing recent work on Greek history, spoke of religion as having 'rejoined the mainstream' after a long period of being studied in isolation.[2] I think this is true. This in turn has at the extreme end had two consequences. First, sometimes things get seen as religious when they are not. Second, there is a scornful reaction against earlier approaches which are said to have been too secular, political, and 'modern'.

As for the first consequence, not everything was religious. The Greeks could distinguish perfectly well between sacred and secular money. And they had the conceptual vocabulary to say religious as opposed to non-religious. Two examples: Herodotus says that Spartans value the things of the gods more than the things of men; and Thucydides that the Spartans march to the sound of pipes not for the sake of the divine, τοῦ θείου χάριν, but to keep in step.[3]

As for the second point (political versus religious?), my paper deals with a case study of this. But let us not make the mistake of thinking the difficulty is confined to ancient Greece, or that it would be resolved by, say, a miraculous quadrupling of the quantity of our evidence. Take the English Civil War of the seventeenth century AD, a richly documented period of history compared to what we are used to. The English Marxist Christopher Hill unexpectedly praises the royalist historian Clarendon for seeing the struggle in terms of class struggle and of regional interests. Other modern historians like Firth have been severe on Clarendon precisely for his neglect of the religious factor.[4] These intellectual divides are, I suggest, not the kind of dispute which can be settled by further accumulation of data—or even settled at all.

Having insisted that Greeks knew how to say 'this is not religious, it is something else', we might still want to say that 'political' is not the right word for that something else. When talking of the Panhellenic sanctuaries, it might be better to talk of enhancing the prestige there of one's own community, rather than of control, manipulation and so on. But when the quest for prestige and reduction or denial of the prestige of others leads to exiling of individuals or factions, to *stasis*, even to war, it seems legitimate to use the word political in the hard sense, and that is what I shall argue for as well. But let us not forget that, as F. Lefèvre said in 1998 in his monumental and authoritative study, the organization of the Pythian games was the most prestigious of the tasks of the amphiktiony.[5] Success in that sphere conferred more than just athletic, equestrian or musical prestige.[6] Pindar in his Fourth Pythian ode says τῷ μὲν Ἀπόλλων ἅ τε Πυθὼ κῦδος ἐξ Ἀμφικτιόνων ἔπορεν ἱπποδρομίας (Apollo and Pytho granted glory [to King Arkesilas of Kyrene] at the hands of the Amphiktions in horse-racing).[7] We might call this the soft sense of 'political'. Soft, but nonetheless very important, and I shall return to this importance at the end. Implicit in the soft political sense is, I concede, what we might call a zero-sum model: my prestige goes up as yours goes down. That may, however, be too crude. A similar problem was discussed a century or so ago with reference to the alleged neutrality of Hellenistic Delos. Tarn denied this neutrality; his critics pointed to the co-existence of royal festivals there. If the matter was simply political (they argued), one king would have abolished the festival of his rival. That, replied Tarn, would have been sacrilege, so it did not happen; but this restraint did not stop the festivals having political meaning.[8]

The Fourth Century BC

Introduction and Background: The Situation after 370 BC

I now close in on classical Delphi and its amphiktiony, starting with the fourth century, and working back to the hazier fifth century in the second section. Here are a few statements of the orthodoxy about the post-370 position. First, Beloch: 'After Jason's death [in 370], Boiotian influence in the Delphic amphiktiony was dominant, and Thebes did not hesitate to exploit this influence for its own political ends.' Second, Rhodes in the *Cambridge Ancient History*: 'More extensive [than the Panionion] was the Amphictyony of Anthela and Delphi, which Boeotia exploited as a body of political importance after Leuctra [371]'; in his new history of the classical Greek world he says that the Thebans made the amphiktiony an instrument of power after Leuktra. Third, Arnush: 'Thebes, Thessaly, Philip and Alexander of Macedon, and the Aetolian *koinon* all managed to wrest control of the amphiktiony temporarily from its members, assisted by the acquisition of seats on the *synedrion*.'[9] This is what I call maximalism. My own view of the fourth-century amphiktiony and the Third Sacred War was formed by reading a superb but never published 1961 Oxford B.Litt. thesis on maximalist lines by Christopher Ehrhardt, David Lewis's first graduate student[10] (1961 is not so long ago, but it is a sad fact that all four people involved are now dead. That is, the supervisor, Lewis, the examiners, Peter Brunt and Guy Griffith; and now the candidate himself in Otago, New Zealand, two years ago).

Now this orthodoxy has been challenged, first by Buckler in 1985[11] and then in a more extreme minimalist form by Bowden in 2003.[12] Buckler sought to reduce the responsibility of the Thebans for the war, and to deny Pomtow's theory, as he called it, of Boiotian-Phokian enmity, as reflected in Delphian politics. But he was far from denying a political role for the amphiktiony. On the contrary, he said 'it was in strict terms a religious organization, but politics always played an important part in its deliberations and operations'.[13]

Bowden goes much further. His article concludes 'attempts to claim a larger, political, role for the Amphictyony in this period fail to convince because the evidence cannot support them'.[14] The 'period' he refers to is the classical period down to 346. I shall argue that, on the contrary, the evidence does support the more usual view. In particular I shall re-examine some key items, events and episodes of the period 373–356—the 'long 360s', as I call it—which Bowden puzzlingly omits altogether from his account. He does this by passing from the Second Sacred War to the Third, a startling jump of nearly a century, 450 to 356. But the turbulent decade of the 360s cannot so be ignored, the main decade of what Pomtow called the 'magna stasis Delphorum'. That was in the third edition of the *Sylloge inscriptionum graecarum*, but he was reprising a fundamental article of 1906.[15]

Before I get on to the facts, I make a concession. What do we mean by 'role'? We might hold that a very short usurpation (loaded word) of harmless religious functions by a malignantly disposed individual (like Jason) or group or *ethnos* or polis says nothing about the normality. (Note Arnush's 'wresting *temporary* control'.) To that

I reply: it depends how brief or lengthy was the usurpation or period of control leverage or whatever. Nobody would dispute that in the period of Aitolian domination (most of the third century BC) the amphiktiony's role was political. Not everyone liked it; Sparta under Areus tried to dislodge them in 281; the Greek allies denounced it in 220 (the Social War), as Polybius tells us: 'They also added a clause in the decree engaging to recover for the Amphiktyonic council its ancient laws, and its authority over the Delphic temple, of which it had been deprived by the Aetolians, who wished to control the affairs of the temple themselves' (trans. Paton).[16] But political the Aitolian takeover and domination was. To say this is not to deny other aspects. It would, for instance, be wrong to refuse all religious significance to the Soteria festival reorganized by the third-century Aitolians.

Excellent recent francophone work on the amphiktiony—two books by Lefèvre, and one book by Sánchez,[17]—shows no consensus on detail;[18] but the shared trend is minimalist and anti-historicist. On the fifth century, I shall be mainly replying to Lefèvre *CID* IV (2002) who, in a seven-page appendix, argues with the Simon Hornblower of 1991/92. But I do not want to go over old ground in detail, or just to restate my 1991 position. I do not see things quite as I did fifteen years ago, or I would not have written this paper.

Now for my story. Two important events occurred in the latter years of the 370s. First, the temple of Apollo burned down in 373. Second, the Thebans defeated the Spartans at the battle of Leuktra in 371. I assume these events were unconnected, though one could write an entertaining conspiracy theory novel starting from the opposite premise. Ancient Greek temples burnt down remarkably often. All those oil-soaked wool fillets, *stemmata*, were a terrible fire risk,[19] though another tradition says that Delphi in 373 was rocked by an earthquake. Anyway, we think of archaic Delphi, the Argive Heraion in the 420s and the temple of Athena Lindia at Rhodes in 392.[20] One consequence was that records might be damaged. If we ask why we know so much more about fourth-century Delphi than about the Delphi of the fifth or sixth centuries, the answer may in part be the fire of 373 (cf. Lindos). The fragmentary new amphiktionic decree published in 1994 was re-engraved before the Third Sacred War of 355, and is thought to have been 'originally engraved on the temple, before its destruction in 373'.[21] Not every document will have been so lucky. In 1998 John Davies, in a fine, meticulous study of Delphic finance administration and realpolitik (his words) acutely noted that without the Herodotean tradition 'we would know nothing about the Alkmaionid rebuild [in the late sixth century] beyond the surviving physical fragments'.[22] One result of the 373 fire and the rebuilding programme was probably the appointment of a new board of amphiktionic naopoioi. They started work in 367/66. That this rebuilding was itself a prestige point is indicated by the inscription recording Athenian negotiations with Dionysius I of Sicily in 369/68, who is there related to have written to Athens 'about the building of the temple and the peace'.[23] Ehrhardt himself spoke of the amphiktiony reviving after 371, 'quite suddenly'.[24] This may be right, but we should allow for lost evidence before that.

But the bigger world event was surely Leuktra, and not only Leuktra but its consequences in Thessaly in the 360s. By their alliances with the Magnesians and Phthiotic Achaians in 364, after the battle of Kynoskephalai, the Thebans had achieved a remarkable position within the amphiktiony: of the twenty-four votes, twenty-two were now exercised by themselves or their allies, leaving only the votes of Athens and of the Dorians of the Peloponnese. Even this last vote was sometimes in the hands of one of Thebes' Peloponnesian allies. The normal view is that the amphiktiony now started to operate in a political, and specifically pro-Theban, way. Two pieces of evidence, which Bowden does not mention at all, have always played a central part in the argument.

The (Original) Amphiktionic Condemnation of the Spartans

The first item is found in Diodorus book 16, in his account of the outbreak of the Third Sacred War in the mid-350s. The aspect which concerns me is a retrospective mention, an analepsis. Diodorus' sources for the war itself (a war which Bowden does of course discuss; see below) are an old problem.[25] I have not space here to say more than that the Diodoran material is of high quality, not to be discarded with Bowden, and that, like Ehrhardt, I accept Hammond's demonstration, in one of his best articles, that there is no major narrative doublet.[26] What Diodorus says (16.23, cf. 29) is that the Thebans, after defeating the Spartans in 'the Leuctrian war', brought suit against them before the Amphiktions. The charge was that Phoibidas the Spartan had seized the Kadmeia, and the Amphiktions assessed a fine of 500 talents. Now this Spartan outrage, the seizure of the Theban acropolis, took place as long ago as 382. When did the amphiktionic vote condemning the Spartans occur? '*At the time of*' (ch. 29) or '*after*' (ch. 23) the Leuktrian war are both vague temporal indicators. How much later than 371 and the battle itself? In 1957, Sordi made a good case for 366 (but see below).[27] And 361 is certainly possible. Buckler prefers the 350s, just before the Phokian seizure of the sanctuary.[28] But this cannot be right. The 'Leuktrian war' cannot be extended beyond 362 and the battle of Mantineia. On Buckler's late dating, Diodorus' source would not have bothered to mention 'the war' as a chronological indicator at all. We surely need a date in the 360s.

Buckler treats Diodorus as saying, at both chs. 23 and 29, that the fine was first imposed 'after' the Leuktrian War which ended in 362. That is certainly what Diodorus says at ch. 23. (paras. 2–3: Λακεδαιμονίων πρὸς Βοιωτοὺς διαπολεμησάντων τὸν Λευκτρικὸν πόλεμον καί καταπολεμηθέντων οἱ μὲν Θηβαῖοι διὰ τὴν κατάληψιν τῆς Καδμείας δίκας μεγάλας ἐπαγάγοντες τοῖς Λακεδαιμονίοις ἐν᾽Αμφικτύοσι κατεδίκασαν αὑτοὺς πολλοῖς χρήμασιν, οἱ δὲ Φωκεῖς ἐπεργασάμενοι πολλὴν τῆς ἱερᾶς χώρας τῆς ὀνομαζομένης Κιρραίας δίκας ὑπέσχον ἐν Ἀμφικτύοσι καὶ πολλοῖς ταλάντοις κατεκρίθησαν).

But ch. 29.2 is not so easy. Diodorus seems to say there that the condemnation happened '*in* the Leuktrian War' (ἐν τῷ Λευκτρικῷ πολέμῳ), which would allow us to go some years further back. The translation has been disputed, so the point must be elaborated. The Greek is ἐν τῷ Λευκτρικῷ πολέμῳ

Θηβαῖοι καταπολεμήσαντες τοὺς πολεμίους δίκην ἐπήνεγκαν εἰς Ἀμφικτύονας κατὰ τῶν Σπαρτιατῶν. Here the word order surely indicates that καταπολεμήσαντες τοὺς πολεμίους is in parenthesis. Sherman's Loeb translation inserts commas in the English, to make this clear: 'In the Leuctrian War the Thebans, after defeating the enemy, brought suit before the Amphictyons against the Spartans.' Here Diodorus seems, if we may expand him, to mean that 'during the Leuktrian War, the Thebans, after defeating the Spartans [namely, at the battle of Leuktra in 371], brought suit', and so forth. Buckler[29] ignores the word order when he says that 'the aorist participle καταπολεμήσαντες with the aorist main verb points to a time after the war, not during it, for the Theban prosecution of Sparta'.

In any case, Diodorus is explicit that the condemnation of 356 repeated an older one. The Spartans, says Diodorus, had not paid the fine—which was a colossal 500 talents—'within the period laid down by the laws', οὐκ ἐκτινόντων κατὰ τὸν ὡρισμένον ἐκ τῶν νόμων καιρόν. So the Thebans in 356 brought suit against them for double that amount. We have no idea how long was the period 'laid down by the laws'. Was there a standard period within which you had to pay any fine, large or small? Surely not. Common sense would indicate that an amount as large as 500 talents would need years to collect, so that ὡρισμένον ἐκ τῶν νόμων καιρόν would mean 'the period which had been stipulated in accordance with the laws'.[30]

There is a small difficulty, noted by Ehrhardt.[31] The Spartan state paid contributions to the naopoioi in autumn 360 and autumn 358,[32] and individual Spartans (including a woman, Philostratis) in spring 360 and spring 358.[33] Would the Spartans, collectively or individually, have paid contributions like this when under condemnation for a heavy and in their opinion unjust fine? Ehrhardt rightly refuses to see this as evidence for a very late date, at or after autumn 358, for the condemnation. ('Though this theory cannot be disproved, it seems improbable, for the offence had by then been committed twenty-four years earlier, and there had been ample opportunity to condemn it in previous years.') Instead, he offers an ingenious speculation of his own. The first levy (ὁ πρῶτος ὀβολός), he conjectures, was assessed on the assumption that Sparta would pay the fine, and that their 500 talents would, together with the proceeds of the first levy, meet the costs of the temple. Levy and condemnation were both about 366. When the Spartans did not pay, and it was recognized that they could not be forced to, the second levy was raised (first collected in 361). The Spartans paid this on the understanding that the fine would be ignored. But it was then remembered at the time of the condemnation of the Phokians in 356.

Most recently, Lefèvre[34] wonders if 371 itself is possible, straight after Leuktra, when Cicero says (perhaps from Ephorus?) that the Spartans accused the Thebans before the Amphiktiony of erecting a trophy on the battlefield (Cic. *De inv.* 2.69). The Theban prosecution will then have been a reply. But then we would have expected μάχη, not πόλεμος, 'battle' (of Leuktra) not 'war'. However that may be, the longer the interval we posit since the original capture in 382, the more the vote will look like political point-scoring. To be sure, some amphiktions may, like Xenophon, have felt religious outrage at Sparta's occupation of the Kadmeia. Specifically, it was perpetrated during, and was thus a violation of, Demeter's festival the Thesmophoria, and Demeter Amphiktionis was patron goddess of the amphiktiony.[35] But then why did these pious folk wait at least twelve years before issuing the condemnation? This (to return to the general topic raised at the beginning of this paper), is a test-case: is this politics? Or 'just' religion? On an extreme view, the amphiktionic condemnation of the Spartans was simply and solely generated by the offence against Demeter. I do not think this view will do on its own.

To conclude and sum up: the *first* condemnation of the Spartans did not take place either as late as the 350s, or as early as 371. It belongs in the 360s somewhere, the period of Boiotian preponderance in the amphiktiony.[36] Certainty is not possible. There is a good case for 366, if we *either* press the possible implication of Diodorus 16.29 and put the condemnation 'in' not 'after' the Leuktrian war, *or* if (as many scholars have done without saying so explicitly)[37] we think that Diodorus' source wanted to say that the condemnation was after, but not very soon after, the actual battle of Leuktra in 371. The alternative date will be 361, after the battle of Mantineia, and two years after the exiling of Astykrates, to which I now turn.

The Amphiktionic Condemnation of Astykrates and Co.

The second item is epigraphic. It is a remarkable Athenian inscription securely dated to 363. I have used Michael Osborne's 1981 text. The fourth-century Athenian decrees are currently being re-edited by Angelos Matthaiou as part of the new *IG* 2^3 project, and though he has not yet properly dealt with this text, he has looked at the stone for me, and he has kindly offered me some comments on details.[38]

The Athenian boule successfully recommends to the assembly that Astykrates of Delphi be given citizenship. He and his associates have been exiled from Delphi on a proposal of a named Thessalian hieromnemon and have had their property confiscated (lines 20–21). The Athenians (17) say this exile was 'contrary to the laws of the amphiktions', and pronounce the condemnation null and void, τὰς δίκας ἀτελεῖς εἶναι (23–25). This last part is amazing, but has been little discussed in works on Athens and its democracy: how could the Athenian assembly purport to cancel an amphiktionic decision? We can also ask, why go through the amphiktiony? Ehrhardt thought the accusers could not rely on the Delphic courts to bring in the verdict they wanted, so used the 'rather cumbrous and dubiously legal Amphiktyonic procedure'.[39]

Astykrates' later career is fortunately well attested from other Delphic inscriptions, through which we can follow what happened to his sold-up property and that of his associates.[40] His fortunes went up and down spectacularly. He and his associates returned to Delphi from exile after 356. This has been doubted, most recently by Sánchez, but the clinching onomastic item is the rare[41] Delphian name Hag/Hegesarchos. Prosopographically, he is the key to the affair. He was one of those banished with Astykrates in 363 (see line 58 of the Athenian inscription), but was then a naopoios in 353, during the period of Phokian control of the sanctuary.[42] What happened to the group after 346 is hard to establish precisely, but their property seems to have been confiscated again.[43] From Pomtow to M. Osborne, Astykrates has been regarded as leader of an Phokian faction, and his exile as political. I am aware that some of the pro-Theban activity at Delphi in this period was the work of the Delphians rather than the amphiktiony. But the difficulty of disentangling the two bodies was remarked already by Busolt-Swoboda in what remains one of the most useful accounts of the amphiktiony and what it did; and their close relationship has been explored in more detail recently by Arnush.[44] Thus Busolt-Swoboda noted that grants by Delphi of promanteia — the right of privileged oracular consultation — were often accompanied by a similar and simultaneous gesture by the amphiktiony (though the amphiktiony did not, as far as we can see from inscriptions, actually grant promanteia itself).[45] This being so, we should not be too pedantic about denying an amphiktionic aspect to the promanteia granted to the Thebans in precisely summer 363.[46] This recalls that granted to Kroisos and the Lydians in the sixth century.[47]

The Astykrates inscription from Athens is of great importance. I do not know what Bowden means[48] when he speaks of the 'near-invisibility of the Amphiktiony in accounts of the history of Greece' between 479 and 356. I think this means modern histories, and even this is simply not true. All histories of the fourth century have a fair amount to say about the Amphiktiony when describing how the Thebans threw their weight about in central Greece in and after the 360s.

Why was Astykrates exiled? To be sure, we are not told in detail, but his associates were accused of wronging, ἀδικεῖν, some Delphians (l.26). This is unspecific, but it does not obviously look as if Astykrates was a religious offender. Indeed Bousquet in *CID* 1989, despite his disparagement of Pomtow, was sure that the exiles were banished for purely political reasons, since the Athenians would not have taken them in if they were ἐναγεῖς, under a curse.[49] Sánchez disagreed, on the grounds that Astykrates' property passed to Apollo not the amphiktiony, as ἱερά not κοινὰ χρήματα. In other words they were ἐναγεῖς in all but name.[50] But I think Bousquet and now Lefèvre[51] are right, and that the Athenians were not likely to have said that his exiling was contrary to amphiktionic law if his offence had been religious. The natural assumption is that the exile was political. The Athenians will then be saying that the amphiktiony should not behave like that. So you can either stress that some people thought the amphiktiony *should* not play politics, or you can stress the fact that it *did* (we may recall Polybius on the declaration of 220 about the Aitolians, above). Personally I put more weight on what it did than on what people believed or pretended to believe it was supposed to do. But ideally the case must be cumulative and extend over time.

The Outbreak of the Third Sacred War

With this I turn to the outbreak of the Third Sacred War. True, it does not without circularity *confirm* the above picture. From the active–passive switch in Diodorus (the Thebans indicted the Spartans for seizing the Kadmeia, but the Phokians *were indicted* in the amphiktiony for cultivating sacred land)[52] Buckler concludes: 'Thebes did not indict the Phokians.' The linguistic observation is shrewd and, as far as it goes, correct (though it does not settle the question, 'what exactly did Diodorus' *source* say?'). But the historical conclusion does not necessarily follow, not least because Diodorus is not quite the only source: see below. Buckler suggests the accusers were the Delphians, Sánchez the Thessalians; both go beyond Diodorus.[53] Whoever was the technical accuser, it is surely very plausible to think that the Thebans and their Thessalian allies were *behind* the prosecution and its success. We cannot in analysing the causes of the war ignore Astykrates, Athenian–Theban rivalry of the 360s, and Theban resentment of Sparta as shown in the amphiktionic condemnation for the Kadmeia episode. The war did not happen one day out of nothing earlier. Wars usually do not. The Thebans are usually—and I think rightly—regarded as wanting to punish the Phokians for refusing to join Epaminondas' last invasion of the Peloponnese (Mantinea campaign, 362). This refusal is reported with careful explicitness by Xenophon near the end of his *Hellenica*.[54] Buckler seeks to get rid of this by the argument that the Phokians were strictly within their treaty rights.[55] This misses the point, which has to do with perceptions—how the Thebans *viewed* their 'sulky' Phokian ally.[56] There is also a little matter of narrative technique. *Why* does Xenophon go out of his way to mention the Phokian refusal like this? His *Hellenica* stops in 362, but from the mention of Tisiphonus of Pherai (6.5.52) we know he was aware of events of the mid-50s.[57] I suggest that his emphasis on Phokian–Theban tension in book 7 is what Irene de Jong calls a 'seed', an item whose true significance will become clear only later[58]—in this case some years later than the end of the narrative proper. I refer to the outbreak of the Third Sacred War. We should recall also the long-standing hostility between the Phokians and the Thessalians (Hdt. 8.30; Paus. 10.2.1), with whom the Thebans were at this time hand in amphiktionic glove.[59]

Finally, in this section, I look at some literary sources other than Diodorus. Their evidence is bittier than his, but not necessarily valueless on that account. Several of them point to Theban–Phokian hostility in one way or another.

I begin with an item which Buckler does not even think worth mentioning. It is a fragment of Duris (*FgrHist* 76 F2 = Athenaeus 560B). Athenaeus says that the greatest wars started because of women. He starts, inevitably, with the *Iliad*, and goes on to say that 'the so-called Sacred War happened because of another married woman (γαμετή), as Duris says in the second book of his *Histories*, when a Theban woman called Theano was abducted by a certain Phokian. This war too lasted ten years and was ended by Philip in the tenth year. For that is when the Thebans took Phokis'. Of this Jacoby in his commentary remarks that 'Frauenraub', theft of women, may indeed have featured among the offences alleged against individual Phokians.[60] Why should we discount this item completely? Buckler evidently does discount stories of this general sort. He discusses Aristotle's story (1304a 10–13) that an internal Phokian quarrel over an *epikleros* (heiress) gave rise to the Sacred War—no mention of Thebans in Aristotle, to be sure. Buckler says that Aristotle has 'blown both incidents out of proportion'. The other incident he has in mind here ('both') is the Mytilene revolt of 427, where Aristotle says the precipitating *stasis* arose from trouble over heiresses

(Ib. 4–7). But Thucydides' Mytilene narrative (says Buckler, citing 3.1.2–30) 'deals only with political causes'.[61] This is too simple a dichotomy. True, Thucydides did not mention Mytilenean heiresses. But he was well aware that quarrels over marriage matters could lead to great conflicts: he says just this about the origins of the disastrous Athenian involvement in Sicily of 415–13 BC, which began with a quarrel between the Egestans and Selinuntines of western Sicily about 'marriage matters and disputed land' (6.6.2: περί τε γαμικῶν τινῶν καὶ περὶ γῆς ἀμφισβητήτου). Valuable title to land and property might go along with a betrothal or marriage—in other words, Thucydides' expression may be a hendiadys. There is also the matter of honour, in a touchy Mediterranean society:[62] it was not at all absurd of Homer and Herodotus to suggest that violent male reaction to the abduction of women might escalate into war. To return to Duris and Theano, this episode may be only a small part of the whole story, but it would be arrogant to assert that it is baseless, and as far as it goes, it supports the idea of Theban–Phokian tension.

Isokrates (5.54), listing the crimes of the Thebans, includes the fact that they began war against the Phokians. The other 'crimes' can all be related to known historical events, as the Loeb edition footnotes show. Why not this?

Justin, whom it is usual to dismiss as rhetorical and worthless, says plainly in his account of the war that the Thebans accused both the Spartans and the Phokians in the common council of the Greeks, 'apud commune Graeciae concilium' (8.1.4–5), the latter for raiding ('quod Boeotiam populati essent'). Justin is seen these days as a 'creative writer' with oratorical interests,[63] concerned above all to provide pleasure and moral *exempla*. He introduces his account of the Third Sacred War by announcing that the Greek states each wanted hegemony, but because of this they all ended up losing it together, and rushed to mutual destruction, 'in mutuum exitium' (8.1.1). The Thebans were the cause of this evil, 'causa et origo huius mali' (para. 4). We may scoff at Justin's programme and at the corners he cuts in the execution of it, but many ancient historians wrote to entertain and to instruct, and this does not entitle us to throw out all their assertions of fact. Theban actions against Sparta and Phokis illustrated Justin's moralizing point nicely. It does not follow that he invented those actions. Even modern historians are, quite reasonably, tempted to moral comment by the outcome of the Third Sacred War. 'Singularly destructive in physical, intellectual and political terms alike ... marked the bankruptcy of the polis as a political institution', says Buckler in his Epilogue.[64] He goes on to quote, precisely, Justin (8.1) on the way 'the Greek states ruined one another, only to fall prey to Philip'. Buckler had on an earlier page characterized the same Justin as 'chaotic and unreliable', when rejecting his assignment of blame to the Thebans for the war.[65] Justin's account is simple and severe; but not obviously chaotic, and not necessarily false.[66]

As for Pausanias, he wondered if the Thessalians were giving expression to ancient hatred. But he was frankly unable to determine the cause of the war (10.2.1).

The Duris, Isokrates and Justin items tend to be dismissed; they tend to lead marginal lives in the footnotes of modern accounts. But they seem to be quite independent of each other, and to be usable additional evidence for the Theban–Phokian hostility which Pomtow, rightly in my view, detected in the 'long 360s'. It is not at all implausible that there should have been minor irritants (raids, abductions, and so on) which accompanied the deeper causes of this war, just as there were for the Peloponnesian War. I leave unanswered the question: was the truest cause, the ἀληθεστάτη πρόφασις, the alleged Phokian cultivation of sacred land, or was it the hostility between the Thebans (and Thessalians) and the Phokians?

The Fifth Century

In the remaining part of this paper, I want to go back to the fifth century. The evidence is much less full; even so I shall have to be selective. But there is a plausible case for saying that the Spartans after 480 are the equivalents of the Thebans after 371: they tried to maximize their influence with the amphiktiony. I argued this in 1991/92, and it is this which has now prompted a seven-page reply by Lefèvre in 2002, in an appendix to *CID* IV.[67]

The starting point is Georges Daux's 1957 demonstration, accepted by Roux, Lefèvre, and Sánchez, that fifth-century Sparta was represented in the amphiktiony through and only through the so-called 'Dorians of the metropolis'.[68] This small community in central Greece, Doris, thus had an importance for the Spartans which had little to do with strategic considerations but much to do with the kind of kinship relationship so well investigated recently by scholars like C. P. Jones.[69] It is remarkable that Thucydides uses identical language to carefully flag the metropolis relationship on the only two occasions when he mentions Spartan military help to Doris, first in 457 (the Tanagra campaign against Phokis), and then in 426. His Greek is ἐς Δωριῆς τὴν Λακεδαιμονίων μητρόπολιν (i. 107, iii. 92). This is the Homeric technique of using repeated phraseology to make a comparison. The designation is amphiktionic. Might the Spartans or rather some Spartans (as always, different Spartans will have been differently motivated) have been worried about their precarious toe-hold in the amphiktiony? So I suggested fifteen years ago, precisely on the evidence of Thucydides. He is not a crude writer and I think that, far from suppressing the amphiktionic aspect here, he is in his subtle way actually hinting at it. In 1992 I noted that without little Thucydidean touches of this sort it would not be possible for us to draw the slightly different picture which I believe it possible to draw. Spartan action against the Phokians and in favour of Doris in 458/57 is very relevant to the Second Sacred War of 450, also fought against the Phokians. This did not happen, with a bang, in 450. Like the Third War, it flowed out of earlier developments. As for the 426 intervention, it led to the founding of Herakleia-in-Trachis, a new city which would have a vote in the amphiktiony by 343. Kip in 1911 argued that the Spartans at some time between 426 and 343 got one of the old Malian votes in effect transferred to the Oitaians, that is to Herakleia, to augment Spartan influence in the amphiktiony. I agreed. (The Oitaians, I should explain, are regularly designated by their city ethnic of Herakleia.) Béquignon, whom I followed in 1991, suggested the date was 426 itself.[70] Anyone who dislikes that date must argue for another in the period 426–343. That Sparta tried to manipulate the amphiktiony as early as 478 is explicitly said by Plutarch: 'The Spartans tried to expel medizers so as to control the votes completely and carry out their own wishes'—striking words, but Themistokles stopped them.[71] Some moderns disbelieve this, fancying in their modest way they know more about Delphi than did Plutarch, a Delphic expert, an amphiktionic representative of Boiotia, and an attested epimelete and agonothete.[72] Lefèvre, who accepts the story, says against me that it was one thing to try to take over the entire amphiktiony as in 478, another to exert huge energy on behalf of a single vote.[73] True. But the failure of the bigger attempt helps explain the urgency of the lesser project.

Almost finally, two a priori considerations, of similar sorts. First, if the Hellenistic Aitolians were not merely taking further what classical predecessors like Spartans and Thebans had tried to do, that makes them improbably innovative. Second, why should Delphi have been so different from that other Panhellenic place Nemea? David Lewis's elegant and convincing account of the First Peloponnesian War, of 460–446, was in terms of competition between Corinthians and Argives for control of Kleonai and the Nemean games and festival.[74] Delphi's organization was different (more complex); the basic idea is not.

I want to put aside detail now and address the fundamental grounds of disagreement between myself and Lefèvre. He quotes a 1992 sentence of mine and says he agrees with the first half but not with the second. What I had said was that the amphiktiony was *active and important* in the fifth century, and that we should not conclude otherwise from Thucydides' silence about it. He thinks I am right against Sordi that it was active; in 1957 she concluded from Thucydides' silence that it went to sleep in the fifth century. But he thinks I was wrong to say it was important.[75] But what does 'important' mean in this context? Lefèvre goes on to argue forcefully that the amphiktiony was not an instrument of direct political hegemony, though oddly he accepts Plutarch's startling formulation in his *Life of Themistokles*, saying that the amphiktiony was not an instrument of power, it conferred only prestige: 'Il n'est pas un instrument du pouvoir, mais ne confère que du prestige.'[76] It is the 'only' ('ne ... que') which I object to there. I would now want, as a result of the work I did for my book *Thucydides and Pindar*, to emphasize this, what I called earlier the 'soft' political aspect, much more than I once did. To that extent I no longer subscribe, if I ever did, to the 'historicizing' tendency Lefèvre attributes to me—at least as far as the fifth century goes (the 360s were, I have tried to show, rougher and more direct). Lefèvre, as we saw, said in 1998 that the organization of the games was the most prestigious thing the amphiktiony did.[77] I can only applaud this. But he then devotes only two pages to that topic, and treats it in isolation. His short list of the 'rare appearances' ('rares apparitions') of the amphiktions in the fifth century[78] includes the Pythia of 462, when Arkesilas was crowned by the amphiktions, as Pindar tells us in *Pythian Four*. But Pindar could have said exactly this about each and every one of the many fifth-century Pythian victories he celebrated![79] Pindar makes a meal of the detail of Arkesilas' chariot victory (we may conjecture) only because the poem was much longer than usual, and he was no doubt being paid much more for writing it.

The Pythia, we know for certain, were musical as well as equestrian and athletic. Perhaps we can go further. Did the amphiktions go in for theatre too? Lefèvre's list of amphiktionic 'appearances' includes, as well as Pindar, some later fifth-century literary references such as the fragmentary *Amphiktions* of the comic poet Teleclides, and the famous 'Panhellenic' passage in Aristophanes, *Lysistrata* (1129–1131: 411 BC). But there is another, tragic, passage which has been intriguingly exploited recently. Oliver Taplin has suggested that a choral lyric in Sophocles' *Trachiniai* may be evidence for theatrical performances at the Pylaia, and wonders, cautiously, if there was a 'political or diplomatic dimension' to this. The chorus addresses 'you who inhabit the harbour by the rocky hot springs, and the spurs of Oita, and you by the inner Malian gulf, and the shore of the maiden-goddess of the golden shuttle, where the celebrated gatherings of the Hellenes at the Pylai are held', ἔνθ' Ἑλλάνων ἀγοραὶ Πυλάτιδες κλέονται (lines 633–39).[80] Taplin's suggestion is appealing.

What I tried to do in my own recent book (*Thucydides and Pindar*) was to integrate agonistic festival success with other kinds of what Pindar calls *kudos*. Influence with the organizing body was worth having and intriguing for. The fifth-century Spartans were the greatest military power in Greece, but they also thought it worthwhile to win equestrian success at Panhellenic games. Conversely, a quarrel with the organizers could lead to humiliating exclusions, like that of the Spartans by the Eleans from the Olympic games of 420 (Thuc. 5.49–50). Was Delphi so different? Then there is Thessaly and Jason, who is said to have planned in 370 to usurp some of the amphiktiony's functions. Xenophon was not sure what he had in mind for the 'sacred monies'. But he does say he planned to arrange the festival and games himself, note the explicit αὐτός, implying a contrast with the amphiktiony.[81] As for the Thebans, the boy victor in the running-race at the revived Pythian games of 346, presided over by Philip, had a resonantly historical and Pindaric name. He was Aioladas, surely a direct descendant of Pagondas son of Aioladas, who won the battle of Delion in 424, and for whose family Pindar wrote a daphnephorikon. Like that royal victor Arkesilas more than a hundred years earlier, this interesting young man will have received his crown from the hands of the amphiktions.[82]

Acknowledgements

I thank the organizers of the Networks conference for their hospitality, the audience at Rethymno for their reactions, Peter Rhodes for comments on a subsequent written draft, and Angelos Matthaiou for re-examining a crucial Athenian inscription for me (see below, note 38), and giving me the benefit of his comments on the text. These should however be regarded as provisional, pending his republication of it in the new edition of *IG* ii.

Notes

[1] In the notes below I frequently cite the important recent works by Lefèvre, *L'amphictionie pyléo-delphique: histoire et institutions*, and Sánchez, *L'amphictionie des Pyles et de Delphes*, but I would mention here a much shorter and very lively treatment: Forrest, 'The Pre-polis Polis'.

[2] Davies, 'A Discipline in Transformation', 234f.

[3] Money: M. H. Hansen, *Inventory*, 130; Hdt. 5.63, and Thuc. 5.71.

[4] On this aspect of Clarendon, see the references in my forthcoming paper in the Festschrift for O. Murray.

[5] Lefèvre, *L'amphictionie pyléo-delphique: histoire et institutions*, 236.

[6] Hornblower, *Thucydides and Pindar*, ch. 1.

[7] Pind. *Pyth.* 4. 66–67. Race is surely right against B. Snell and H. Maehler Pindari carmina (Leipzig: Teubner, 1987) and W.J. Slater's *Lexicon to Pindar* (Berlin: de Gruyter, 1969) 43 to spell Αμφικτιόνων with a capital: not just 'neighbours'. Cf. Posidippos AB 74 line 4, ἀμφικτύοσιν,Φοῖβ', ἐν ἀγωνοθέταις. (In a paper delivered at the Victory Ode conference at UCL/ICS, July 2006, I suggested that in Pind. F94b line 43 (*daphnephorikon* for the family of Pagondas), the reference to ἀμφικτιόνεσσιν might also be an allusion to a Pythian victory. The poem goes on, after all, to mention an Olympic success ('at Pisa') as well as local victories in between. But G. B. d'Alessio afterwards convinced me this is wrong. For one thing, the order Delphi–local–Olympia is odd. For another, the language at *Isthm.* 4. 8 is similar, and there Delphi is not in question.)

[8] Tarn, *Antigonos Gonatas*, 429–32, with references.

[9] Beloch, *Griechische Geschichte*, 246; Rhodes, *CAH*, 6^2 (1994), 587, and *History of the Classical Greek World*, 201, cf. 254–55 and 302; Arnush, 'Argead and Aetolian Relations', 296.

[10] Ehrhardt, *The Third Sacred War.*

[11] Buckler, 'Thebes, Delphoi, and the Outbreak of the Third Sacred War', and *Philip II and the Sacred War.*

[12] Bowden, 'The Functions of the Delphic Amphictyony before 346 BCE'.

[13] Buckler, *Philip II and the Sacred War*, 7.

[14] Bowden, 'The Functions of the Delphic Amphictyony before 346 BCE', 82.

[15] *Syll.*^{3} vol. 1, 239; Pomtow, 'Eine delphische στάσις im Jahre 363 v. Chr'.

[16] Polyb. 4.28.5; I overlooked this important passage by mistake in 1991, as Lefèvre, *CID* IV 2002, 438 n. 13 rightly points out.

[17] Lefèvre, *L'amphictionie pyléo-delphique: histoire et institution* and *CID* IV 2002; Sánchez, *L'amphictionie des Pyles et de Delphes.*

[18] Note the long section 'points de vue divergents' (sc. from Sánchez, *L'amphictionie des Pyles et de Delphes*) in Lefèvre *CID* IV 2002, 462–67. His section 'points de vue convergents' is at pp. 459–62, and is thus shorter.

[19] See Thuc. 4.133.2 for the στέμματα which caught fire and caused the Heraion at Argos to burn down in 423 BC.

[20] Higbie, *Lindian Chronicle*, 8f.

[21] *SEG* 44. 425 (whence the quotation) = *CID* IV 2002, no. 2.

[22] Davies, 'Finance, Administration and *Realpolitik*', 9.

[23] Rhodes and Osborne, *Greek Historical Inscriptions*, no. 33 lines 9–10.

[24] Ehrhardt, *The Third Sacred War*, 7 and 10.

[25] The most recent investigation is Markle, 'Diodorus' Sources for the Sacred War': a very complex multi-source theory.

[26] Hammond, 'Diodorus' Narrative of the Sacred War'.

[27] Sordi, 'La fondation du collège des naopes et le renouveau politique de l'amphictionie au ive siècle', 49–52.

[28] Buckler, 'Thebes, Delphoi, and the Outbreak of the Third Sacred War', 241; cf. Sánchez, *L'amphictionie des Pyles et de Delphes*, 181 n. 145 (non-committal). Buckler, *Philip II and the Sacred War*, 15, n. 12 says 'Delphic inscriptions give the correct date', and refers to his own 1985 treatment at 242–43. But this is misleading. Delphic inscriptions do not date the condemnation.

[29] Buckler, 'Thebes, Delphoi, and the Outbreak of the Third Sacred War', 242.

[30] Sordi, 'La fondation du collège des naopes et le renouveau politique de l'amphictionie au ive siècle', 49 ('le délai légalement fixé') and 51 ('ce délai avait été fixé conformément aux lois'). The closest parallel I can find for the Greek is Dem. 24.28, ἀνελὼν τὸν ἐκ τῶν νόμων χρόνον. But admittedly Wayte in his edition of the speech (1882) glosses this as 'the time prescribed by law'.

[31] Ehrhardt, *The Third Sacred War*, pp. 91–93, Appendix I, 'The Date of the Amphictyonic Condemnation of Sparta for the Seizure of the Cadmea'.

[32] *CID* II (1989) no. 4, col. II, lines 48-54; no. 5, col. II lines 28–31.

[33] *CID* II (1989) no. 4 col. I lines 33–34 and 55–56 (Philostratis); no. 5 col. I lines 15–18.

[34] Lefèvre, *CID* IV 2002, 445 note 55.

[35] Xenophon: *Hell.* 5.4.1 and 6.4.3. Demeter Amphiktionis: Hdt. 7.200 and Strabo 420, with Sánchez, *L'amphictionie des Pyles et de Delphes*, 29, 32.

[36] This preponderance is admitted by Lefèvre, *L'amphictionie pyléo-delphique: histoire et institutions*, 73. But at the same time he says (note 346) that Buckler has 'justement minoré' Thebes' part in these affairs. He promises to return to these questions in his forthcoming Budé edition of Diodorus 16, to which we look eagerly forward.

[37] This is probably true of all those who, like Sordi or Ehrhardt, have opted for a 366 date, not bothering about the difference between 'Leuktrian war' and 'battle of Leuktra'.

[38] *IG* ii^{2} 109 [Ditt. *Syll*^{2} 100, ^{3} 175 (Pomtow); Hicks, *Manual*, 91, Hicks and Hill, 116]; Osborne, *Naturalization* 1, 49–51 no. D11. But Angelos Matthaiou (per epist., 25 May 2006) worries about the restoration of the penalty word ἀ[ειφυγίαν] at line 18: we expect an accusative of person after εἰσήγαγεν, or at least some word like γραφήν or δίκην. The restoration ἀ[ειφυγίας] (gen.) is excluded because Astykrates is in the genitive in line 19. But Matthaiou concedes that the sense ἀ[ειφυγίαν] is required by the unavoidable [ὥστε φυγ]αδεῦσαι at lines 19–20 (and also, we may add, by ἐκπεπτωκόσι at line 55, cf. 38), and he can think of nothing better than the usual text at line 18. He notes the remarkable singular of ἀφείλετο (Andronikos not the amphiktiony) at line 21. He rejects Osborne's [ὁλοκλήρους] at lines 21–22. and notes Rangabé's θ[εωρούντων].

[39] For which see Bonner and Smith, 'Administration of Justice in the Delphic Amphictyony'; Ehrhardt, *The Third Sacred War*, 13.

[40] The relevant inscriptions, mentioning the leased-out properties of the exiles, are now *CID* II (1989) nos. 67–72; note the helpful table at p. 133. For the property of Astykrates and Hagesarchos, see for example no. 68 lines 1 and 3. See also Rousset, *Territoire de Delphes et terre sacrée d'Apollon*. In these texts, Astykrates' name has sometimes been deliberately erased: see Pomtow, 'Neues zur delphischen στάσις im Jahre 363 v. Chr.', 400. *Syll*^{3} 178 prints parts of nos. 68 and 71. *CID* II no. 73 (*Syll*^{3} 177), by contrast, seems to be a list of anti-Phokians, who perhaps returned to Delphi in 346: see Bousquet, *CID* II p. 143, who reports Pomtow's reconstruction, while saying it is closer to a historical novel than to history (but see his p. 131, very close to Pomtow).

[41] Sánchez, *L'amphictionie des Pyles et de Delphes*, 170, concedes this rareness. *Lexicon of Greek Personal Names* IIIB (entry under Ἀγήσαρχος) lists four Delphians, of whom no. (3) is second-century BC and no. (4) is first-century BC. Nos. (1) and (2), listed separately in *LGPN*, are both attested at mid-fourth-century Delphi, and it is reasonable to identify them both with each other, and with the Ἡγήσαρχος of *Syll*^{3} 175. *LGPN* rightly identifies the latter with no. (1) in its entry.

[42] *CID* II (1989) no. 31 line 34.
[43] Pomtow, 'Delphische στάσις, 114, 126, 404; cf. Sánchez, *L'amphictionie des Pyles et de Delphes*, 169.
[44] Arnush, 'Argead and Aetolian Relations'.
[45] Busolt and Swoboda, *Griechische Staatskunde*, 1303 n. 2 (*promanteia* point).
[46] *Syll*^{3} 177; cf. no. 7, actually = Hdt. 1.54.2.
[47] *Syll*^{3} 7, actually not an inscription in the usual sense at all, but reconstructed from Hdt. 1.54.2.
[48] Bowden, 'The Functions of the Delphic Amphictyony before 346 BCE', 67.
[49] Bousquet, *CID* II (1989), 131.
[50] Sánchez, *L'amphictionie des Pyles et de Delphes*, 171 and 143.
[51] Lefèvre, *CID* IV 2002, 466
[52] Diod. 16.23.3.
[53] Buckler, 'Thebes, Delphoi, and the Outbreak of the Third Sacred War', 245, and *Philip II and the Sacred War*, 16f.; Sánchez, *L'amphictionie des Pyles et de Delphes*, 177.
[54] Xen. *Hell.* 7.5.4. Cf. Hdt. 8.30 for long-standing Phokian–Thessalian hostility.
[55] Buckler, *Philip II and the Sacred War*, 10.
[56] 'Sulky': Ehrhardt, *The Third Sacred War*, 19.
[57] And he surely alludes to the Third Sacred War at *Poroi* (*Revenues*), 5. 9.
[58] De Jong, *Narratological Commentary on the Odyssey*, xvii–xviii.
[59] Schaefer, *Demosthenes*, 489 and n. 3.
[60] See also Schaefer, *Demosthenes*, 89 and n. 2.
[61] Buckler, *Philip II and the Sacred War*, 19.
[62] See Horden and Purcell, *The Corrupting Sea*, ch. XII.
[63] Yardley and Heckel, *Justin: Epitome of Trogus*, 11–12, 17f.; cf. Justin, preface para. 4.
[64] Buckler, *Philip II and the Sacred War*, 142. Cf., from a different angle but in the same moral vein, Ehrhardt *The Third Sacred War*, 90: 'Thebes, by wanton aggression for her own selfish ends, was responsible for crippling Greece and finally destroying her freedom.'
[65] Buckler, *Philip II and the Sacred War*, 16 n. 13.
[66] Note the scholiast to Dem. 19. 20: ὅτι Θηβαίων καὶ Φωκέων πολεμούντων πρὸς ἑαυτοὺς διὰ τὴν παραποταμίαν γῆν, with Schaefer (above, n. 59), 489 n. 2.
[67] Hornblower. 'The Religious Dimension to the Peloponnesian War', and *Commentary on Thucydides* on i. 107, 112, iii. 92. Criticized at length by Lefèvre, *CID* IV, 2002, 436–44, cf. 452: 'un certain excès d'"historicisme".'
[68] Daux, 'Remarques sur la composition du conseil amphictionique'; Roux, *L'amphictionie, Delphes et le temple d'Apollon au IVe siècle*, 5 and n. 2; Lefèvre, *L'amphictionie pyléo-delphique: histoire et institutions*, 53.
[69] Jones, *Kinship Diplomacy in the Ancient World*.
[70] Kip, *Thessalische Studien*, 19; Béquignon, *La vallée du Spercheios*, 350 n. 1; cf. Flacelière, *Les Aitoliens à Delphes*, 40 n. 2.
[71] Plut. *Them.* 20: [Θεμιστοκλῆς] φοβηθεὶς μή....[οἱ Λακεδαιμόνιοι] παντελῶς ἐπικρατήσωσι τῶν ψήφων καὶ γένηται τὸ δοκοῦν ἐκείνοις.
[72] *Syll*^{3} 829A (cf. Jones, *Plutarch and Rome*, 26).
[73] Lefèvre, *CID* IV, 2002, 443 n. 42.
[74] Lewis, 'The Origins of the First Peloponnesian War'; cf. Hornblower, *Greek World*, 27–30, cf. 21. I do not here try to bring the intriguing but fragmentary Athenian decree *IG* i^3 9 into the mid-century story, though I note that Lefèvre, *CID* IV 2002, 463, para. 3, thinks (against Sánchez, *L'amphictionie des Pyles et de Delphes*, 109–11) it may be 'abusif de dénier à l'amphictionie toute dimension symmachique'.
[75] Lefèvre, *CID* IV 2002, 438.
[76] Lefèvre, *CID* IV 2002, 449.

[77] Lefèvre, *L'amphictionie pyléo-delphique: histoire et institutions*, 236
[78] Lefèvre, *CID* IV 2002, 441.
[79] See note 7 above; see Paus. 10.7, etc.
[80] Taplin, 'Spreading the Word Through Performance', 46ff. The passage is cited by Lefèvre, *L'amphictionie pyléo-delphique: histoire et institutions*, 193 n. 149, but merely as one of the literary testimonia for the location of the sessions of the amphiktiony. Similarly Sánchez, *L'amphictionie des Pyles et de Delphes*, 32, 62.
[81] Xen. *Hell.* 6 4.30, πανήγυριν τῷ θεῷ καὶ τοὺς ἀγῶνας αὐτὸς διατιθέναι.
[82] Paus. 10.7.8. See Hornblower and Morgan, 35–39.

References

Arnush, Michael. 'Argead and Aetolian Relations with the Delphic Polis in the Late Fourth Century BC'. In *Alternatives to Athens*, edited by R. Brock and S. Hodkinson. Oxford: Oxford University Press, 2000.

Beloch, Julius K. *Griechische Geschichte* 3^2 1. Berlin: W. de Gruyter, 1967. (Orig. pub. Berlin, 1922).

Béquignon, Yves. *La vallée du Spercheios*. Paris: Editions de Boccard, 1937.

Bonner, R. J., and G. Smith. 'Administration of Justice in the Delphic Amphictyony'. *Classical Philology* 38 (1943): 1–12.

Bowden, Hugh. 'The Functions of the Delphic Amphictyony before 346 BCE'. *Scripta Classica Israelica* 22 (2003): 67–83.

Buckler, John. 'Thebes, Delphoi, and the Outbreak of the Third Sacred War'. In *La Béotie antique*, edited by P. Roesch and G. Argoud. Paris: Editions du Centre national dela recherche scientifique, 1985.

Buckler, John. *Philip II and the Sacred War*. Leiden: E. J. Brill, 1989.

Busolt Georg, and H. Swoboda. *Griechische Staatskunde*, Vol. 2. 1292–1310, 'Die pylaeisch-delphische Amphiktionie'. München: C.H. Beck, 1926.

Cambridge Ancient History. London: Cambridge University Press, 1970–2005.

CID. See Lefévre; Rougemont.

Davies, John. K. 'Finance, Administration and *Realpolitik*: The Case of Fourth-century Delphi'. In *Modus operandi: Essays in Honour of Geoffrey Rickman*, edited by M. Austin *et al.* London: Institute of Classical Studies, School of Advanced Study, University of London, 1998.

———. 'A Discipline in Transformation'. In *Classics in Progress*, edited by T. P. Wiseman. Oxford and New York: Oxford University Press, 2002.

Daux, Georges. 'Remarques sur la composition du conseil amphictionique'. *BCH* 81 (1957): 95–120.

De Jong, Irene. *A Narratological Commentary on the Odyssey*. Cambridge: Cambridge University Press, 2001.

Ehrhardt, Christopher T. H. R. 'The Third Sacred War'. B.Litt. thesis, University of Oxford, 1961.

Flacelière, Robert. *Les Aitoliens à Delphes*. Paris: Editions de Boccard, 1937.

Forrest, W. George. 'The Pre-polis Polis'. In *Alternatives to Athens*, edited by R. Brock and S. Hodkinson. Oxford: Oxford University Press, 2000.

Hammond, Nicholas G. L. 'Diodorus' Narrative of the Sacred War'. *JHS* 57 (1937): 44–78.

Hansen, Mogens H., and Thomas. H. Nielsen, eds. *An Inventory of Archaic and Classical Greek Poleis*. Oxford: Oxford University Press, 2004.

Higbie, Carolyn. *The Lindian Chronicle and the Greek Creation of their Past*. Oxford: Oxford University Press, 2003.

Horden, Peregrine, and Nicholas Purcell. *The Corrupting Sea: A Study of Mediterranean History*. Malden: Blackwell, 2000.

Hornblower, Simon. *Commentary on Thucydides*. Vol. 1, Books I–III. Oxford: Clarendon Press, 1991.

———. 'The Religious Dimension to the Peloponnesian War, Or, What Thucydides Does Not Tell Us'. *HSCP* 94 (1992): 169–97.

———. *The Greek World 479–323 BC*^{3}. London: Routledge, 2002.

———. *Thucydides and Pindar: Historical Narrative and the World of Epinikian Poetry.* Oxford: Oxford University Press, 2004.

Hornblower, Simon, and Catherine Morgan. Introduction to *Pindar's Poetry,* edited by Simon Hornblower and Catherin Morgan. Oxford: Oxford University Press, 2007.

Jones, Christopher P. *Plutarch and Rome.* Oxford: Clarendon Press, 1971.

———. *Kinship Diplomacy in the Ancient World.* Cambridge, MA: Harvard University Press, 1999.

Kip G. 'Thessalische Studien.' diss., Halle, 1910.

Lefèvre, François. *L'amphictionie pyléo-delphique: histoire et institutions.* Athens: École française d'Athènes; Paris: De Boccard édition-diffusion, 1998.

Lefèvre 2002 = *CID* IV 2002.

Lewis, David M. 'The Origins of the First Peloponnesian War'. In *Selected Papers,* edited by P. J. Rhodes. Cambridge: Cambridge University Press, 1997. (Orig. pub. 1981).

Markle, M. M. 'Diodorus' Sources for the Sacred War in Book 16'. In *Ventures into Greek history,* edited by I. Worthington. Oxford: Clarendon Press, 1994.

Osborne, Michael J. *Naturalization in Athens 1.* Brussels: AWLSK, 1981.

Pomtow, Hans R. 'Eine delphische στάσις im Jahre 363 v. Chr.'; 'Neues zur delphischen στάσις vom Jahre 363 v. Chr.' *Klio* 6 (1906): 89–126, 400–19.

Rhodes, P. J., and R. G. Osborne. *Greek Historical Inscriptions 404–323 BC.* Oxford: Oxford University Press, 2003.

Rhodes, Peter. J. *A History of the Classical Greek World.* Malden, Mass. and Oxford: Blackwell Publications, 2006.

Rougemont, G., J. Bousquet, and A. Bélis, eds. *Corpus des inscriptions de Delphes* (*CID*), I–IV (so far). Paris: De Boccard édition-diffusion, 1977–2002.

Rousset, Denis. *Territoire de Delphes et terre sacrée d'Apollon.* Athens: École française d'Athènes; Paris: De Boccard édition-diffusion, 2002.

Roux, Georges. *L'amphictionie, Delphes et le temple d'Apollon au IV*[e] *siècle.* Lyon: Maison D'orient, 1979.

Sánchez, Pierre. *L'amphictionie des Pyles et de Delphes.* Stuttgart: F. Steiner, 2001.

Schaefer, Arnold. *Demosthenes und seine Zeit*^{2}, vol. 1. Leipzig: B.G. Teubner, 1885.

Sordi, Margareta. 'La fondation du collège des naopes et le renouveau politique de l'amphictionie au ive siècle'. *BCH* 81 (1957): 38–75.

Taplin, Oliver. 'Spreading the Word Through Performance'. In *Performance Culture and Athenian Democracy,* edited by S. Goldhill and R. Osborne. Cambridge: Cambridge University Press, 1999.

Tarn, William. W. *Antigonos Gonatas.* Oxford: Clarendon Press, 1913.

Yardley, J. C., and W. Heckel. *Justin: Epitome of the Philippic History of Pompeius Trogus vol. 1, Books 11–12, The History of Alexander the Great.* Oxford and New York: Clarendon Press, 1997.

Pythios and *Pythion*: The Spread of a Cult Title

J.K. Davies

It is a cliché of scholarship that the terms *Pythios* and *Pythion* derive from *Pytho*, the earliest attested name for the cultic area later known as Delphoi,[1] and that they came to refer unequivocally to Apollo as the principal god of the upper sanctuary there. They therefore belong among the 'locational' epithets of the god, parallel to *Klarios* or *Didymeus*, not among those derived from function or from association or identification with animals.[2] As such, they pose a problem, for it was unusual for locational epithets to spread far beyond their primary sanctuary, still less to show such widespread attestation throughout much of the Greek-speaking world: Zeus Olympios provides perhaps the only true comparison, with even Eleusinian Demeter well behind. Their distribution therefore needs an explanation, and one which is couched not just in terms of geography or of the 'profile' of the god but also, and primarily, in terms of human needs and of the ways in which a specific cult could be transferred, or replicated/cloned, elsewhere. What follows here began as an attempt to provide such an explanation, though the reader will find that it does little more than canvass possibilities. Equally, I leave open the question of whether the attested distribution amounts to a 'network', in terms of the examples and theoretical models which were offered at the Rethymno colloquium, for even at a wholly pragmatic level three models need to be distinguished. The first, discussed in more detail below, is that of the formal establishment in territory B of a cult which has its primary location in territory A, itself a model which may or may not include periodic ceremonial *theoriai* from B to A. The second, attested only for the major agonistic festival sanctuaries, is the complement of the first, comprising the periodic circuits of Greek states undertaken by *theoroi* as envoys of such sanctuaries in order to announce the date of a forthcoming festival, to invite competitors to present themselves, and to proclaim a sacred truce. A third model, Procrustean in form and varying continually in space and in intensity, consists of the community of all participants in this or that cultic activity carried out at a particular sanctuary, whether as dedicants or as competitors, or as suppliants, or as enquirers of an oracle, or as manumitters of slaves, or as participants in feasts. Since my own study of the third model is still very much work in progress, while the second model has recently been excellently mapped (if only for Peloponnese) by Paula Perlman's *City and Sanctuary*, what follows here will focus on the first model, in the hope that some preliminary progress can be made.

One must start with the data. That is easier said than done, for the only systematic assemblages of all the evidence for all the epithets of Apollo and his sanctuaries, including Pythios and Pythion, remain those of Wernicke and Farnell, now over a century old.[3] As might be expected, their lists cannot be used in their raw state in the search for attestations of Apollo Pythios (A.P.), for they now need both expansion and pruning. The latter is needed because there had clearly been a temptation to draw the net too widely and to create an over-synchronic picture, which is the very opposite of what is needed for present purposes. For example, attestations of the epithet in the areas of the post-Alexander diaspora, e.g., in Egypt or the cities of Syria, must stem from a period of secondary or tertiary diffusion, tell us little about the processes that generated earlier stages, and went on for centuries: the Pythian festival that came to be celebrated in Thessalonike was (remarkably) a creation of the 240s CE.[4] Again, perhaps paradoxically, it may be wise to leave out of the account the cities that are known from literary evidence (mostly Pausanias) to have erected treasuries at Delphi. Wernicke included them all, but even apart from the well-known difficulties of attribution, the hidden assumption—namely, that all such cities must already have imported and given civic space to the cult of A.P.—does not necessarily hold without other evidence. The same is true for those cities that Plutarch and Strabo name as having made 'golden summer' offerings to Delphi,[5] for, accepting the normal interpretation of such gifts as one-off thank-offerings,[6] all that need be postulated is a community's desire to advertise its good fortune by clothing it in suitable garb of piety. Caution requires instead that only the direct attestation—preferably datable, and preferably in epigraphic or reliable antiquarian form—of an altar, sacrifice, sanctuary, or temple explicitly devoted to Apollo Pythios can reliably be used as evidence.

On the other hand, new evidence, largely epigraphic and archaeological, has so much increased the amount of robust relevant data that, for reasons of space, it has proved impossible to retain them within the framework of the present paper. The only expedient available was to consign the detailed assemblage of data to a separate paper, which will be published elsewhere,[7] and to present here a simple, largely un-annotated working summary that omits the dubious categories instanced above, and allows the argument to go forward. The expedient is less than satisfactory, not least because this present paper has had, *hysteron proteron*, to take precedence in reaching editors' desks; I can only beg for the reader's understanding. For convenience, the following summary follows the geographical order of *IG* I–XIV (non-existent volumes being flagged [thus]), thence (though it scarcely applies) the order used by *SEG*.

Athens and Attika (*IG* I–III) offer a very substantial harvest of evidence. Not only did the city itself show two shrines, the temple of A(pollo) P(ythios) on the Ilissos attributed to the Peisistratids by Thucydides (2.15.3) and that of A. Hyperakraios/Pythios on the north slope of the Akropolis, but shrines or altars (for present purposes the distinction is unimportant) of A.P. are known at Daphne, Eleusis, Ikarion, Marathon, Oinoe, Thorikos, and in the Kephisos valley. To them one must add both the curious set of cultic consultants called the *exegetai pythochrestoi*,[8] and the evidence of a pilgrimage called the Pythais,[9] which came to have great symbolic importance after 146. Evidence from the Korinthia (IV), in contrast, is very slim indeed, unless one is prepared either to take the Treasuries of Korinth and Sikyon at Delphi into account, or to short-circuit the endless debate about the deity housed so monumentally on Temple Hill at Korinth.[10] The Argolid (IV) offers far more evidence, with attestations of various kinds from Aigina, Argos, Asine, Epidauros, Hermione, Kynouria, Mykenai, and Troizen, but also with evidence for the parallel cult of Pythaieus, which is discussed below. Lakonia and Messenia (V,1), on the other hand, offer very little apart from a dedication to A. Pythaieus at Prasiai, a fifth-century verse dedication to A.P. in the sanctuary of Athene Chalkioikos, an image of A.P. at Sellasia, and the Kings' sacred remembrancers called the *Pythioi* (Hdt. 6.57.2 and 57.4). Arkadia (V,2) offers evidence only at Mt Lykaios, Pheneos, and Tegea, if numismatic evidence of the Severan period from Kaphyai and Orchomenos is to be discounted; while Olympia and Elis ([VI]) offer nothing except an altar of Apollo Pythios before the *proedria* at Olympia (Paus. 5.15.4). Boiotia and the Megarid (VII) offer problematic evidence at Megara (for Pythaieus, as well as for Pythios) and Thebes, and explicit evidence at Tripodiskos, which is discussed in more detail below. Delphi itself ([VIII]), of course, is full of the epithet, while in north-western Greece (IX^2, 1), sites in Aitolia, Akarnania, and western Lokris offer virtually nothing except at mid-fifth-century Naupaktos, and in the post-279 decrees for the Soteria, Kerkyra alone preserving a fifth-century [*o*]*rwos Puthaios* (*IG* IX^2 1,863). Thessaly (IX, 2), in contrast, shows a remarkably rich haul,[11] with explicit attestations in Argissa, Atrax, Azoros, Gonnoi, Halos, Larisa, Melitaia, Pythion itself (hardly surprisingly), and Tempe. Further north, however, Makedonia and the Black Sea yield virtually nothing until late in the Hellenistic period, unless the third-century dedications made at Delphi by the

Paionian royal family can be taken to reflect a daughter cult of A.P. established in Bylazora and not (as is rather more likely) an anti-Makedonian statement of independence and Greekness pinned onto the national notice-board.[12]

If we return to the Aegean, the picture changes dramatically, with evidence for altars, sanctuaries, and temples of A.P. widespread throughout the region: Delos, of course (XI), Rhodes with its oscillation between Pythios and Pythaieus (XII,1), Syme, Anaphe, and Thera (XII, 3), Halasarna (XII, 4), Andros, Ios, Keos, Paros, Sikinos, and perhaps Tenos (XII, 5), Samos (XII, 6), Amorgos (XII, 7), Thasos (XII, 8), and Chalkis on Euboia (XII, 9). Even that list, with its interesting absences noted below, pales besides Krete ([XIII]), where the cult is attested at Allaria, Dreros, Eleutherna, Gortyn, Hierapytna, Itanos, Lato, Lyttos, Malla, Phaistos, Praisos, and Tylisos, with perhaps Aptera, Axos, Chersonesos, Kydonia, Lappa, Rhithymna, and Sybrita on the more fragile basis of numismatic evidence. In contrast, apart from the Treasuries, western Greece offers thin pickings, with firm early evidence for an installed cult only being available for Kroton and (remarkably) Rome. Lastly, Asia Minor and the eastern Mediterranean, where the only installations of the cult which may pre-date the post-Alexander expansion seem to be those of Mylasa, Ephesos, Erythrai, and Klaros.

For fairly obvious reasons these data must be used with great caution. Areas with a pronounced epigraphic habit, *imprimis* Athens–Attika, show a very rich picture, while Magna Graecia is correspondingly under-represented, as also are areas which fell outside the purview of Pausanias. More radically, cult titles are not necessarily mutually exclusive. In this regard Travlos[13] cited the statement of Demosthenes (18.141) that Apollo Pythios was honoured in Athens as Patroios and that on the evidence of Euripides, *Ion* 283–85, the Apollo who was worshipped as Hypoakraios ('Below the crags') on the North Slope of the Akropolis was also the Pythios whose shrine figures in the Panathenaic procession.[14] Likewise there can be little doubt that the Apollo Archegetes whose altar on the shore of Sicilian Naxos is reported by Thucydides (6.3.1) was the Apollo of Delphi in his capacity as mythic leader of the colony—though that does not allow us to list it as a cult of A.P. However, on other occasions aspects of a god are explicitly differentiated. For example, Philochoros reports (*FGrH* 328 F 75) that in the temple of Apollo Delios at Marathon it was the altar of A.P., not by implication the main altar, which was used for sacrifices when the Pythais was being sent to Delphi. So, too, oaths to be sworn by the citizens of Cretan communities distinguish, for example, Apollo Delphinios and Apollo Pythios at Dreros, or even more elaborately at Eleutherna give equal and separate billing to A. Lykeios, A. Delphinios, A. Pythios, A. Karneios, A. Bilkonios, A. Styrakritas, and A. Sasthraios.[15] That will especially have been the case when, as at these two Cretan towns, a community has inserted into its cultic idiosystem two or more aspects of the god. As usual, this is clearest at Athens, where besides A. Pythios/Patroios there was also A. Delphinios, whose profile, especially in Krete, closely parallels that of Apollo Pythios, not least in the use of their sanctuaries as public archives. In his exemplary paper of 1979 Fritz Graf argued that Apollo Delphinios is not a god of the seaways, as the (itself very dubious) etymological association with dolphins might suggest, for in contrast to Poseidon by no means all his sanctuaries are on the coast. Nor, a few brief allusions in the *Homeric Hymn to Apollo* notwithstanding (400; 493–96), has he, or his epithet, any discernible link with Delphi, where (a) A. Delphinios is not attested, and (b) the form of the epithet should be Delphaios, not Delphinios. Instead, Graf argued, Apollo Delphinios had specific roles in the public life of the community, not least in respect of male bonding groups and the protection of ephebes—this latter aspect explaining, via the myth of the return of Theseus from Krete, the existence and role of the Delphinion law-court in Athens.

If one overlap between Pythios and another epithet of Apollo can be mapped thus, there remains another overlap, that with Apollo Pythaieus. Here too the essentials of this aspect of Apollo have been clarified, this time by Marcel Piérart. He has shown that it is attested above all in the Argive Kulturgebiet, had an oracle at Argos as its central component, and came to be seen as an offshoot of the Delphian shrine.[16] However, and notwithstanding the oscillation between the forms 'Pythios' and 'Pythaieus' which the Rhodian texts show,[17] there is a clear tendency in the texts to use the word 'Pythaieus' on its own, as if it were not so much an epithet as a self-standing proper name. Though not conclusive in itself,[18] that hint gains substance further when Pausanias (2.35.2) quotes Telesilla for the statement that Pythaieus was the son of Apollo. That suggests strongly that the overlap between 'Pythaieus' and 'Pythios' reflects a different pattern, namely that whereby it becomes convenient to merge with Apollo an originally distinct divine entity.[19] There is a precisely parallel oscillation in northern Boiotia, where the hero Ptoios merged with Apollo but could also rate as his son,[20] and another in eastern Argolid, where the hero Maleas eventually merged with Apollo and/or with Asklepios,[21] not to mention the even more complex processes which approximated Diktynna-Britomartis, Aphaia, Artemis, and Saronia to each other.[22] In this way, though it is not the primary focus of this paper, one can see how an adventitious similarity of name (Pythios/Pythaieus), and the social need for a particular function—here, oracular—to be naturalized in a community's territory, combined to generate an untidy convergence and to give a god a more complex and a more widespread profile. Whether this amounts to a 'network' is in part a matter of terminology, but in part also a function of the degree to which such rapprochements came to be used as the basis of institutionalized practices such as *theoriai*.

After these preliminaries, the argument can now at last address the principal question: can we account for the distribution of the cult epithet 'Pythios', and what connotations, or reflections of function, did it carry? There are two main ways of approaching the question: procedurally, by considering likely mechanisms of transfer or colonization, or psychologically, by reviewing likely motives on the part of communities. If I first consider 'mechanisms' of transfer, that word is not meant to imply that the diffusion of this, as of any, cult was a mechanistic process. Since there is a whole spectrum of them, helpfully illustrated for Athens by Garland and Parker and in more general terms by Burkert,[23] it will be useful, with no pretence of completeness or originality, to set them out summarily.

(1) At one extreme stands the systematic, 'top-down' promulgation of a cult on the part of a powerful state outside its own boundaries for political purposes, as has been claimed for the third-century plantation of the cult of Serapis and Isis in the Aegean by the Ptolemaic regime of Egypt.[24] (2) Closely allied thereto is the process of building up local deities or heroes as symbols of national identity, in a way well illustrated by the magnification of Theseus before and after the Graeco-Persian Wars or by that of Athene herself within Athens and Attika.[25] (3) Also closely allied is the picture—drawn above all by Defradas in 1954 for Delphi itself, but applicable in principle to other sanctuaries—of a cult centre actively promulgating itself and its deity. If the list now moves from visibly human action to initiatives which are deemed to be due to the deities themselves, (4) a straightforward version is that of a divine command, enunciated by an oracle to a community, to establish a specific cult, two examples among many being the injunction in the Great Rhetra to establish cults of Zeus and Athene at Sparta and the coded instruction to the Athenians in 480 to incorporate a cult of Boreas.[26] (5) A variant of (4) comprises those establishments of cult where an individual is reported to have witnessed an epiphany or received an instruction but where the community, then or later, gives the cult formal recognition and space. Well-known examples include the creation of the public cult of Pan after the epiphany reported by Pheidippides, or the reception of Asklepios in Athens, with Telemachos of Acharnai as the driving force.[27] (6) More informal again are establishments of cult by an enthusiast, prophet, or guru, where the individual takes the initiative but the cult remains 'private' in the sense of not being taken into the formal cultic calendar of the community (whether city, canton, tribe, or deme), the classic example being the cult of the Nymphs in the Vari cave set up by the *nympholeptos* Archedemos.[28] (7) A variant of (6) comprises those cults of their homeland deities which are set up abroad by displaced persons such as slaves, freedmen, metics, or mercenaries.[29] (8) Lastly, though perhaps to be left out of account as unlikely to be applicable to the present problem, are those cultic practices which are reported to have been a response to an untoward or unexpected natural or human event treated as an omen oblativum: examples are a lightning strike, a plague, a sudden violent nervous breakdown with tragic consequences, or an unexpected success or salvation.[30]

Such mechanisms of diffusion plainly reflect a community's need to harness or avert the power of a particular deity. In the case of A.P., the question therefore becomes that of asking which aspect of his ascribed powers (obvious possibilities are oracular, purificatory, and legislative) formed the main attraction. Fortunately, whatever the preferred answer, one can use a fairly clear chronological horizon for its initial diffusion, for although recent reassessment of the archaeological evidence has sharply reduced the length of the period in the tenth century BCE, when Delphi may have been a ghost town, or nearly so, and has instead given it a status in the ninth and eighth centuries BCE as already a significant settlement, nevertheless there is absolutely no reliable evidence to suggest that it was a cult centre of any consequence until the dedications begin in the late ninth or early eighth century BCE.[31] We must therefore be dealing with a process which does not begin to unfold until the Late Geometric period, perhaps even towards the end of it. Indeed one might cautiously go further, for the Kretan evidence in particular, recently and most helpfully assembled and reviewed,[32] shows that sanctuaries which at least were later called 'Pythion' were in place in Gortyn and elsewhere by the late seventh century. In consequence, the process of diffusion must have been in motion well before then.[33]

That being so, various possible explanations of the process need to be canvassed, though the reader should be warned that the quest leads rather to *aporia* than to tidy answers. The simplest hypothesis is institutional, namely that the diffusion reflects the crystallization of the Pylian Amphiktyony and its re-orientation towards Delphi at the expense of Anthela.[34] However, although the strong showing of A.P. in Thessaly supports that hypothesis, his equally strong profile in Krete and the Aegean is intractable counter-evidence, for participation in the 'Ionian' ethnos-constituency never extended further east than Euboia, while participation in the 'Dorian' constituency came eventually to include Sparta and the Dorian states of north-eastern Peloponnese, but never embraced Krete. In any case, while a close relationship between Krete and Delphi can indeed be documented for the seventh century BCE, alike via the Kretan artefacts found at Delphi[35] and via the 'Pythian' sanctuaries detectable on the island, Guarducci's case that the relationship weakened from the late seventh century onwards, for whatever reason, still holds good,[36] and carries with it the awkward consequence that the Kretan 'presence' at Delphi tailed off before the period when, on conventional understandings of the 'Sacred War', Amphiktyonic influence began to make itself felt there. While therefore it would be foolish to deny that the crystallization of the Amphiktyony around the Apollo sanctuary assisted the spread of the cult, that cannot be the whole story.

Nor can the known distribution of the cult be explained in purely geographical terms, for closely adjacent and easily accessible regions such as Achaia, Aitolia, Akarnania, Elis, and even Boiotia and the Korinthiad show little trace of it. The hypothesis that the cult did not take root where there were strong local oracles would explain its absence from Boiotia or from north-western Greece, and perhaps also from Euboia if the 'most reliable' oracle of Selinountian Apollo at Histiaia (Strabo 10.1.3, 445C) offered an effective alternative. However, that explanation falters for the Aegean, for whose communities the oracles of Apollo at Klaros and Didyma were far more accessible. Nor, thirdly, can the diffusion of the cult of A.P. be seen as that of Apollo as patron saint of colonial foundations, tempting though the *termini ante* and *post quem non* render it. That conclusion is firm, not so much because he would then be *Archegetes*, as on the seashore of Sicilian Naxos, or because the recent 'colonization debate' has argued for seeing most such claims of divine patronage as *ex post facto* rationalizations,[37] as because the cult is widespread in regions which had little or nothing to do with colonization (Krete itself in particular) but is absent or near-absent in communities such as Megara and above all Korinth, which tradition regarded as being among the most active colonizers. Nor, fourthly, at least in respect of the Archaic and Classical periods, should we see the cult as a reflection of Apollo as the patron saint of musical or athletic contexts. Again, the argument rests partly on the existence of another, more appropriate epithet, Nomios, but mainly on the fact that on the conventional chronology of the Pythian Games, that aspect of Apollo does not gather force until well after the earliest Pythian sanctuaries had been established, again especially in Krete.

Even so, certain other possibilities remain open. If one thinks of the two main roles which are chartered for Apollo in the sixth-century hymns of Alkaios and 'Homer', that is, to speak the will of Zeus and to utter *themistes*, there is a plausible case for both: for him as the utterer of *themistes* such as the Great Rhetra because of the role as repositories of public documentation which Pythian sanctuaries came to play, and for him as oracular mouthpiece, both because that is what we should expect in any case, and particularly because, as noted above, A.P. is noticeably sparse or absent in areas such as Boiotia or north-west Greece, which had active oracles of their own. However, again there are difficulties. To invoke his oracular function in general, whether for private or public guidance, fails to explain why it should be the 'Pythian' oracle of Apollo at Delphi which was being favoured, rather than one of his other oracles in (say) Boiotia or Asia Minor: we need to be looking for some specialized but nonetheless widely needed sub-set of oracular pronouncements, wherein Delphi was deemed to have special authority.

One remaining possibility is therefore the need for purification, for that is a perpetually recurring theme in the oracles attributed to Pythian Apollo, and its prevalence has to be taken seriously even if every single example is marked down, by Fontenrose's or other criteria, as a *falsum*. In this context the ritual purifications which are widely attested in *leges sacrae* for pollution incurred by birth or death matter less than homicide, the pollution which it engenders, and the purifications which are deemed to neutralize such pollution. Over twenty years ago Robert Parker made a very plausible case (a) for seeing the package of ideas and phobias comprised in the concept of pollution as 'a kind of shadowy spiritual Doppelgänger of the law', and (b) for linking them with the versions of homicide law which crystallized in the seventh century BCE, such laws being part of the growth in a community's need and capacity not so much to control this basic breakdown in proper human behaviour, as to manage its consequences.[38]

The trouble is that though in principle Parker's analysis offers an attractive explanation for why the cult of A.P. came to be invoked and to take root outside Delphi, the supporting evidence is frail. Two texts do indeed illustrate a direct link, but both are taken from Pausanias, with all the dangerous baggage of antiquarian elaboration through the centuries which such paternity implies (though also with the high marks for reporter's accuracy which recent work has awarded him). The first concerns the crime of Krotopous[39] in killing Poiné, whom Apollo had sent to punish the Argives because the sheepdogs of Krotopous had killed the child which Apollo himself had fathered on Krotopous's daughter Psamathe. Krotopous himself sorts out the ensuing inversion of moral order (plague) by submitting himself to Delphi, where he was forbidden to return to Argos and ordered instead to take a tripod away and to live, and to build a temple, at the place where it eventually slipped out of his hands—

which was Tripodiskoi in western Megarid. Now of course this is 'pure' aetiological myth to account for the place-name Tripodiskos—of course it is Apollo taking vengeance for the death of his child, but it is also an explanation of why there was a temple of Apollo at that spot, and more importantly still an example of how to redeem a murderer by chartering a legitimate residence for him in exile. The second text,[40] the foundation myth of the Athenian cultic lineage of the Kephalidai,[41] contains closely similar structural motifs. Again a murder, this time of his wife Prokris by Kephalos; again an exile, to Thebes before leaving to found Kephallenia; again a desire to return, this time on the part of two descendants of his; again a consultation of Delphi; again an injunction to do so but to found a temple at a place where certain conditions would be fulfilled, in this instance the Apollo sanctuary which Travlos locates underneath the eleventh-century monastery church at Dafni.[42]

The trouble is that even if such myths could be reliably deemed to have crystallized in the Archaic period, and to reflect community preoccupations of that epoch accurately, they scarcely begin to match the widespread distribution of the cult of A.P. In theory, indeed, it is easy to accept that its diffusion could reflect a specific aspect of the role of oracular Apollo, namely that of offering practical ways in which homicide could be negotiated, and its destructive effects mitigated, in the interests alike of the community, of the murderer himself, or his relatives, and of the relatives of his victim. Such a pattern of diffusion would suit the horizon of the eighth and seventh centuries BCE, when population increase—surely real, whatever we make of the quantitative arguments—will have fuelled tensions over land and inheritance, and will have generated a need for clear rules about homicide. It could therefore have made the oracular shrine which showed an interest in such matters into a source of authority and legitimation, both of one-off solutions for each individual murderer, and of codified general solutions for those polities which experienced significant disruption. That much said, the case is fragile, for only a clear correlation between regions that adopted the cult of A.P., and regions that were particularly troubled by the need to negotiate homicide, would provide proof. That will scarcely be forthcoming, for though one may take comfort from the fact that Athens–Attika—plainly very active in adopting the cult—was one community for which a preoccupation with homicide can be firmly documented for the seventh century BCE,[43] while somewhat later Gortyn on Krete was another,[44] neither they nor such few other early homicide laws as are reported or attested[45] show themselves much preoccupied with purification: only in the case of Kleonai is the link explicit.[46]

The argument therefore ends, disconcertingly and disappointingly, in bafflement. On the one hand we have a 'network' of attestations of the cult of Apollo Pythios, widely though not uniformly spread throughout the pre-Alexander Greek world, and plainly in some sense reflecting the importance of the Apolline sanctuary of Delphi. On the other hand no single explanation goes very far to explain its general diffusion, still less its specific distribution pattern; while to invoke all suggested explanations in combination is a sad abdication of investigative scholarship. One might, with strictly guarded optimism, hope that the discovery and excavation of further sanctuaries identifiable as Pythia would increase the number of regional *termini post quos non*, and perhaps allow different explanations to have predominated at different epochs. Until then, however, all I can do is to pose the problem and to leave it open to others.

Acknowledgements

My warmest thanks for their assistance go to Dr Mili (n. 11 above), and to Professor Robert Parker, whose speed, kindness, precision and general helpfulness in reading a draft of this paper were palmary. She and he bear no responsibility for any remaining *bêtises*. My thanks also to the editors, and to the suggestions made by two anonymous referees, which I have gratefully incorporated.

Notes

[1] *Iliad* II 509 and IX 405; *Od.* VIII 80 and XI 581; Hes. *Theog.* 499; Hdt. 1.54.1.
[2] Details in Davies, 'The Moral Dimension of Pythian Apollo', 50–51.
[3] The survey of A.P. by Vogel, *Apollon Onos*, 89–122, adds little.
[4] *IG* X 2,1, nos. 38, 178, and 214, with *SEG* L 638.
[5] Myrina on Lesbos, and Apollonia (but which Apollonia is not clear) (Plut. *De Pyth. Or.* 16 = *Mor.* 402A); Metapontion (Strabo 6.1.15, 264C). The human *ver sacrum* dedications made by Eretrians and Magnesians and cited by Plutarch (ibid.), even if historical, should likewise be left on one side.
[6] Thus at least F. Hiller v. Gaertringen, "Χρυσουν θέρος", *RE* III, 2 (1898): 2522.
[7] Davies, 'Attestations of the Cult of Apollo Pythios'.
[8] Jacoby, *Atthis*, 40–41 and 272 n. 225; Parker, *Miasma*, 131; Parker, *Athenian Religion*, 220 with n. 10. For the word *pythochrestos* see also *IG* XII 3, 329 (Thera, as epithet of the hero Anthister).
[9] Boethius, *Die Pythais*, passim, with brief updates in Parker, *Athenian Religion*, 247 and 250 n. 111. The post-146 renewal of the pilgrimage has attracted much more recent discussion (e.g., Daux, *Delphes au IIe et au Ier Siècle*, 521–83; Bélis, *Les hymnes à Apollon*, 133–42; Habicht, *Athens from Alexander to Antony*, 275–79), but Boethius's remains the best discussion of earlier epochs.
[10] Jost, *Sanctuaires et cultes*, 490–91, reads Paus. 2.7.7–9 as implying a temple of A.P. in the *agora* of Sikyon, but the myth linking the historical temple of Peitho with the need for Artemis and Apollo to be purified after killing Python does not justify the inference.
[11] I am exceptionally grateful to Dr Maria Mili, who has allowed me to use the evidence she has collected, for Apollo Pythios as for other cults of Apollo, in her 'Studies in Thessalian Religion'.
[12] See Moretti, *ISE* II 79 with commentary.
[13] *Bildlexikon*, 91.
[14] Robert Parker reminds me that Travlos's inference has been challenged by Wycherley, *Stones of Athens*, 167 n. 29 and 177 n. 9. Further debate must await Steven Glowacki's forthcoming book on the North Slope.
[15] *SEG* XLI 743, with further commentary by Chaniotis, *Die Verträge*, 190–95 no. 6.
[16] Piérart, 'Un oracle', citing Paus. 2.24.1.
[17] Ibid., 324 n. 30.
[18] See Graf, *Nordionische Kulte*, 38.
[19] The idea goes back at least to Maass, cited by Wernicke, 'Apollo', 65.
[20] Schachter, *Cults of Boeotia* I, 52–73, especially 55–58.
[21] References in *Kleine Pauly*, s.v. *Maleatas*.
[22] References in *Neue Pauly*, and more fully in *Kleine Pauly*, s.v. Diktynna.

[23] Garland, *Introducing New Gods*; Parker, *Athenian Religion*, 152–98; Burkert, 'Greek Temple-Builders'.
[24] Thus Roesch, 'Les cultes égyptiens'; Fraser, 'Two Studies', was more cautious.
[25] Garland, *Introducing New Gods*, chs. 4–5; Parker, *Athenian Religion*, 153f.
[26] Plut. *Lykourgos* 6.2; Hdt. 7.189, with Parker, *Athenian Religion*, 156–57. A referee helpfully cites as a further parallel the oracle given to Mnesiepes to establish a cult of Archilochos on Paros (*SEG* XV 517).
[27] Garland, *Introducing New Gods*, 47–63 and 116–35; Clinton, 'The Epidauria'.
[28] See Burkert, 'Greek Temple-Builders', 27.
[29] See the cults attested in the silver-mining region round Laureion (Lauffer, *Bergwerkssklaven*, 277–85, or the examples cited by Chaniotis, *War in the Hellenistic World*, 149–54.
[30] Examples in Burkert, 'Greek Temple-Builders', 25–28: add Paus. 6.9.6–8 (Kleomedes at Astypalaia).
[31] Morgan, *Athletes and Oracles*, 106–47; Piérart, 'Un oracle', 323 n. 22; Davies, 'The Origins of the Festivals', 54.
[32] Cf. Sporn, *Heiligtümer und Kulte Kretas*, 321 and 380–81.
[33] It is tempting to use the extreme rarity of the month-name Pythoios (attested only at Halos and Meliteia in Thessaly: Samuel, *Greek and Roman Chronology*, 81) as an index of late diffusion, but given the evidence in Linear B of month-names of recognizable later form (Ventris and Chadwick, *Documents*, 304–5) and the accompanying likelihood that calendars had crystallized long before the late seventh century, month-names unfortunately tell us little. Likewise, the known distribution of names in Πυθι- emerges too late, and is subject to too many contributory factors, to be a useful indicator. My thanks, nonetheless, to the referee who raised these points.
[34] The process is set out at length by Sánchez, *L'Amphictionie*, 31–80, sketchily by Lefèvre, *L'Amphictionie*, 13–16.
[35] Morgan, *Athletes and Oracles*, 142ff., and cf. the artefact evidence from Delphi summarized by Rolley, *Les trépieds*, 103, and Rolley, *Les statuettes*, 112–13.
[36] Guarducci, 'Creta e Delfi', correcting the Bronze Age inclinations of Aly, *Der kretische Apollonkult*, and Swindler, 'Cretan Elements'. The whole issue needs close re-examination in the light of the work of Sporn, *Heiligtümer und Kulte Kretas*; Prent, *Cretan Sanctuaries and Cults*; and Skounaki, *Proimes latreies*.
[37] See the papers in Hurst and Owen, *Ancient Colonizations*.
[38] Parker, *Miasma*, 104–43.
[39] Paus. 1.43.7–8. Structuralists will enjoy the *noms parlants*.
[40] Boethius, *Die Pythais* 74–75, ingeniously combined Paus. 8.15.5 with Plut. *De sero numinis vindicto* 16 (Mor. 557C) to yield a third case, but cf. Jost, *Sanctuaires et cultes*, 489–91, for a wholly different analysis.
[41] Paus. 1.37.6–7, with Parker, *Athenian Religion*, 300. For much further detail on Kephalos, see Kearns, *Heroes of Attica*, 70 and 177.
[42] Travlos, *Bildlexikon*, 177.
[43] *IG* I^3 104 = *ML* 86 = *Nomima* I, 02.
[44] *IC* IV 8 and 9 (= *Nomima* II, 78).
[45] *Nomima* I, 01 = *SEG* XXXIX 13 = *IGT* 86*, of debated paternity from south-eastern Sicily, ?525–500; ML 20 = *Nomima* I, 43 (Naupaktos, mid-fifth-century BCE, an *andreophonikos tethmos*). Ad hominem measures such as the curses of Teos (ML 30 = *Nomima* I, 104, c. 475), the judgement against sacrilegious persons at Mantineia (*IG* V 2, 262 = *IGT* 34 = *Nomima* II, 2), or the Alkmaionid curse (Thuc. 1.126–27) are a different matter.
[46] *Nomima* II, 79 = *IGT* no. 32.

Abbreviations

FGrH Jacoby, Felix (ed.) *Die Fragmente der griechischen Historiker.* Berlin: Weidmann, later Leiden: Brill, 1922.
IG *Inscriptiones Graecae.* Berlin: De Gruyter, 1873.
SEG *Supplementum Epigraphicum Graecum.* Leiden: A.W. Sijthoff, later E.J. Brill, 1923.

References

Aly, Wolfgang. *Der kretische Apollonkult. Vorstudie zu einer analyse der kretischen Götterkult* (Habilitationsschrift, Freiburg i.B.) Tübingen: H. Laupp, 1908.

Bélis, Annie. *Les Hymnes à Apollon.* (*Corpus des Inscriptions de Delphes*, III). Paris: École Française d'Athènes, 1992.

Boethius, Axel. *Die Pythais. Studien zur Geschichte der Verbindungen zwischen Athen und Delphi.* Uppsala: Almquist & Wiksells, 1918.

Bourboulis, Photeine P. *Apollo Delphinios.* Thessalonike: N. Nikolaides, 1949.

Bowden, Hugh. *Classical Athens and the Delphic Oracle. Divination and Democracy.* Cambridge: Cambridge University Press, 2005.

Burkert, Walter. 'Apellai und Apollon'. *Rheinisches Museum* 118 (1975): 1–21.

———.'Greek Temple-Builders: Who, Where, and Why?'. In *The Role of Religion in the Early Greek Polis. Proc. 3rd* INT. SEMINAR *on Ancient Greek Cult, Athens Oct. 1992*, edited by R. Hägg. Stockholm: Svenska Institutet I Athen, 1996.

Chaniotis, Angelos. *Die Verträge zwischen kretischen Poleis in der hellenistischen Zeit.* Stuttgart: Franz Steiner Verlag, 1996.

———. *War in the Hellenistic World.* Oxford: Blackwell, 2005.

Clinton, Kevin. 'The Epidauria and the Arrival of Asclepius in Athens'. In *Ancient Greek Cult Practice from the Epigraphical Evidence. Proc. 2nd int. Seminar on Ancient Greek Cult, Athens Nov. 1991*, edited by R. Hägg. Stockholm: Svenska Institutet I Athen, 1994.

Colin, G. *Le culte d'Apollon Pythien à Athènes.* Paris: A. Fontemoing, 1905.

Daux, Georges. *Delphes au IIe et au Ier Siècle.* Paris: De Boccard, 1936.

Davies, John K. 'The Moral Dimension of Pythian Apollo'. In *What Is a God? Studies in the Nature of Greek Divinity*, edited by Alan B. Lloyd. London: Duckworth with The Classical Press of Wales, 1997.

———.'The Origins of the Festivals, Especially Delphi and the Pythia'. In *Pindar's Poetry, Patrons and Festivals*, edited by S. Hornblower and Cathy Morgan. Oxford: Oxford University Press, 2007.

———.'Attestations of the Cult of Apollo Pythios.' Forthcoming.

Defradas, Jean. *Les thèmes de la propagande delphique.* Paris: Klincksieck, 1954. 2nd ed. Paris: Les Belles Lettres, 1972.

Detienne, Marcel. *Apollon le couteau à la main.* Paris: Gallimard, 1998.

Farnell, Lewis R. *Cults of the Greek States,* IV. Oxford: Clarendon Press, 1907.

Fontenrose, Joseph. *Python. A Study of Delphian Myth and its Origins.* Berkeley, Los Angeles and London: University of California Press, 1959.

Fraser, Peter M. 'Two Studies on the Cult of Sarapis in the Hellenistic World'. *Opuscula Atheniensia* 3 (1960): 1–54.

Garland, Robert. *Introducing New Gods. The Politics of Athenian Religion.* London: Duckworth, 1992.

Graf, Fritz. 'Apollo Delphinios'. *Museum Helveticum* 36 (1979): 2–22.

———. *Nordionische Kulte: religionsgeschichtliche und epigraphische Untersuchungen zu den Kulten von Chios, Erythrai, Klazomenai und Phokaia.* (Bibl. Helv. Rom. 21.). Rom: Schweizerische Institut in Rom, 1985.

Guarducci, Margharita. 'Creta e Delphi'. *Studi e Materiali* 19–20 (1943/46): 85–114.

Habicht, Christian. *Athens from Alexander to Antony.* Translated by D.L. Schneider. Cambridge, MA and London: Harvard University Press, 1997.

Hurst, H., and S. Owen, eds. *Ancient Colonizations. Analogy, Similarity, and Difference.* London: Duckworth, 2005.

Jacoby, Felix. *Atthis. The local chronicles of Ancient Athens.* Oxford: Clarendon Press, 1949.

Jost, Madeleine. *Sanctuaries et cultes de l'Arcadie* (*Études Péloponnésiennes,* 9.). Paris: Librairie Philosophique J. Vrin, 1985.

Kearns, Emily. *Heroes of Attica.* (BICS, Suppl. 57.). London: Institute of Classical Studies, 1989.

Lauffer, Siegfried. *Die Bergwerkssklaven von Laureion.* 2nd ed. Wiesbaden: F. Steiner Verlag, 1979.

Lefèvre, François. *L'amphictionie Pyléo-delphique: Histoire et institutions.* (*Bibliothèque des écoles françaises d'Athènes et de Rome,* 298). Athens: École française d'Athènes, and Paris: De Bocard, 1998.

Mili, Maria. 'Studies in Thessalian Religion'. D.Phil. thesis, University of Oxford, 2005.

Morgan, Catherine A. *Athletes and Oracles. The Transformation of Olympia and Delphi in the Eighth Century BC.* Cambridge: Cambridge University Press, 1990.

Parker, Robert. *Miasma. Pollution and Purification in Early Greek Religion.* Oxford: Clarendon Press, 1983.

———. *Athenian Religion: A History.* Oxford: Clarendon Press, 1996.

Perlman, Paula. *City and Sanctuary in Ancient Greece: The Theododokia in the Peloponnese.* Göttingen: Vandenhoeck & Ruprecht, 2000.

Piérart, Marcel. 'Un oracle d'Apollon à Argos'. *Kernos* 3 (1990): 319–31.

Prent, Mieke. *Cretan Sanctuaries and Cults. Continuity and Change from Late Minoan IIIC to the Archaic Period.* Leiden: E.J. Brill, 2005.

Roesch, Paul. 'Les cultes égyptiennes en Béotie'. In *Egitto e la storia antica dell'ellenismo all'età araba. Bilancio di un confronto* (*Atti del Colloquio internazionale, Bologna 1987*), edited by L. Criscuolo, and G. Geraci. Bologna: Editrice CLUEB, 1989.

Rolley, Claude. *Les statuettes de bronze. Fouilles de Delphes.* V. Paris: De Boccard, 1969.

———. *Les trépieds à cuve clouée. Fouilles de Delphes.* V, 3. Paris: De Boccard, 1977.

Samuel, Alan E. *Greek and Roman Chronology: Calendars and Years in Classical Antiquity.* München: Verlag C.H. Beck, 1972.

Sánchez, Pierre. *L'Amphictionie des Pyles et de Delphes. Recherches sur son role historique, des origins au IIe siècle de notre ère,* (*Historia, Einzelschrift* 148.). Stuttgart: Franz Steiner Verlag, 2001.

Schachter, Albert. *Cults of Boeotia, I: Acheloos to Hera.* (BICS, Suppl. 38, no. 1.) London: Institute of Classical Studies, 1981.

Skounaki, Ioulia. *Proimes latreies tou Apollona. I Apollonia triadha.* (*Themata klasikis arhiologias,* 25). Rethimno, 2005.

Solomon, J., ed. *Apollo: Origins and Influences.* Tucson and London: University of Arizona Press, 1994.

Sporn, Katja. *Heiligtümer und Kulte Kretas in klassischer und hellenistischer Zeit.* (*Studien zu antiken Heiligtümern,* 3). Heidelberg: Verlag Archäologie und Geschichte, 2002.

Swindler, Mary. 'Cretan Elements in the Cults and Rituals of Apollo'. Ph.D. thesis, Bryn Mawr College 1913.

Travlos, Ioannis. *Bildlexikon zur Topographie des antiken Athen.* Tübingen: Verlag Ernst Wasmuth, 1971.

———. *Bildlexikon zur Topographie des antiken Attika.* Tübingen: Verlag Ernst Wasmuth, 1988.

Ventris, M., and J. Chadwick. *Documents in Mycenaean Greek.* Cambridge: Cambridge University Press, 1959.

Vogel, Martin. *Apollon Onos.* Bonn: Orpheus Verlag GmbH, Verlag für systematische Musikwissenschaft GmbH, 2003.

Wernicke, Konrad. 'Apollo'. RE 2. (1895): 1–111.

Wycherley, Richard E. *The Stones of Athens.* Princeton: Princeton University Press, 1978.

Cults of Demeter Eleusinia and the Transmission of Religious Ideas

Hugh Bowden

Network Theory is being increasingly applied in the study of both ancient and modern religion,[1] as it allows scholars to examine the way religious practices, vocabulary, and ideas can spread, how new forms of worship are able to grow, and indeed how and why established religions can decline. But theory cannot replace information: hypotheses can only be tested if there exists reliable evidence. While sociologists of religion in the contemporary world can use interviews and questionnaires in their investigations, ancient historians are reliant on whatever can be extracted from the historical record. In examining the connections between the various cults of Demeter Eleusinia in the Greek world, my concern is to ask as much about the nature of the evidence, as about how the cult might have spread.

In his account of the battle of Mykale in 479 BCE, Herodotos remarks that 'the divine aspect of things is clear from many proofs' (9.100.2), noting first that a rumour of the victory at Plataia reached the Greeks at Mykale, even though the battles happened on the same day; he also refers to a *thyrsos* found mysteriously lying on the beach, dropped by some divine messenger, as the reader is encouraged to believe. The historian then points out that both battles took place where there were sanctuaries (*temenea*) dedicated to Demeter Eleusinia (9.101.1). This comment takes the reader back to Herodotos's account of the battle of Plataia, where he notes that 'although the battle was right by the grove (*alsos*) of Demeter, there was no sign that any Persian had been killed in the *temenos* or entered into it; most of them fell near the temple in unconsecrated ground'. And here Herodotus makes one of his least equivocal comments about divine involvement in human affairs, adding, 'I think—if it is necessary to decide about divine matters—that the goddess herself denied them entry, since they had burnt her temple, the shrine at Eleusis' (9.65.2). As he understands it, a crime committed against Demeter in one of her sanctuaries is punished within sight of two other sanctuaries dedicated to her.[2] It would seem that some kind of network is present here.

Mykale and Plataia were not the only places with sanctuaries dedicated to the goddess.[3] Table 1 lists other cases known from our sources, either literary or epigraphic.[4] The list raises some questions: Can we trace the origins of these sanctuaries? Were they formally connected? Did they, in other words, form a wider Eleusinian network? As we will see, these questions are not as straightforward as they might seem. Before we address them, we need to look at the evidence a little more closely.

Herodotos tells us something of the origins of the cult at Mykale. It was founded by 'Philistos son of Pasikles when he went with Nileus son of Kodros to the founding of Miletos' (9.97), that is, as part of the Ionian migration. Interestingly, a parallel explanation is given for the presence of the cult at Ephesos by Strabo, who says that the priest of Demeter Eleusinia was drawn from the *genos* of the Basileis, who claimed descent from Androklos, son of Kodros, founder of the city (14.1.3). This is supported by inscription from the imperial period describing someone as καὶ αὑτὸς βασιλεὺς [καὶ ἱερεὺς Δήμητρος διὰ] γένọụς (IEph 2018.4–5). But there appears to be virtually no mention in the epigraphic record of the cult title 'Eleusinia' either in Ephesos or in Mykale or Miletos.[5] Instead, the cult titles that we find for Demeter at Miletos and Ephesos are usually Karpophoros and Thesmophoros; indeed at Ephesos inscriptions refer to μυστήρια Δήμητρι Καρποφόρῳ καὶ Θεσμοφόρῳ.

There is evidence for 'Eleusinian elements' in the cult of Demeter in these *poleis*, including a dedication at Ephesos to the combination of Plouton, Kore, and Demeter Karpophoros, which might strike one as Eleusinian. Another Asian *polis* with a cult of Demeter *Karpophoros kai Thesmophoros*, Pergamon, is reported to have had cult personnel, including a Hierophant, a Dadouchos, and *ho epi bomon*—all figures paralleled at Eleusis. Herodotos and Strabo had their reasons for their identifications, therefore. But the relationship between these cults and that at Eleusis itself is not necessarily straightforward.

There are two more observations that can be made about the Ionian cults of Demeter Eleusinia. The first is that one might reasonably assume that Eleusinian cults would be associated with the Eleusinian *gene*, that is, the priestly clans of the

Table 1 Evidence for Cults of Demeter Eleusinia

	Evidence for title Eleusinia	*Cult including Mysteries*		*Start of cult*
Attika				
Eleusis		Yes	*Homeric Hymn to Demeter* etc.	Before 600 BCE
Phlya		Yes	Paus. 1.31.4, 9.27.2, 9.30.12, cf. 4.1.5; Plut., *Them.* 1.3	Before 480 BCE
Boiotia				
Hysiaia (Plataia)	Literary		Hdt. 9.57.1–2, 62.2, 65.2, 69.1; Plut. *Arist.* 11.3–8; Paus. 9.4.3	Before 479 BCE
Arkadia				
Basilis	Literary		Paus. 8.29.5, Athen. 13.609e–f	Possibly before 371 BCE
Lykosoura			Paus. 8.37.1–10	By mid C2 CE
Mantineia			IG v^2, 265–6	
Megalopolis		Yes	Paus. 8.31.7–9; IG v.2 517 (Late C2/Early C3 CE)	Possibly ca. 300 BCE
Onkeion		Yes	Paus. 8.25.4–7	By mid C2 CE
Pheneos	Literary	Yes	Paus. 8.15.1–4	By mid C2 CE
Thelpousa	Literary		Paus. 8.25.2–3	Probably not before C4 BCE
Trapezos			Paus. 8.29.1	By mid C2 CE
Argolid				
Epidauros			IG iv^2.1 83 (40-2 CE); IG iv^2.1 126 (post 117 CE)	Probably by 42 CE
Keleai (near Phlious)		Yes	Paus. 2.14.1–4	By mid C2 CE
Lerna		Yes	Paus. 2.36.7, 37.3	By mid C2 CE
Messene				
Andania		Yes	Paus. 4.1.5–9, 4.26.6–8, 4.33.4–6; IG v.1 1390 = LSCG 65 = Syll^3 736 (92/1 BCE); Syll^3 735 (93/2 BCE)	Probably ca. 370 BCE; certainly by 91 BCE
Lakonia				

(*continued*)

Table 1 – *Continued*

	Evidence for title Eleusinia	*Cult including Mysteries*		*Start of cult*
Therai	Literary, Epigraphic		Paus. 3.20.5–7, IG v.1 607 (C2–3 CE)	By mid C2 CE
Aegean Islands				
Delos			ID 2475	
Ionia				
Ephesos	Literary	Yes	Strab. 14.1.3 (633); GIBM 481 (83/4 CE); I Eph 1270 (imperial);	By 25 CE
Erythrai	Epigraphic		LSAM 26 (189–150 BCE)	By 150 BCE
Miletos (Mykale)	Literary		Hdt. 9.97	By 479 BCE
Teos	Epigraphic		GIBM 1032 (= SEG 4, 598)	
Mysia				
Pergamon		Yes	$Syll^3$ 694	By 129 BCE
Karia				
Panamara			IStr 147 (dedication)	
Stratonikeia			IStr 1124 (dedication)	
Pisidia				
Termessos			TAM III 870	
Egypt				
Alexandria	Literary	Yes	Tac. Hist. 4.83, Schol. Callim. In Cer. 1	By ca. 300 BCE

Eumolpidai and the Kerykes. Thus, for example, according to Tacitus, a certain Timotheos (a Eumolpid) was brought to Alexandria by Ptolemy Soter, apparently to establish the cult of Demeter Eleusinia there (*Hist.* 4.83).[6] But in Ephesos and Miletos the cult of Demeter is associated with the sons of Kodros, not with the Eumolpidai; indeed the *genos* in charge of the rites at Ephesos is explicitly descended from the kings of Athens, and not from an Eleusinian ancestor.

My second observation is to note that the Ionian epigraphic evidence is not completely silent about the title Eleusinia. Accounts for sacrificial victims from Erythrai from the mid-second century BCE do include sacrifices to Demeter Eleusinia there, and a decree from Teos mentions τῶν θεῶν] τ̣[ῶν] Ἐ̣λ̣ε̣υσινίων, though without providing a clear context (SEG 4, 598). But at what date these cults were introduced, and what form they took, we cannot say.

Table 1 indicates another part of the Greek world in which the cult of Demeter Eleusinia seems to have been well established, and that is the Peloponnese, and above all Arkadia.[7] For information about these cults we must rely heavily on Pausanias, and unfortunately where we do have other evidence it does not necessarily fit well with Pausanias's statements. This is most evident in the case of the mysteries at Andania, to which I will return later. Even in cases where there is no contradictory evidence, the example of Ephesos and Mykale should make us cautious.

Pausanias identifies four sanctuaries in the Peloponnese as belonging to Demeter Eleusinia, and a further five instances in which he claims, or implies, that the cult practices were either introduced from, or modelled upon, the mysteries at Eleusis. In most cases, he tells the reader a little about the sanctuary, or about the cult. For the cults that he identifies as having the epithet 'Eleusinia', we learn that in Lakonia there is a procession of the cult statue of Kore from Helos to the sanctuary at Therai (3.20.7); that at Pheneos in Arkadia during the so-called *telete meizon*, secret texts are brought out and read to initiates (8.15.2); and that at Thelpousa there are large statues of Demeter, Kore, and Dionysos (8.25.3). From Athenaios comes the additional fact that the sanctuary at Basilis was the location for an annual beauty contest (13.609e–f). For none of these features do we know of parallels at Eleusis itself. Nor is it clear that mysteries were celebrated at any of these sanctuaries other than Pheneos. Conversely, in the case of Thelpousa, Pausanias implies that mysteries were associated with the nearby sanctuary of Demeter Lousia and Demeter Erinys (8.25.7).

In two cases, Pausanias states that Peloponnesian cults were founded directly from Eleusis. The cult at Keleai, near Phlious, was supposedly founded by Dysaules, brother of Keleus, the king of Eleusis who, according to the *Homeric Hymn to Demeter*, reigned when Demeter appeared there (2.14.2). The cult observed at Pheneos in Arkadia, where the greater mysteries included the reading out of secret texts, was founded by Naus, grandson of Eumolpos, the eponym of the *genos* of the Eumolpidai (8.15.1). It is worth noting briefly that, whereas Ionian cults are associated with kings of Athens, these Peloponnesian foundations are associated with rulers of Eleusis. We cannot know whether these ancient 'pedigrees' belong to a long-established tradition, or are of recent antiquarianism.

There is no evidence in any of these cases—whether in the Peloponnese or in Asia Minor—of a formal connection with the sanctuary of Demeter at Eleusis. We have no records of *theoriai*, or of dedications from communities that observed cults related to Demeter Eleusinia. This notwithstanding, there were at times good reasons to claim such a connection. In the fifth century BCE the Athenians instituted, or re-instituted, the festival of the First Fruits, encouraging all the Greek cities to send offerings of wheat and barley to the sanctuary at Eleusis, and promising good harvests to those who did so, thereby remaining on good terms with the Athenians (IG i^3 78).[8] Those cities already connected to Athens and Eleusis by a common cult would have been more likely to join in the venture. In the second century CE the creation under Hadrian of the Panhellenion had significant effects on Eleusis. The fabric of the sanctuary underwent major refurbishment, and it has been suggested that the Panhellenes made dedications at Eleusis in imitation of the First Fruits dedications of the fifth century.[9] The significance of Eleusis in the world of the Panhellenion can also be seen in dedications made by Panhellenes in their home towns. We have a dedication by Eurykles in Aizanoi in northern Phrygia from 157 CE to various deities, including Antoninus Pius, the divine Hadrian Panhellenius, the Eleusinian goddesses, Athena Polias, Poseidon, and Amphitrite (SEG 42:1191). Nonetheless, there appears to be no correlation between members of the Panhellenion and the *poleis* with cults of Demeter Eleusinia. Admittedly, the evidence for all of these matters is very limited, but it seems that there is no connection between the Panhellenion and the cult of Demeter outside Attika.

In spite of the lack of evidence of ongoing contact, a number of attempts have been made to explain why the apparently explicitly Attic cult title 'Eleusinia' should appear in the Peloponnese at all. It has been suggested that the cult, and presumably therefore also the mysteries, were introduced from Athens by the Dorians when they moved into the Peloponnese. Alternatively, it has been suggested that the cult title was not related to Eleusis in Attika, but is a version of 'Eleithyia', and that the connection with Demeter is a later development.[10] Such generalizing explanations depart from the assumption that both cult and title go back to before the historical period, an assumption that is supported by a number of stories mentioned by Pausanias, but which is open to question.

An alternative view, put forward by Jost for Arkadia in particular, points to an existing indigenous cult of Demeter and her daughter (cf. Paus. 8.37.9)—not in this case Kore—onto which an Eleusinian form of cult was later grafted.[11] The lack of any precise evidence for when cults were actually established makes this claim difficult to substantiate, and consequently any discussion of origins must remain purely hypothetical. In any case it is difficult to separate discussion of cults of Demeter Eleusinia from discussion of cults of Demeter, which are said to derive from Eleusis, and even from cults of divinities who may be associated with Eleusis, for example cults of *Megalai Theai*.[12]

Of course, one factor that should not be ignored is the long-term spread of interest in matters Eleusinian, as initiates from outside Attika returned to their native communities, and as Athenian initiates settled abroad. Initiates were of course forbidden to discuss the secret elements of the Mysteries, but the fact that they had been initiated was not a secret. Awareness of the significance of Eleusis itself would also have been increased through the work of *spondophoroi* (heralds) dispatched through the Greek world to invite people to the mysteries.[13] Although the effect of these things cannot be easily measured, it is likely that the name of Eleusis, and the cult associated with it, was widely known and generally respected in the Greek world. But the case needs further exploration before one can infer that local cults were themselves related to Eleusis.

Recent work on 'ritual dynamics', in particular by Angelos Chaniotis, has brought new attention to the way in which ritual practice, and the interpretation of it, evolved over time.[14] This may often have been the result of political changes—and the period covered in this paper witnessed more than one fundamental transformation of the political landscape of the Greek world—but there could be other forces at work. As Chaniotis notes:

> A factor of enormous importance for the evolution of rituals—no matter whether we are dealing with revival or transmission, amalgamation or syncopation, aesthetic or ideological transformation—is the part played by individuals, their idiosyncrasy, personal piety, social position, education, or even political agenda.[15]

Given this, I wish now to consider the role of one particular category of individuals, who might be described as 'religious experts', and the effect they might have had on mystery cults. 'Religious experts' come in a variety of types. While we are familiar with prominent men like the Athenian Lampon, who are called upon by the *polis* to consider religious matters,[16] other such figures are less visible among the historical records, but are nevertheless referred to; for example, Epimenides of Crete (Pl., *Leg.* 642d, D.L. 1.110), or Methapos, whom we will meet shortly. Homer and Hesiod too are treated by later generations as sources of religious knowledge (e.g., Ar. *Ran.* 1032–35). But we should not underestimate the influence of some of the writers whose work has survived, in particular Herodotos and Pausanias. It is certainly the case that, by the early second century CE, Herodotos had become an authoritative source of information about Delphi for the officials at the sanctuary, if Plutarch's categorizing him as an oracle collector has any meaning (*Mor.* 403e). Herodotos is presented in his history as more than ready to talk to religious officials, in Egypt (2.143), in Dodona (2.52.1), and at Delphi. He was, it would appear, an initiate at Samothrake (2.51) and probably at Eleusis, and in his work he is quite ready to make comparisons between different cult practices (e.g., 2.81).[17] Pausanias too was ready to talk to religious officials. He specifically mentions the *dadouchos* (torch-bearer) of the mysteries at Phlya, as being someone he has discussed things with (9.27.2). He was himself an initiate at Eleusis (1.38.7) and at Andania (4.33.5), and at Lykosoura (8.37.9), and quite probably at Megalopolis and Keleai. There have been recent studies of Pausanias, particularly in his role as observer and learner;[18] and he may also be considered as an actor and teacher.

As has already been noted, between the time of Herodotos and that of Pausanias, the Greek world had undergone major transformations. The independent *poleis* of Herodotos's time had been made subordinate first to the Macedonian kings and their various Hellenistic successors, and then to Rome. Given the close connection between the religious and the political affairs of cities, this must have had an impact on how the citizens of the *poleis* understood their own cults and festivals.[19] Nonetheless, the *polis* remained the basic organizational unit of the Greek world throughout this period,[20] and, as we will see, while festivals whose interpretation was a matter of public concern might have to be reinterpreted in the light of new political circumstances, festivals in which significant elements were not allowed to be publicly discussed might be less affected by these changes. So in this respect the world of Pausanias was not necessarily completely different from the world of Herodotos.[21] And for the purpose of this study, the two aforementioned historian-writers are taken to be representative of a larger phenomenon, and belonging to a class of educated travellers who can be identified from Solon (Hdt. 1.29) onwards.

Herodotos and Pausanias then were well-travelled, literate men interested in discussing matters of cult. From the perspective of most small communities in the Greek world they would be exceptional individuals, who would have a perspective quite probably far superior to that of the cult personnel in places like, say, Phlious. We can get a picture of Pausanias's method perhaps from what he has to say about two mystery cults, those at Keleai, and at Andania. Although he avers that he holds the mysteries at Andania as second in importance only to Eleusis, it would seem that he obtained most of his information about it while in Attika. He mentions a conversation with a torch-bearer of the Lykomidai, an official at the mysteries at Phlya (9.27.2), and a statue base in a sanctuary there, dedicated by Methapos, who claimed to have made some changes to the cult at Andania (4.1.8). It would seem to be on the basis of this evidence, supported by a toponym from the poetry of Rhianos of Crete (4.1.6),[22] that Pausanias reconstructs the early history of the mysteries at Andania. According to this account, the mysteries were introduced to Messenia at its creation by Kaukon, grandson of Phlyos, the eponymous founder of Phlya, who brought the rites from Eleusis. They were then modified by Lykos son of Pandion, and further improved by Methapos (4.1.5–9).[23]

But Pausanias tells another story about the (re)foundation of the mysteries, in which Epaminondas and Epiteles, at the time of the fourth-century founding of Messene, were led by a dream to discover a jar containing an inscribed piece of tin foil, which had been buried by Aristomenes at the end of the Messenian revolt (4.20.4): the foil related the secrets of the Andanian mysteries (4.26.6–8). This latter story is more likely to have been told in Messenia. Here Pausanias not only provides two distinct stories about the mysteries, but he also makes links between them. We might ask whether the Messenians who told him the second story heard the first one from him, in return, and if so, what they made of it.

Pausanias's comments about Keleai suggest that some kind of debate might have followed. He notes that the rite at Keleai is a *mimesis* of the *teletai* at Eleusis, and adds that the Phliasians agree (*homologousi*—perhaps 'concede' catches the tone better) that they imitate *ta en Eleusini drômena* (2.14.1). He then reports the Phliasian claim that the mysteries at Keleai were founded by Dysaules, brother of Keleus, king of Eleusis, who fled there after defeat by the Athenians under Xouthos (2.14.2), and goes on to point out that there was no actual battle in the war with Xouthos, and to suggest that Dysaules could not have been a significant Eleusinian, since he is not named in the *Homeric Hymn to Demeter* (2.14.2–4). What his Phliasian interlocutors thought about their own traditions after this barrage of erudition is not revealed, but the stories they told after this exchange with Pausanias may have differed from those they had been telling before.

We should note that there is an issue of authority here. We might suppose that people directly involved with the cults gave precedence to local traditions, because they were passed down from generation to generation. However, so far there is no good evidence to believe that this is how things actually worked. In a society where a reference in the catalogue of ships could be taken as a historical charter (see Plut. *Sol.* 10.1), the foreigner who knew his Homer could probably win out over the native who did not.

Such a supposition leads to some odd results. Herodotos, who was initiated at Samothrake, considered that the *Theoi Megaloi* of the island were the Kabeiroi (2.51.1–2), despite the fact that the name is never used in any document on Samothrake. Even more bizarrely, Pausanias, who was initiated at Andania, considered that it was a cult of Demeter and Kore, the *Theai Megalai*, when the one inscription we have from the site refers to the mysteries of the *Theoi Megaloi* ($Syll^3$ 736). There have been numerous attempts to explain this startling confusion,[24] but it seems to me that here there is a clash between the authority of Pausanias's literary sources on the one hand, and practice on the ground on the other. The strength of the former seems to have been enough to distract Pausanias's attention from the gender of the word endings.

Is it possible to attribute a role to men like Herodotos and Pausanias in the creation of links between cults? In truth, all ritual activity is made up of basic elements, most of which are shared between cults over a wide area. There is likely to have been a standard form of animal sacrifice, for instance, with its accompanying purifications, libations and so on, shared by most divine cults across the Greek world and beyond, whereas other ritual acts might be associated with particular divinities. The daily interactions between individuals and communities could therefore have spread something approaching a shared understanding of these acts, and as such ritual can be seen as a sort of language conveying information about gods and mortals, and about the relationships between them.

With mystery cults this inter-communication is necessarily much more limited. While the public aspects of such cults are likely to be understood well enough, the secret parts—those known only to initiates—will inevitably be more obscure to participants. As it was rare for individuals to be initiated into a large number of cults, there will have been comparably little opportunity to draw parallels between practices.

For most participants of course this need not have mattered, but for some it might. What was actually secret seems to have varied from cult to cult, but often it appears to have played a minor role in the general practice of worship. The mysteries of Demeter seem generally to be related to the story of the abduction of Persephone, which must in part reflect the wide circulation of the *Homeric Hymn to Demeter*, but which also might relate to the importance of the seasonal cycle to cults of Demeter. But in itself that single story can generate a range of secrets. At Lykosoura and Onkeion in Arkadia, for example, the name of Despoina, the daughter of Demeter and Poseidon, conceived while Demeter was searching for Kore, was a secret known only to initiates; whereas at Pheneos and probably at Andania some 'secret' texts were apparently read out loud to the worshippers present.

Consequently, just how much the actual ritual of the mysteries varied from cult to cult is difficult to know—after all, we only have the vaguest idea of what happened at Eleusis itself. Nor do we know how exactly ritual actions were copied and therefore transmitted within cults. The assumption that this was a crucial element of ancient religion should therefore perhaps be questioned. If there was a certain fluidity in the transmission of ritual, then the contribution of learned outsiders could be quite significant. If Pausanias were to tell you that your mysteries were obviously modelled on those of Eleusis, but that they differed in one small respect, would you keep doing the different thing to assert your distinctiveness, or would you modify your practice to be closer to the more prestigious cult? The latter possibility might well be more likely—a case perhaps of the theologians taking over the Pantheon.

As to how far such influence actually travelled, the question is of course unanswerable, given that there must have been other factors affecting the interpretation of cult activity. It does, however, give us an idea of how the process might have operated. If we return to our starting point, we have Herodotos's reference to a sanctuary dedicated to Demeter Eleusinia at Mykale, a title otherwise unattested by epigraphic evidence; and a similar case with Strabo's discussion of Ephesos. We might explain this away as the difference between the official title on the one hand, and a more colloquial one on the other, but this may not be the whole truth. Although the usual title of the cult of Demeter in Miletos and elsewhere is 'Karpophoros', we do have one reference in an inscription to Demeter Karpotokos (IEph 1305), which presumably refers to the same cult. But the confusion goes a little further: when Apollo at Didyma was asked about the erection of an altar to Kore next to the altar of Demeter *Karpophoros*, he replied that it would indeed be acceptable to erect the altar 'next to the altar of Demeter *Karpotrophos*' (IDid 504. 11–2.24)—a reply that suggests that the deity himself was a little confused about his aunt's title.

The appellations Karpotrophos, Karpotokos, and Karpophoros are closely related in meaning, and Demeter Karpophoros is sometimes also called 'Thesmophoros'. Perhaps it is better not to think of them as titles at all, but simply as descriptions that help to distinguish between different sanctuaries and cults. But descriptions can also be used to point up similarities, and writers like Pausanias may have used them to designate such connections. From this evidence, to distinguish between links made by scholarly tourists, and those which result from the movement of ideas across networks, may not always be possible.

Notes

[1] Ancient: e.g., Collar, this volume; Rutherford, this volume. Modern: e.g., Hirst, 'Social Networks'.

[2] For this as a feature of Herodotos's approach, see Bowden, 'Herakles, Herodotos'.

[3] Farnell, *Cults*, 3.198–213; Sfameni Gaspasso, *Misteri e culti*.

[4] The absence of two areas of the Greek world from Table 1 deserves explanation. At least two Cretan *poleis*, Biannos and Olos, as well as the island of Thera, named their months *Eleusunios* or *Eleusinios*. This makes it highly likely that these communities celebrated a festival called the Eleusinia or Eleusunia. It is, however, not known whether they had sanctuaries dedicated to Demeter Eleusinia. On this, see Parker, 'Demeter'. Magna Graecia is an area particularly associated with the cult of Demeter and Kore. However, there is no evidence at all for the cult title 'Eleusinia', or indeed for any sanctuary-based initiatory cult of the goddesses (Hinz, *Der Kult von Demeter*), and no other evidence linking the sanctuaries specifically with Eleusis. In Attica there was an Eleusinion at Phaleron, as well as one in the city and the sanctuary at Eleusis. It is clear that these were all part of the same cult complex, as they were administered together (IG i^3, 32). Inscriptions from Marathon, Paiania and Phrearrhioi refer to an Eleusinion, but there is no other evidence for local sites of that name, and all the inscriptions could be referring to the city Eleusinion. Although nothing is certain, there seems no good reason for supposing the existence of local cults in addition to that at Eleusis. On Attica in general see Parker, *Polytheism*, 327–68, esp. 332–33.

[5] The fact that an inscription from Milesian colonies in the Black Sea includes the name 'Eleusinios' has been taken as evidence for a cult of Eleusinian gods from an early date in Miletos itself (Graf, *Nordionische Kulte*, 274 n. 18; *LGPN* IV s.v. Ἐλευσίνιος). However, it is not safe to assume that, even if Eleusinios were a direct descendent of one of the original colonists, his ancestor was a Milesian citizen: Milesian 'colonies' would have had a Milesian *oikist*, and no doubt the blessing of Apollo at Didyma, but the settlers themselves may have come from further afield—thus enabling Miletos to be metropolis of so many settlements in a relatively short time.

[6] 'Timotheum Atheniensem e gente Eumolpidarum, quem ut antistitem caerimoniarum Eleusine exciverat.' Cf. Schol. Callim. *In Cer.* 1.

[7] Cf. Jost, *Sanctuaires et cultes*; 'Mystery Cults'.

[8] Discussion in Bowden, *Classical Athens*, 125–29; Cavanaugh, *Eleusis and Athens*, 29–95.

[9] Spawforth and Walker, 'World of Panhellenion', 100.

[10] Discussed in Graf, *Nordionische Kulte*, 274–77. See also Parker, 'Demeter', 101–2.

[11] Jost, 'Mystery Cults'.

[12] Jost, 'Grandes Déesses'; 'Nouveau regard'.

[13] Mylonas, *Eleusis*, 244.

[14] Chaniotis, 'Ritual Dynamics: Boiotian Festival'; 'Ritual Dynamics in the Eastern Mediterranean'; 'Statusänderung'.

[15] Chaniotis, 'Ritual Dynamics in the Eastern Mediterranean', 162.

[16] Bowden, 'Oracles for Sale'.

[17] On religion in Herodotus in general: Harrison, *Divinity and History*; Mikalson, *Herodotus and Religion*.

[18] Rutherford, 'Tourism'; Elsner, 'Pausanias'.

[19] Chaniotis, 'Ritual Dynamics: Boiotian Festival', 39–40.

[20] As witnessed for example by Jones, *Greek City*.
[21] Pausanias himself reinforces this perception through his imitation of Herodotos. See e.g., Hutton, *Describing Greece*, 190–213.
[22] For Pausanias's reliance on Rhianos see 4.6.1–3. On his method in Book 4 see Musti and Torelli, *Pausania*.
[23] Cf. Ogden, *Aristomenes*, 89–103.
[24] Guarducci, 'I culti'; Robertson, 'Melanthus, Codrus'; Zunino, *Hiera Messeniaka*; Graf, 'Lesser Mysteries'.

References

Bowden, Hugh. 'Oracles for Sale'. In *Herodotus and his World: Essays from a Conference in Memory of George Forrest*, edited by Peter Derow and Robert Parker. Oxford: Oxford University Press, 2003.

———. *Classical Athens and the Delphic Oracle: Divination and Democracy*. Cambridge: Cambridge University Press, 2005.

———.'Herakles, Herodotos and the Persian Wars'. In *Herakles and Hercules: Exploring a Graeco-Roman Divinity*, edited by Louis Rawlings and Hugh Bowden. Swansea: Classical Press of Wales, 2005.

Cavanaugh, Maureen B. *Eleusis and Athens: Documents in Finance, Religion and Politics in the Fifth Century B.C.* Atlanta: Scholars Press, 1996.

Chaniotis, Angelos. 'Ritual Dynamics: the Boiotian Festival of the Daidala'. In *KYKEON. Studies in Honour of H.S. Versnel. Religions in the Graeco-Roman World 142*, edited by H. F. J. Horstmanshoff, H. W. Singor, F. T. van Straten, and J. H. M. Strubbe. Leiden: Brill, 2002.

———.'Ritual Dynamics in the Eastern Mediterranean: Case Studies in Ancient Greece and Asia Minor'. In *Rethinking the Mediterranean*, edited by William V. Harris. Oxford: Oxford University Press, 2005.

———.'Griechische Rituale der Statusänderung und ihre Dynamik'. In *Investitur- und Krönungsrituale. Herrschaftseinsetzungen im kulturellen Vergleich*, edited by Marion Steinicke and Stefan Weinfurter. Köln: Böhlau, 2005.

Elsner, John (Jas). 'Pausanias: A Greek Pilgrim in the Roman World'. *Past and Present* 135 (1992): 3–29.

Farnell, Lewis R. *The Cults of the Greek States*. 5 vols. Oxford: Clarendon Press, 1896–1909.

Graf, Fritz. *Nordionische Kulte: religionsgeschichtliche und epigraphische Untersuchungen zu den Kulten von Chios, Erythrai, Klazomenai und Phokaia*. Rome: Schweizerisches Institut in Rom, 1985.

———.'Lesser Mysteries – Not Less Mysterious'. In *Greek Mysteries: The Archaeology and Ritual of Ancient Greek Secret Cults*, edited by Michael B. Cosmopoulos. London and New York: Routledge, 2003.

Guarducci, M. 'I culti di Andania'. *Studie materiali di Storia delle religioni* 10 (1934): 174–204.

Harrison, Thomas. *Divinity and History: The Religion of Herodotus*. Oxford: Oxford University Press, 2000.

Hinz, Valentina. *Der Kult von Demeter und Kore auf Sizilien und in der Magna Graecia*. Wiesbaden: Reichert, 1998.

Hirst, Rob. 'Social Networks and Personal Beliefs: An Example from Modern Britain'. In *Predicting Religion: Secular and Alternative Futures*, edited by Grace Davie, Paul Heelas, and Linda Woodhead. Aldershot: Ashgate, 2003.

Hutton, William. *Describing Greece: Landscape and Literature in the* Periegesis *of Pausanias*. Cambridge: Cambridge University Press, 2005.

Jones, Arnold H. M. *The Greek City from Alexander to Justinian*. Oxford: Clarendon Press, 1940.

Jost, Madeleine. 'Les Grandes Déesses d'Arcadie'. *Revue des Études anciennes* 77 (1970): 138–51.
———. *Sanctuaires et cultes d'Arcadie*. Paris: J. Vrin, 1985.
———.'Nouveau regard sur les Grandes Déesses de Mégalopolis: influences, emprunts, syncrétismes religieux'. *Kernos* 7 (1994): 119–29.
———.'Mystery Cults in Arcadia'. In *Greek Mysteries: The Archaeology and Ritual of Ancient Greek Secret Cults*, edited by Michael B. Cosmopoulos. London and New York: Routledge, 2003.
Mikalson, Jon D. *Herodotus and Religion in the Persian Wars*. Chapel Hill: University of North Carolina Press, 2003.
Musti, Domenico, and Mario Torelli, eds. *Pausania, Guida della Grecia, IV, La Messenia*. Milan: Mondadori, 1991.
Mylonas, George E. *Eleusis and the Eleusinian Mysteries*. Princeton: Princeton University Press, 1961.
Ogden, Daniel. *Aristomenes of Messene: Legends of Sparta's Nemesis*. Swansea: The Classical Press of Wales, 2004.
Parker, Robert. 'Demeter, Dionysus and the Spartan Pantheon'. In *Early Greek Cult Practice*, edited by Robin Hägg, Nanno Marinatos, and G. C. Nordquist. Stockholm: Paul Åströms Vörlag, 1988.
———. *Polytheism and Society at Athens*. Oxford: Oxford University Press, 2005.
Robertson, Noel. 'Melanthus, Codrus, Neleus, Caucon: Ritual Myth as Athenian History'. *Greek Roman and Byzantine Studies* 29 (1988): 201–61.
Rutherford, Ian. 'Tourism and the Sacred: Pausanias and the Traditions of Greek Pilgrimage'. In *Pausanias: Travel and Memory in Roman Greece*, edited by Susan E. Alcock, John F. Cherry, and Jas Elsner. New York: Oxford University Press, 2001.
Sfameni Gasparro, Giulia. *Misteri e culti mistici di Demetra*. Rome: 'L'Erma' di Bretschneider, 1986.
Spawforth, A. J., and Walker, Susan. 'The World of the Panhellenion I: Athens and Eleusis.' *Journal of Roman Studies* 75 (1985): 78–104.
Zunino, Maddalena L. *Hiera Messeniaka. La storia religiosa della Messenia dall' età micenea all'età ellenistica*. Udine: Forum, 1997.

What Travelled with Greek Pottery?

Robin Osborne

During the sixth and fifth centuries BC very large amounts of Athenian black- and red-figure pottery were transmitted round the Mediterranean. The nature of the exchange relations underlying this pottery distribution have long been a topic for discussion, and in two earlier papers I have argued both that Athenian potters responded to very specific orders from particular Italian markets, and that, taken as a group, Italian markets consumed voraciously whatever Athenian potters produced, with the iconographic initiative resting at least predominantly with the Athenians.[1] In this paper I sidestep the issue of the direction of that initiative, and revisit the question of what types of pots ended up in which places, to investigate what *sort* of information flowed along the network created by the exchange of pottery. Building on work by other scholars designed primarily to answer slightly different questions, I ask: did Corinthian and Athenian pots export a lifestyle and an outlook on life? By looking on the one hand at the find contexts of Greek pottery outside mainland Greece, and on the other at the images found on that pottery, I ask whether Greek pots served to construct a network of shared cultural practice and knowledge, or were simply parasitic upon, a (changing) cultural *koine*.

In the physical sense, all sorts of things travelled with Greek pottery and, vice versa, Greek pottery travelled with all sorts of things. One of the things that has become clear from the excavation of wrecks is that some ships carried only small amounts of pottery, their main cargo being something quite different, while other ships carried very significant amounts of pottery. The Giglio wreck, dated to shortly after 600 BC, yielded just fifty pieces of fine pottery; whereas the Pointe Lequin 1A wreck, dated to ca. 515 BC, is estimated to have had 800 Attic cups, 1,600 'Ionian' cups, and 150 further fine vessels, mostly Attic.[2] Some fine pottery seems to have travelled with transport amphorae from the same origin, others to have circulated independently of commercial amphorae. Rouillard has suggested that Corinthian and Etruscan fine pottery and amphorae seem largely to travel together, but noted that SOS amphorae disappear from the picture when Attic fine pottery begins to be exported in quantity.[3] Again, we have very direct confirmation from a wreck of the way fine pottery and amphorae from the same origin travelled together in the Ecueil de Miet wreck with 100 Etruscan amphorae and bucchero kantharoi.[4]

But I am only marginally concerned here with the physical sense. What I am really interested in is what consumers outside the place of production gained when they acquired fine Greek pottery. Did they simply acquire the *kudos* of owning an exotic item, or did they acquire some form of social or cultural knowledge? Did acquiring a particular type of drinking vessel mean also assuming certain drinking habits? Or did the use of an imported perfume vessel affect habits of bodily hygiene or social intercourse? Or did a vessel with figured decoration afford knowledge of the life of the imaginary (whether that be the imaginary of myth or the imaginary of warrior or athletic or gender ideology)? Was there a network of those who 'knew' Greek, not in the sense of knowing the Greek language (though language acquisition cannot be entirely divorced from these questions), but in the sense of knowing, and being able to replicate by behaviour or in discourse, Greek culture?

At one level these are unanswerable questions. For most Greek artefacts found abroad we are never going to know whether the person into whose hands, by whatever means, they came knew what they were for or what they embodied in cultural terms. When pots shaped specifically for use in the symposium end up in graves, this does not mean that those who acquired them were unaware of their original intended use. When vessels showing warriors end up in a warrior grave, it is safe enough to assume that those who put them there realized that what they showed were warriors, but one cannot infer from this that they knew anything about hoplite warfare, let alone deduce how they themselves fought.[5] Acquiring a pot with an image of the sacrifice of Polyxena did not mean that the owner could relate the story of the burial of Achilles.

But if these questions cannot be answered at the level of the individual consumer or, often, at the level of the individual artefact, this does not rule out altogether reasonably secure answers. Patterns of assemblage give a fair idea of the company that Greek pots kept in their new homes, and images produced by non-Greeks showing Greek artefacts, while hardly qualifying as snapshots, provide some fairly strong contextual information. Context also offers some clues about the interpretation of figural imagery, while the reproduction of such imagery by those who acquired it often reveals their degree of understanding. While it is impossible to show what was always the case, we can have some idea of what was sometimes the case.

I begin with issues of shape. Shape is a particularly good indicator of lifestyle, since some styles of life are impossible without suitable shapes of pot. For instance, the presence of aryballoi does not necessarily indicate that Greek athletic activity is present, but their absence is more telling, simply for the reason that one cannot oil oneself in the gymnasium using an amphora or a cup. When a full repertoire of sympotic shapes is found in a given context (above all kraters and cups), we can reckon that there is at least some chance that the institutions of the symposium have also arrived. In the absence of kraters, some sign is needed that large open vessels of some other kind were available for the mixing of wine before one leaps from the presence of drinking vessels to the presence of the symposium. Consequently, the fact that at Ensérune between 600 and 450, 93 per cent of the Greek material consists of cups, and most of the fine Etruscan wares of the sixth century consist of *kantharoi*, is insufficient evidence for sympotic practice there—indeed, it might be taken to indicate the opposite.[6] What is more, the pattern of ceramic imports at the Greek settlement at Massalia is not so very different; here again, cups and Etruscan *kantharoi* dominate the sixth-century assemblage.[7] Literary sources going back to the Aristotelian *Constitution of Massalia* record the Greek settlers as bringing the vine and the olive to the south of France, but it is far from clear whether they also brought the symposium with them, or whether they themselves adopted the drinking practices of the local population.[8] Early imports to the Greek settlements were dominated by Etruscan amphorae and bucchero kantharoi, but when local ceramic production was developed—the so-called Pseudo-Ionic and grey monochrome wares—most local products took the form of cups and shapes derived from the native repertoire.[9] Equally, to import a krater does not mean using it for mixing wine. Higher up the Rhone valley we find a different pattern of Greek ceramic imports, with the krater a significant presence, along with some exceptionally fine cups. There is some reason to believe, however, that kraters in general—like the great bronze Vix krater—were markers of elite status, rather than functional vessels for mixing wine and water in a sympotic context.[10] Similarly, at Los Nietos in Spain, excavations have uncovered eight fourth-century red-figure kraters along with amphorae, but no cups.[11] In various Spanish contexts the *chous* seems to have become a libation vessel,[12] and there are indications that in Andalusia the use of Greek vases for ritual purposes was a reflection of their prestige status, not the acquisition of Greek habits.[13]

How does this compare with what we can deduce about the use of perfume vessels, exported above all from Corinth a century earlier? Roughly half of all known Corinthian seventh-century pottery has been found in Greek settlements abroad or in native Italian cemeteries. For the earlier part of the century (i.e., Protocorinthian) the cemeteries of Greek settlements abroad themselves account for more than half of all known Corinthian pots; for the second half of the century, those cemeteries remain the largest single source, but Italian native cemeteries have become larger consumers of Corinthian pots than cemeteries on the Greek mainland or in the Aegean.[14] In themselves, the number of finds tells us little about usage. The pots in question are partly sympotic (jugs, cups) and partly perfume vessels (aryballoi, alabastra). Residue analysis supports the idea that the aryballoi were not simply vessels for perfume, but they were vessels sold with perfume in them, in some cases apparently rather pungent perfume. Depositing such vessels in graves by Greek settlers abroad might merely mark a conservative repetition of the practices of the cities from which they originated, but the variety of funerary practices attested in the Greek settlements of Sicily, and the numerous ways in which they differ from the practices of 'mother cities', suggests that when we get perfume vessels it is most likely because they played an active role in funerary ceremonial.[15] There is, of course, nothing about the shape of the aryballos or alabastron itself which demands its funerary use. The use of such vessels in cemeteries by the native population of Italy cannot have been learned from the pots alone, but must have been either a product of observation of what the Greeks did with those pots or a result of these pots providing new and convenient containers for what the Italian peoples already employed in funerary ritual.

If for many sites in the western Mediterranean one can only guess how the imported pottery was employed, in Etruria the situation is clearer. From the seventh century on, one finds Etruscan graves that combine drinking vessels of both Greek and of Etruscan type and origin; indeed, a scene on a sixth-century bucchero vase shows both an Etruscan kantharos and a Greek kotyle being used, both by seated men.[16] Greek sympotic vases — kraters, amphorae, jugs, and kylikes — are found too, along with a lyre player, in Etruscan tomb paintings of the end of the sixth and middle of the fifth century.[17] Terracotta architectural reliefs from Etruria and Rome similarly include sympotic scenes. These show the playing of the *aulos* and the kithara, drinkers reclining, kraters, and servants ladling or bringing jugs to refill their drinking cups. But they also show reclining women sharing couches with men. We appear to have all the practices of the Greek symposium, but in a different social framework.[18] This is further emphasized by the peculiarity of the symposium scenes painted on stamnoi (a shape which seems to have been particularly made for the Italian market[19]) by an Athenian workshop, which seems to have been one of few to adapt its imagery to what it took to be Etruscan taste: Alan Shapiro has drawn attention to the way in which the 'Perizoma Group' not only shows athletes in very un-Greek loincloths, but shows women reclining at sympotic couches along with men.[20]

What consideration of pot shapes and functions suggests, therefore, is that Greek pots did not carry Hellenized practices. Simply acquiring Greek pots made for a particular use did not mean acquiring that pattern of use. Non-Greek peoples often took an interest only in such pots as they could employ to do, and do better, the activities in which they were already engaged; in so far as they acquired pots to do something different, it is because they had taken over the different practice from observation of Greek settlers, not because the pots themselves carried that practice with them. Even those of Greek origin and descent, it appears, sometimes adjusted to local patterns of behaviour, rather than insist on acquiring the material commodities that would enable them to carry on behaving as they had in the places from which they came.

If little Greek cultural learning travelled with pot shapes, what about the imagery on pots? Should we imagine that those who acquired figured pottery were paying close attention to the imagery and learning from it a wealth of mythological stories and social practices? It has been tempting for scholars sometimes to suggest, for instance, that the careful labelling not only of human figures but also of objects on the François vase was an educative matter.[21] But a moment of reflection suggests that this will not work. Putting known names next to figures works well as a way of teaching reading to those to whom the name means something, but whose reading abilities are limited. But it makes no sense at all to those for whom the name once deciphered will mean nothing. In this sense, the François vase might have been a great educational tool for Greeks new to writing, but it was a useless tool for those used to reading Etruscan, but who want to learn to read Greek. Only someone with prior knowledge of the stories shown on the vase would have found the writing helpful.

In fact there is quite good reason to think that Greek mythology was sufficiently familiar to Etruscans already in the seventh century for the sophistication of the François vase to be enjoyed, either by those who could or those who could not read Greek. The evidence for this claim is partly archaeological. The Etruscan *pithos* which Anthony Snodgrass has shown to be closer to the literal description of the blinding of Polyphemos than any extant representation on archaic Greek pottery constitutes good evidence that Etruscan artists did not derive their knowledge of Greek myths from the iconography of Greek pottery.[22] Similarly, the Medea named on an Etruscan relief-decorated bucchero *olpe* from the second half of the seventh century does not share an iconography with the Medea of Greek art.[23] Another example dating from the sixth century is the figure of Troilos on sixth-century Etruscan pottery and in the Tomb of the Bulls in Tarquinia, Italy, in which Spivey and Stoddart have shown convincingly that there are features of the iconography that are not found in any Greek version, but which can be explained from features of the story as told which surface in later literary versions.[24] But there is even more powerful linguistic evidence. All the names of Greek gods and heroes in Etruscan derive from the Doric dialect of Corinth. Philologists sometimes refer to this as the 'Demaratean layer', and, whatever we make of the history of Demaratus, this linguistic evidence is evidence that the Etruscans learnt myth (in the seventh century) from the telling of stories, not from looking at pictures.[25]

This independence of knowledge of myth from pictures of myth on pottery is not a phenomenon limited to Etruria. Greek settlements in Sicily and southern Italy rarely produced figured pottery in the Archaic period, and when they did they seem to have done so by importing (at least initially) potters trained elsewhere (as with Chalkidian pottery; and compare the hydriai from Etruscan Caere). But the seventh-century polychrome pottery from Megara Hyblaea shows originality both in its shapes and in its iconography.[26] While from the eighth century on there are certainly cases of potters setting up in Sicily and Italy to produce look-alikes of pottery made on the Greek mainland, as soon as that pottery turns itself to figurative decoration it begins distinctly to diverge.

There is, of course, plenty of evidence for intelligent and discriminating employment of imported Greek vases in accordance with their iconography. Most straightforward are the cases where vases with scenes of warfare are deposited in the tomb of a warrior. Juliet de la Genière has discussed one example from Certosa, Clemente Marconi another from the Contrada Mosè cemetery at Akragas.[27] But neither this nor the peculiar enthusiasm of particular places for particular scenes (for instance the dominance by Vulci of the market for black-figure hydriai with fountain scenes[28]) shows that the pots themselves were transmitting Hellenic culture—these cases may demonstrate, in de la Genière's words, 'l'hellénophilie des Étrusques', but there is no reason to believe that they generated it.

It is particularly interesting in this context to see what emerges from Shanks's analysis of seventh-century Corinthian pottery in the west.[29] Shanks has shown that the Corinthian pots from Greek settlements abroad and from Italian native cemeteries are slightly less inclined to have figured decoration than those found in cemeteries and particularly in sanctuaries on the mainland and in the Aegean. He has also shown that the sorts of figures found in the west are distinct both from those on pots from cemeteries in the mainland and Aegean and from those from sanctuaries there in the first half of the seventh century, but that in the second half they become very close to mainland and Aegean cemeteries, though remaining distinct from sanctuaries.[30] The difference of potentially greatest significance in the early seventh-century pots is the absence of people from the pots in Greek and native cemeteries in Sicily and Italy. This suggests that those carrying the vases began cautiously concentrating on pots whose decoration they thought translated most readily. As they discovered that the market was voracious they gave up being discriminating.

My major conclusion should have become clear. Pottery did not, as far as I can see, serve in any significant fashion to spread Greek cultural knowledge or values, and it did not of itself create a network of people linked by shared cultural knowledge and experience. Of course Greek pots will have excited curiosity. We can expect that many people on seeing some Greek pot shapes for the first time will indeed have asked what that pot was for. Similarly, the scenes on pots will have excited questions both about stories and about Greek life. But all the evidence favours the hypothesis that while pots might reinforce knowledge and habits already acquired, it was people behaving in a particular way who were required to spread habits, and it was stories and histories passed on by word of mouth, or by text, that were required to spread knowledge.

This conclusion goes along, I think, with the view that individual markets might be particularly discriminating, and that far from trade being 'down-the-line' it was mostly 'directed', with merchants setting out, whether on the basis of orders or simply on the basis of their knowledge of the market they were serving, with goods that had been selected to meet particular local taste.[31] That taste was not, in my view, particularly highly selective—overall anything Greek potters produced seems to have found its way to an export market somewhere—but it was preformed.[32] What had formed it was no doubt in part past experience of Greek pottery, but it was also cultural habits acquired independently of any pottery. The common Greek culture of the central Mediterranean in the period from ca. 700 onwards may have been reinforced by Greek pottery exports, but it was not significantly shaped by them.

There are implications here for the question, ignored so far in this paper, of who it was that carried the Greek pottery to non-Greek markets. Nothing about the impact of Greek pots demands that those who sold them on knew about their use or their decoration. There are reasons for thinking that, at least in the eighth and seventh centuries, Corinthian pottery may have been carried around the Mediterranean by Phoenicians as well as by Greeks.[33] The distribution of pottery shapes, including shapes of seventh-century Corinthian pottery, argues against the pots being carried by opportunist traders who reckoned to pay only one, profitable, visit, which is the Homeric picture of the Phoenicians, but it has long been clear that that Homeric picture is an ideological construct, not re-telling of historically familiar experience. We should expect that a lot of non-Greeks, as well as many Greeks, travelled with Greek pots.

I conclude that very little cultural baggage travelled with Greek pottery in the Archaic period. But not none. What the above conclusion overlooks is that what Greek pots gave those who bought them a taste for was things that looked like Greek pots. However much local pot production may differ in detail from Greek painted pottery, there is no doubt at all that Etruscan black- and later red-figure vases broadly imitate the shapes and decorative schemes of Athenian black- and red-figure pottery.[34] Nikosthenes' enterprise of imitating Etruscan shapes, although not without some short-term success (he disposed of as many Etruscanizing pieces as Attic), inspired no emulators. This suggests that his fellow potters in the Kerameikos regarded his imitating Etruscan shapes to be as misguided as the Perizoma group's attempt to produce an Etruscan iconography, and for much the same reason — that they knew the Etruscan market voraciously to consume all Greek iconography and equally voraciously to purchase pots that were of the traditional Athenian shapes.[35] And what is demonstrated in some detail by Etruscan potters is manifest in a simpler way in the various imitations of cups produced at various times and places in the central and western Mediterranean. Whether what they used those pots for bore any relationship to what Greeks of the mainland and Aegean used pots of the same shape for or not, both Greeks and non-Greeks acquired a taste for pots that had a broadly Greek look to them. In the end, the network created by Greek pottery as it travels is a network of aesthetic preferences.

Acknowledgements

I am grateful to Christy Constantakopoulou and Katherina Panayopoulou for their invitation to the most hospitable conference at Rethymnon in May 2006, and to the two sympathetic and helpful readers who substantially reinforced the final product.

Notes

[1] Osborne, 'Pots'; 'Why Did Athenian Pots Appeal?' (with useful corrections of detail in Paléothodoros, 'Pourquoi les Étrusques achetaient-ils?'). The conclusions about voracity square with those offered independently by Reusser, *Vasen für Etrurien.*

[2] Long *et al.*, 'Les épaves archaïques', 205, base this estimate upon the recovery of a minimum number of 1265 'Ionian' cups and more than 500 Attic cups of various sorts (along with 68 transport amphoras. See also Stissi, 'Modern Finds', 354; Parker, *Ancient Shipwrecks*, 192, 323.

[3] Rouillard, 'Le vase attique', 332; for the complex pattern of fine pottery and transport amphoras in archaic wrecks see Long *et al.*, 'Les épaves archaïques', 229. It is not the case that *no* Athenian pottery, other than transport amphoras, is exported in the late eighth and first half of the seventh century, but the quantities are small. For Megara Hyblaea, see De Angelis, *Megara Hyblaea and Selinous*, 89–90.

[4] Hesnard, 'Nouvelles recherches', 237.

[5] Compare the discussions in Marconi, 'Images'; and Osborne, 'Images'.

[6] Dubosse, 'Ensérune'. Open vessels similarly predominate at Saint-Pierre-lès-Martigues: Campenon, 'La céramique grecque'.

[7] Gantès, 'La physionomie'.

[8] Athenaios, *Deipnosophistai*, 576; Justin 43.3–4. See Dietler, 'Driven by Drink'; 'The Cup of Gyptis'.

[9] Dietler, 'The Cup of Gyptis'; 'The Iron Age', 278–79, 303.

[10] Dietler, 'The Cup of Gyptis', 98: 'In the highly stratified societies of the Hallstatt area, Mediterranean imports were valued primarily for their diacritical symbolic value in distinguishing elite consumption rituals ... In the less politically centralized and socially stratified societies of the Lower Rhône basin, wine was valued as an additional element in the traditional arena of commensal politics by which prestige was competed for.' The classic survey of the evidence is Shefton, 'Zum Import und Einfluss'.

[11] Olmos, 'Usos', 428.

[12] Ibid., 433.

[13] Cabrera, 'Comercio'.

[14] Shanks, *Art and the Early Greek State*, fig. 4.2.

[15] Shepherd, 'The Pride of Most Colonials'.

[16] Gran-Aymerich, 'Vases grecques', 449.

[17] Spivey, 'Greek Vases in Etruria', 135–37, discussing the Tomba dei Vasi Dipinti and the Tomba della Nave.

[18] Menichetti, *Archeologia del potere*, 96–98.

[19] Philippaki, *The Attic Stamnos.*

[20] Shapiro, 'Modest Athletes', 330–33. For suggestions that some Dionysiac imagery on *stamnoi* relates to Etruscan rather than Athenian festivals, see de la Genière, 'Vases des Lénéennes?' and 'Images attiques'; and, for counter-arguments, see Osborne, 'The Ecstasy'.

[21] De la Genière, 'Quelques reflexions', 417: 'Quant à l'abondance exceptionnelle des épisodes de la légende grecque qui couvrent le reste du vase, elle répond à l'appétit d'érudition d'une société riche et perméable; et pour le cas où l'usage du grec n'y serait pas parfaitement maîtrisé, le peintre a parfois indiqué le nom des objets représentés.'

[22] Snodgrass, *Homer and the Artists.*

[23] Rizzo, 'Un incunabulo'.

[24] Spivey and Stoddart, *Etruscan Italy*, 99.

[25] I owe my knowledge of the philology to Albio Cassio. On this phenomenon, see de Simone, 'Per la storia', 497–501, 517–88 and, more generally, *Die griechischen Entlehnungen.* For the exception of Odysseus, see Malkin, *The Returns*, 161.

[26] De Angelis, *Megara Hyblaia*, 57–61; Massa-Pairault, *'Megarica'*, 110–19.

[27] De la Genière, 'Quelques reflexions', 420; Marconi, 'Images'.
[28] De la Genière, 'Quelques reflexions', 419.
[29] There are limits to the robustness of Shanks conclusions because his data set does not include fragmentary material and a number of important sites are not represented at all in his data. Nevertheless, it is currently the only such study that is published and it is his data from which I derive my claims here. For what can be achieved with more thorough analysis of the material, see Cooper, *Winged Figures in Corinthian Vase-Painting*.
[30] Shanks, *Art and the Early Greek State*, figs. 4.4, 4.5, 4.6.
[31] That is the view argued in Osborne, 'Pots'.
[32] That is the view argued in Osborne, 'Why Did Athenian Pots Appeal?'.
[33] Morris and Papadopoulos, 'Phoenicians'.
[34] On Etruscan pottery, see Beazley, *Etruscan Vase Painting*; Spivey, *The Micali Painter*; Martelli, *La Ceramica degli Etruschi*.
[35] On Nikosthenes, see Tosto, *The Black-Figure Pottery Signed [NIKOSTHENESEPOIESEN]*.

References

Michel Bats *et al. Marseille grecque et la Gaule (Études Massalietès 3)*. Lattes, 1992.

Beazley, John D. *Etruscan Vase Painting*. Oxford: Clarendon Press, 1947.

Cabrera, Paloma. 'Commercio, intercambios, esferas de interacción: el commercio griego en la Península ibérica'. In *Céramique et peinture grecques: modes d'emploi*, edited by M.-C. Villanueva Puig, F. Lissarrague, P. Rouillard, and A. Rouveret. Paris: Documentation française, 1999.

Campenon, Christine. 'La céramique grecque de Saint-Pierre-lès-Martigues, L'apport des fouilles récentes: quelques remarques préliminaires'. In *Céramique et peinture grecques: modes d'emploi*, edited by M.-C. Villanueva Puig, F. Lissarrague, P. Rouillard, and A. Rouveret. Paris: Documentation française, 1999.

Cooper, Catherine. Winged Figures in Corinthian Vase-Painting: A Contextual Approach to the Use of Iconography on Corinthian Pottery in the Greek World, Ph.D. diss., King's College London, 2007.

De Angelis, Franco. *Megara Hyblaia and Selinous: The Development of Two Greek City-States in Archaic Sicily*. Oxford: Oxford University School of Archaeology Monograph 55, 2003.

De la Genière, Juliette. 'Vases des Lénéennes?'. *Mélanges de l'Ecole française de Rome Antiquité* 99 (1987): 43–61.

———.'Images attiques et religiosité étrusque'. In *Proceedings of the 3rd Symposium on Ancient Greek and Related Pottery*, edited by J. Christiansen, and T. Melander. Copenhagen: Nationalmuseet; Ny Carlsberg Glyptotek; Thorvaldsens Museum, 1988.

———.'Quelques reflexions sur les clients de la céramique attique'. In *Céramique et peinture grecques: modes d'emploi*, edited by M.-C. Villanueva Puig, F. Lissarrague, P. Rouillard, and A. Rouveret. Paris: Documentation française, 1999.

De Simone, Carlo. *Die griechischen Entlehnungen in Etruskischen*. 2 vols. Wiesbaden: O. Harrasowitz, 1968–70.

———.'Per la storia degli imprestiti greci in etrusco'. *Aufstieg und diedergang der römischen Welt* I, no. 2 (1972): 491–521.

Dietler, Michael. 'Driven by Drink: The Role of Drinking in the Political Economy and the Case of Early Iron Age France'. *Journal of Anthropological Archaeology* 9. (1990): 352–406.

———.'The Cup of Gyptis: Rethinking the Colonial Encounter in Early Iron Age Western Europe and the Relevance of World-Systems Models'. *Journal of European Archaeology* 3 (1995): 89–111.

———.'The Iron Age in Mediterranean France: Colonial Encounters, Entanglements, and Transformations'. *Journal of World Prehistory* 11 (1997): 269–357.

Dubosse, Cecile. 'Ensérune: les arrivages de céramiques fines grecques et étrusque (600–450 av. J.-C.'. In *Céramique et peinture grecques: modes d'emploi*, edited by M.-C. Villanueva Puig, F. Lissarrague, P. Rouillard, and A. Rouveret. Paris: Documentation française, 1999.

Gantès, Lucien-François. 'La physionomie de la vaisselle tournée importée à Marseille au VIe siècle av. J.-C.'. In *Céramique et peinture grecques: modes d'emploi*, edited by M.-C. Villanueva Puig, F. Lissarrague, P. Rouillard, and A. Rouveret. Paris: Documentation française, 1999.

Gran-Aymerich, Jean. 'Vases grecs et vases étrusques: les services mixtes'. In *Céramique et peinture grecques: modes d'emploi*, edited by M.-C. Villanueva Puig, F. Lissarrague, P. Rouillard, and A. Rouveret. Paris: Documentation française, 1999.

Hesnard, Antoinette 'Nouvelles recherches sur les épaves préromaines en baie de Marseille'. In *Marseille grecque*, edited by M. Bats *et al.* Aix-en-Provence: ADAM editions: Université de Provence, 1992.

Long, L., J. Miro, G. Volpe. 'Les épaves archaïques de la pointe Lequin (Porquerolles, Hyères, Var). Des données nouvelles sur le commerce de Marseille à la fin du VIe siècle et dans la premières moitié du Ve s. av. J.C.'. In *Marseille grecque*, edited by M. Bats *et al.* Aix-en-Provence: ADAM editions: Université de Provence, 1992.

Malkin, Irad. *The Returns of Odysseus: Colonization and Ethnicity.* Berkeley: University of California Press, 1998.

Marconi, Clemente. 'Images for a Warrior. On a Group of Athenian Vases and Their Public'. In *Greek Vases: Images, Contexts, and Controversies*, edited by C. Marconi. Leiden: E.J. Brill, 2004.

Martelli, M., ed. *La Ceramica degli Etruschi. La pittura vascolare.* Milan: Istituto Geografico de Agostini, 1987.

Massa-Pairault, Françoise. '*Megarica* en hommage à Georges Vallet'. In *La colonisation grecque en Méditerranée occidentale. Actes de la rencontre scientifique en hommage âa Georges Vallet.* Rome: École Française de Rome, 1999.

Menichetti, Mauro. *Archeologia del potere. Re, immagini e miti a Roma e in Etruria in età arcaica.* Milan: Longanesi, 1994.

Morris, S. P., and J. K. Papadopoulos. 'Phoenicians and the Corinthian Pottery Industry'. In *Archäologische Studien in Kontaktzonen der antiken Welt*, Veröffentlichungen der Joachim Jungius-Gesellschaft der Wissenschaften e.V. Hamburg, Vol. 87. Edited by R. Rolle, K. Schmidt, and R. Docter. Göttingen: Vandenhoeck and Ruprecht, 1988.

Olmos, Ricardo. 'Usos y transformaciones de la cerámica griega entre los iberos: los siglos V y IV a.C.'. In *Céramique et peinture grecques: modes d'emploi*, edited by M.-C. Villanueva Puig, F. Lissarrague, P. Rouillard, and A. Rouveret. Paris: Documentation française, 1999.

Osborne, Robin. 'Pots, Trade and the Archaic Greek Economy'. *Antiquity* 70. (1996): 31–44.

———.'The Ecstasy and the Tragedy: Varieties of Religious Experience in Art, Drama, and Society'. In *Greek Tragedy and the Historian*, edited by C. B. R. Pelling. Oxford: Clarendon Press, 1997.

———.'Why Did Athenian Pots Appeal to the Etruscans?'. *World Archaeology* 33, no. 2 (2001): 277–95.

———.'Images of a Warrior. On a Group of Athenian Vases and Their Public'. In *Greek Vases: Images, Contexts, and Controversies*, edited by C. Marconi. Leiden: E.J. Brill, 2004. 41–54.

Paléothodoros, Dimitrios. 'Pourquoi les Étrusqueses achetaient-ils des vases attiques?'. *Les études classiques* 70. (2002): 139–60.

Parker, Anthony J. *Ancient Shipwrecks of the Mediterranean and the Roman Provinces.* BAR International Series 580. Oxford: Tempus Reparatum, 1992.

Philippaki, Barbara. *The Attic Stamnos.* Oxford: Clarendon Press, 1967.

Reusser, Christoph. *Vasen für Etrurien. Verbreitung und Funktionen attischer Keramik im Etrurien des 6. und 5. Jahrhuderts vor Christus.*, 2 vols. Zurich: Akantus, 2002.

Rouillard, Pierre. *Les Grecs et la péninsule Ibérique du VIII au^e IVe siècle av*, Jésus-Christ. Talence: Université de Bordeaux III; Paris: Diffusion, De Boccard, 1991.

———.'Le vase attique: de sa récolte à sa cartographie'. In *Céramique et peinture grecques: modes d'emploi*, edited by M.-C. Villanueva Puig, F. Lissarrague, P. Rouillard, and A. Rouveret. Paris: Documentation française, 1999.

Rizzo, Maria A. 'Un incunabulo del mito greco in Etruria'. *Annuaris della Scuola Archeologica di Atene delle Missioni Italiane in Oriente* 66–67 (1988–89): 7–47.

Shanks, Michael. *Art and the Early Greek State: An Interpretive Archaeology*. Cambridge: Cambridge University Press, 1999.

Shapiro, Alan H. 'Modest Athletes and Liberated Women: Etruscans on Attic Black-Figure Vases'. In *Not the Classical Ideal*, edited by B. Cohen. Leiden: E.J. Brill, 2000.

Shefton, Brian B. 'Zum Import und Einfluss mediterraner Güter in Alteuropa'. *Kölner Jahrbuch für Vor- und Frühgeschichte* 22 (1989): 207–20.

Shepherd, Gillian. 'The Pride of Most Colonials: Burial and Religion in the Sicilian Colonies'. In *Ancient Sicily*, edited by T. Fischer-Hansen. *Acta Hyperborea* 6. (1995): 51–82.

Snodgrass, Anthony M. *Homer and the Artists*. Cambridge: Cambridge University Press, 1998.

Spivey, Nigel. *The Micali Painter and His Followers*. Oxford: Clarendon Press, 1987.

———. 'Greek Vases in Etruria'. In *Looking at Greek Vases*, edited by N. Spivey and T. Rasmussen. Cambridge: Cambridge University Press, 1991.

Spivey, Nigel, and Simon Stoddart. *Etruscan Italy: An Archaeological History*. London: Batsford, 1990.

Stissi, Vladimir. 'Modern Finds and Ancient Distribution'. In *Céramique et peinture grecques: modes d'emploi*, edited by M.-C. Villanueva Puig, F. Lissarrague, P. Rouillard, and A. Rouveret. Paris: Documentation française, 1999.

Tosto, Vincent. *The Black-Figure Pottery Signed [NIKOSTHENESEPOIESEN]*. Amsterdam: Allard Pierson Museum, 1999.

Networks of Commerce and Knowledge in the Iron Age: The Case of the Phoenicians

Michael Sommer

'Unde interrogati rustici nostri quid sint, punice respondentes Chanani.'[1] Thus wrote Saint Augustine as late as in the late fourth century CE, giving expression to his surprise that Punic was still alive as a language, and was spoken in the area surrounding Hippo Regius (modern Annaba), in Numidia. Not only did the peasants speak Punic, but they also regarded themselves as *chanani*, an expression of ethnic identity which links them directly to the Levant in the Iron Age whose inhabitants were called 'Sidonians' or 'Phoenicians' by the Greeks. Philo of Byblus, who wrote in the second century CE, called them *chna* (Canaanites), a term also derived from the western Semitic root *kn'n*.[2] Saint Augustine's *rustici* were by no means an exception: the Roman Emperor Septimius Severus, who grew up in Lepcis Magna, spoke Punic as his mother tongue and apparently felt himself close to the inhabitants of Syria (where his second wife Julia Domna came from), which benefited considerably from his rule. Heliodorus of Emesa, where Julia Domna was born, dubbed himself, in the third or fourth century CE, as a man 'from Emesus [Emesa], a city in Phoenicia, son of Theodosius who fetched his pedigree from the sun'.[3] Another case is offered in the Gospels, which relate that when Jesus came to Tyre, he met a woman 'whose daughter was possessed by an impure spirit',[4] a Greek (*Hellēnis*) and by origin a Syro-Phoenician (*Syrophoinikissa*). We have countless examples of individuals who defined themselves as Phoenicians, or were so defined by others.

The remarkable persistence—or, more accurately, revival—of Phoenician or Punic identities in the Imperial period is a complex matter that deserves some explanation. Certainly, the Phoenicians of the Iron Age were as much 'frogs around a pond' as their Greek neighbours.[5] Furthermore, the Phoenicians' seafaring and commercial activities were likewise impressive, as were their colonial ventures which, like those of the Greeks, encompassed the entire Mediterranean, and even beyond.[6] And yet, in terms of colonial diaspora, the migration of the Phoenicians had a fundamentally different pattern from that of the Greeks. Far fewer people were involved, and rather than an escape from the growing pressures of population, the Phoenician merchant adventurers were seeking new opportunities to make profits, new markets, and new sources of raw materials. Whether in Italy, North Africa, or Spain, the Phoenician settlements overseas were different from the Greek *apoikia* in almost every respect: first, their material culture displays far more 'indigenous' traits than that of Greek colonies;[7] second, their settlements were much smaller in size, at least in the early phases; and third, with the significant exception of Carthage (to which we will return later), they disposed of no hinterland comparable to the Greek *chōra*, although the Phoenician presence undoubtedly influenced the material culture of neighbouring indigenous settlements. To sum up, while the Phoenician model of expansion in the Mediterranean is as specific and characteristic as the Greek one,[8] the cultural element brought in by Phoenicians and Carthaginians is generally not regarded as a constituent influence in the formation of the classical world that took shape while Greeks and Phoenicians settled the Mediterranean coasts.[9]

It is fair to say that the Phoenician model of opening up, settling, and colonizing the Mediterranean differed in many respects from the patterns we know from the Greek world. In this volume, many types of networks set up by Greeks over various centuries are discussed. In order to point the way for a comparison we will have to address a number of questions:

1. Is the 'network' paradigm an appropriate heuristic tool to describe the development and organization of the Phoenician trading diaspora in the Mediterranean? In order to get to the bottom of this, we obviously need a little theory of social networks.

2. If so, what were the driving forces which led to the establishment of trading posts and settlements overseas? What impelled the tradesmen of the Levantine coastal cities to leave the security of their homes in exchange for an uncertain future in distant parts of the Mediterranean?
3. What factors enabled this network to endure over many centuries, surviving various critical turning points in the Mediterranean's history, and why did the network prove so relatively coherent? What kinds of knowledge management were practised, and how were information, commodities and people exchanged within the network? And finally, how did the network established by them develop diachronically?

It goes without saying that the evidence provides only sketchy answers to many of these questions, and the overall image we can draw from our sources is inevitably incomplete. Networks have been the subject of social theorizing for half a century or so, spawning a well-established sub-discipline of system theory, and a sub-sub-discipline of social sciences, that deal with social networks, for which there are now dedicated chairs, journals, organizations, annual conferences, and online discussion groups.[10] There is an abundance of definitions, the most general one claiming that a network is a social structure consisting of vertices or nodes (individuals or collectives) which are linked to each other.[11] The maximum number of vertices within one system is generally given to be 150, beyond which the network becomes dysfunctional.[12] Networks require repeated (and not just occasional) and formalized interaction over longer periods of time. Beyond formal barriers and boundaries, a network comprises various kinds of relationships.[13] Economically, a network is a middle-range phenomenon, half-way between market and organizational hierarchy.[14] Network theorists distinguish between 'open' networks, which tend to be rather large and have relatively loose connections ('weak ties') between the vertices, and 'closed' ones, which feature strong but often redundant edges, and are usually smaller. 'Open' networks with weak ties are generally believed to accelerate the exchange of goods and ideas more than 'closed' ones.[15]

Though networks are by definition multi-polar and decentralized, in most cases they are nevertheless hierarchic. The hierarchic structure of a network is mirrored by the different numbers of connections (edges) that the various vertices have. A high quantity and/or quality of edges means access to a wide range of information, material, and symbolic resources, and accordingly reflects a high status within the system.[16] As in von Thünen's famous ring model, there are nodes with different degrees of complexity and connectivity: central places of various levels with catchment areas of different size.[17]

With this in mind, it is easier to decide whether or not the 'network' paradigm is suitable for analyzing the cluster of entrepôts and settlements established everywhere in the Mediterranean from the Early Iron Age onwards by people coming from the Levant. Of course, this cluster did not come into being overnight. Rather, it was the result of a long process starting probably in the third millennium BCE, when first contacts between the Levant and the Mediterranean west were established. In the second millennium BCE, copper in the form of the famous oxhide-ingots found its way from Sardinia via Crete to the Levant.[18] It is probably stretching the issue to call such occasional contacts a 'network', but the situation changed dramatically after the collapse of the palatial centres shortly after 1200 BCE: after a short period of decline in the trans-Mediterranean trade (in which, however, the exchange of commodities continued to take place on a smaller scale), the commercial contacts between east and west became more intense than ever from the tenth century onwards.[19]

The resurgent production of luxury commodities in the Levantine coastal cities, which we find echoed in the biblical account of the construction of the Temple in Jerusalem, in the Homeric Epics and of course in the findings of precious artefacts of Near Eastern origin in Lefkandi[20] and Crete[21] (Knossos and the Idaean Cave), required access to raw materials and markets. It also required knowledge: knowledge of navigational and nautical techniques, but first of all knowledge of places to procure such raw materials, and where to sell the finished products. The Levantine coastal cities with their relatively high degree of continuity between the Bronze and the Iron Ages could possibly build on the experiences of their forefathers, but more importantly on information acquired by visiting other places and establishing contact with different peoples. Scholars have described this process of gradual rapprochement as 'pre-colonization'.[22] Though the material evidence is sketchy, it provides some notion of the stage that precedes the establishment of full-fledged settlements.

A quite instructive passage is offered by the tale of Eumaios in the *Odyssey*,[23] in which a group of Phoenicians visit the island of Syria to trade luxury items (i.e., prestigious high-value goods of no immediate practical use), and stay for more than a year, exchanging information with the locals, but kidnap the king's son. They bring young Eumaios to Ithaca and sell him to Odysseus's father.[24] The account gives us some idea how the exchange of goods and knowledge in the period of pre-colonization might have worked. The activity of visiting distant coasts and evaluating their potential for subsequent commercial interchange does not appear to require a centralized organization, like that of a royal palace.[25] Rather, it was propelled and performed by small groups of tradesmen operating on a modest scale, possibly individual ship-owners or trading companies sharing one vessel.[26] The apparent informality of relations, coupled with a certain regularity in the exchange of goods and information, makes 'network' a suitable category to describe the Phoenician's trade involvements.

Another story from the Homeric epics affords similar insights into the process: in a running contest, Achilles offered

> a mixing-bowl of silver, richly wrought; six measures it held, and in beauty it was far the goodliest in all the earth, seeing that Sidonians, well skilled in deft handiwork, had wrought it cunningly, and men of the Phoenicians brought it over the murky deep, and landed it in harbour, and gave it as a gift to Thoas; and as a ransom for Lycaon, son of Priam, Jason's son Euneos gave it to the warrior Patroclus. This bowl did Achilles set forth as a prize in honour of his comrade, even for him who so

> should prove fleetest in speed of foot. For the second again he set an ox great and rich with fat; and a half-talent in gold he appointed for the last.[27]

The krater, in this case made by Sidonians, was not only by far the most precious prize awarded at such contests, but as an artefact it significantly also had a long prehistory as a gift and counter-gift. From a Greek perspective, it was regarded as normal that Phoenicians (the traders, the term is not necessarily restricted to people from the Levant) and Sidonians (the originators) as well as Greeks, all mutually formed part of each others' gift exchange system and *proxenia* networks.[28]

This is admittedly an outsider's perspective, but nonetheless a revelatory one. Though Greeks and Phoenicians each conformed to their own model of expansion, it is impossible to draw a neat division between a 'Phoenician' and a 'Greek' network. Knowledge and goods flooded to and fro, quite unimpeded. In its early stages, this large, loose Mediterranean trade network was relatively unhierarchical, though the Levantine traders were obviously in the front line in the scramble for markets and sources of raw materials; whereas the Greeks, making their appearance on the scene later, remained at the periphery of the system. Being intermediate traders, the Phoenicians, as they were called by the Greeks, did the bulk of the transport and hence linked up the nodes within the system.[29]

The relatively sudden revival of trade in the later Dark Ages was prompted by a conjuncture of assorted variables,[30] the focal point of which was the coastal plain of modern Lebanon. According to data contained in the Egyptian account of Wenamun, in the eleventh century the Phoenician cities traded in timber, including cedar and other types of wood from the Lebanon mountain forests.[31] Even though they had skilled labour forces at their disposal, they ostensibly remained mere exporters of raw materials. This changed dramatically in the subsequent centuries. The biblical accounts of Solomon claim that, in joint venture with King Hiram of Tyre, the Israelite king equipped an expedition to Spain (Tarshish) every three years.[32] At the same time, Hiram contributed substantially to the construction of the Temple in Jerusalem.[33] Be this historical fact or not, the rise of Tyre to hegemonic power in the Levant in precisely this period is echoed by other evidence, and trade contacts between the Levant and the Mediterranean west can be traced, for example, in the tombs of Francavilla Marittima in central Italy, in the emulation of Phoenician techniques in early Italic pottery (*impasto rosso*), in a Levantine bronze bowl from the Late Bronze Age hoard in Berzocana (Estremadura), and in the occurrence of visual motifs from the Levant (like the Herzsprung shield on tomb stelae in south-western Spain).[34] While all this is too little to prove the precocious foundation dates proffered for some of the Phoenician colonies in Spain and North Africa (Gadir in 1104/3 BCE, Utica in 1101 BCE), it makes it equally impossible to dismiss the Phoenician presence in the west altogether.[35]

Why the Phoenicians? What made possible the Phoenician coastal cities' sudden breakthrough from being suppliers of raw materials to being protagonists of intercontinental long-distance trade? Were they, like the Greeks, reacting to the growing population pressure and the shortage of arable land?[36] To be sure, the

Lebanese coastal plain is narrow, but it is immensely fertile and, according to the biblical tradition, Tyre imported huge amounts of foodstuffs from nearby Palestine. Were they driven into the sea by the expansion of the Neo-Assyrian Empire towards the Levant from the ninth century onwards, just as the Phokaians were expelled by the Persians much later (as some scholars believe)?[37] Or was it just the urge for profit-making that attracted them towards unknown strands?

Though they cannot be quantified as such, all the above factors figured in the genesis of the Phoenician colonial diaspora, but not necessarily directly. While land shortage was a problem in Phoenicia at all times, turning peoples' attention towards the sea and towards other means of subsistence, especially craftsmanship and trade, it did not urge the Phoenicians to leave their home cities and to settle elsewhere; nor did the expanding Assyrian Empire, though this was a source of substantial political pressure. In the long run, the Assyrians proved unable to assert more than a fairly indirect rule over the Levant, and this fact left the Phoenicians plenty of leeway to do their own business, both economically and politically. Nevertheless, the military expansion of the Assyrian Empire was decisive for the commercial growth of the Phoenician cities. It resulted in a precarious symbiotic balance: avid for luxury commodities, the imperial elite provided an insatiable market for the Phoenician centres of production; the Phoenician long-distance trade, in turn, was vital for the social mechanics of the Assyrian Empire. Decisive, however, for the quick expansion and initial openness of the Phoenician-dominated trade network of the Early Iron Age was the absence of highly centralized power structures. Though the institutions of kingship and royal palace survived in the Phoenician cities, they were merely weak replicas of palace centres such as Ugarit or Byblos.[38]

The absence of centralized control was only transitional, however. While the ninth century BCE saw the resurgence of imperial power on a large scale, Tyre became, almost simultaneously, the hegemonic power within the Levant's regional system. Tyre integrated its neighbour and major rival city, Sidon, in one territorial state denominated the 'Kingdom of the Sidonians', stretching from the Nahr el-Kelb north of Beirut to present-day Israel to the south and asserting indirect rule over Cyprus.[39] Within this state, virtually everything was devised to serve Tyre's trade and its merchant classes. The kingdom featured a sophisticated settlement hierarchy, with Tyre sitting like a spider in its web and dedicated especially to long-distance trade, surrounded by major centres of production such as Sidon[40] (where the dyeing industry occupied large parts of the town) and Sarepta[41] (a centre of pottery production) and with rural centres at the periphery. The transformation of the Levantine system had direct repercussions on the Mediterranean trade network. The open network of the Early Iron Age turned into a full-fledged world-system, focused exclusively on Tyre. The vertical relationship between Tyre and the various parts of the world-system are echoed in the famous 'Lament over Tyre' in the book of Ezekiel:

> The inhabitants of Sidon and Arvad were your rowers; your wise men, O Tyre, were aboard; they were your pilots. The elders of Gebal and her wise men were with you

> repairing your seams; all the ships of the sea and their sailors were with you in order to deal in your merchandise. Persia and Lud and Put were in your army, your men of war. They hung shield and helmet in you; they set forth your splendour. The sons of Arvad and your army were on your walls, all around, and the Gammadim were in your towers. They hung their shields on your walls all around; they perfected your beauty. Tarshish was your customer because of the abundance of all kinds of wealth; with silver, iron, tin and lead they paid for your wares. Javan, Tubal and Meshech, they were your traders; with the lives of men and vessels of bronze they paid for your merchandise.[42]

This passage—in all likelihood a text[43] originating in Tyre herself and dating back to the eighth century BCE—clearly illustrates the city's central position in the Mediterranean trade network, whereas the role of other Phoenician cities (Sidon and Arvad) was largely to support Tyre's trade and provide soldiers for its army; the same did Persia, Lud (Lydia) and Put (probably Libya); Tarshish (Southern Spain[44]) imported luxury commodities from Phoenicia and in turn supplied metals; Javan (Ionia), Tubal (the Iberian peninsula?) and Meshech (Cappadocia?) were Tyre's 'traders', an odd expression which could imply that Tyrians were resident there to perform their trade, or alternatively, that Tyrian commodities were traded via these places. Although the 'Lament over Tyre', for obvious reasons, exaggerated the city's importance, it cannot be dismissed that in the eighth century BCE the Phoenician metropolis was a veritable global player, the one focal point of the Mediterranean network that maintained direct links to most of the more peripheral nodes, much like a spider in its web, as noted before.

The links between Tyre and its colonies remained close. According to Diodorus Siculus (20.14), the Carthaginians sent a delegation to the sanctuary of the Tyrian Melqart year by year, for centuries. Diodorus claims that the sacrificial offerings brought to Melqart accounted for one-tenth of the Carthaginian fiscal budget, although the proportion diminished with time. If this is true, the Carthaginians were, in sharp contrast to Greek colonists, liable to the payment of tribute. Nothing prevents us from assuming that other Phoenician colonies did the same, thus making the Melqart sanctuary not only a steady source of revenue for the mother city, but also a marketplace of information and knowledge. The paramount importance of Melqart throughout the Phoenician colonial diaspora, especially his association with seafaring, suggests that the god played a key role in keeping the network alive—different from, but analogous to, the oracle of Delphi.[45]

From the early seventh century BCE onwards, however, the patterns of the network changed dramatically. Old peripheries gradually developed into new centres, like North Syria and the Aegean; little by little, Tyre lost its focal position, and the network became polycentric once again. In the same period of time, what we call the 'Punic' civilization took shape in the western Mediterranean. The colonial diaspora in this area now featured a material culture strongly influenced by, but clearly distinct from, Tyre and the other centres in the Levant.[46] One element which set apart the Punic world from the rest of the Mediterranean (including Phoenicia proper[47]) was the *tofet*, a sanctuary whose precise function remains unknown, but which is generally associated with human sacrifice.[48] Also, other features of material culture (like the two-spouted lamps, whereas the ones used in Phoenicia proper were one-spouted[49]), the language used in inscriptions—and, apparently, the religious and ethnic identity—contrasted with what we know from the Levant.[50] Less evident in the material evidence, but certainly more decisive for Mediterranean history, was the gradual transformation of the Punic world into a powerful albeit informal empire controlled by Carthage.

Why did the balance in the Phoenician world shift towards the west? And why to Carthage? Why did the Punic world display such cultural homogeneity, though most of the settlements in the western Mediterranean were no foundations of Carthage, but rather fellow colonies established by Phoenicians from the Levant? Many scholars tended to believe that the big shift followed a master plan devised to deal with political decline in the east; that the foundation of Carthage and the western colonies was a long-term strategy designed by far-sighted statesmen in Tyre.[51]

Flattering as the attribution of such a political vision may be for the Phoenicians, the idea of a master plan seems a post-festum notion contrived by modern scholars. First, the political and economic decline of the Phoenician centres in the east cannot be traced in the evidence. To be sure, there was political pressure, but documents such as the so-called treaty between Baal and Asarhaddon, dated to the 670s BCE, rather prove the persistent efficacy of the symbiotic balance between Assyria and Phoenicia.[52] And not even the Chaldaeans, who put Tyre under siege a century later (581–68), succeeded in imposing a lasting supremacy over Phoenicia.[53] Achaemenid rule was a period of prosperity for the coastal cities, though Tyre's hegemony in terms of realpolitik now clearly belonged to the past.[54] There was probably no *political* reason to shift the centre towards the west.

Secondly, Carthage's assumption of leadership in the western Mediterranean was by no means instantaneous. As pointed out earlier, until the late fifth century BCE, the general image is rather diffuse. There was, obviously, some notion of belonging together, but the links between the various nodes had weakened again. It was within the renewed openness of the Mediterranean network that a distinct 'Punic' identity took shape, of which a common language, religious practices, along with participation in the long-distance trade and the growing political power of Carthage, were constituent elements. By no means foreseen by wise politicians, this process was totally contingent, though its ingeniously chosen location at the crossroads of the Mediterranean and trans-African trade routes, and at the heart of the most fertile coastal strip of present-day Tunisia, contributed substantially to Carthage's later success. The Carthaginian empire was essentially the fruit of the dynamism inherent in an open network: the free flow of information, people, and material reinforced existing ties and created new ones. The result was a complete reshuffle of the Mediterranean system, from which the west largely benefited—not only Carthage, but in the long run also Rome.

To sum up, the 'network' paradigm is most apt for explaining the genesis of the Phoenician commercial diaspora in the Mediterranean, and its gradual transformation into an informal empire dominated by Carthage. It has become evident, however, that network links did not necessarily connect existing cultural, linguistic, or ethnic identity groups. In this respect, we should not speak of a 'Phoenician' or a 'Greek' network, but rather of a composite Mediterranean network to which Phoenicians, Greeks and 'indigenous' populations, settled and mobile groups, traders and producers of commodities, mercenaries and slaves each contributed to a specific extent. Finally, far better than other models, the network paradigm explains the shift from east to west, from Tyre to Carthage, that took place gradually from the seventh century onwards. Sketchy as this outline had to be, due to the evidence available it has hopefully managed to provide a paradigm for comparison, a 'non-Greek model of expansion' that fits into the aforementioned single, composite Mediterranean network, whose many facets are the subject of this volume.

Acknowledgements

This is the revised version of my conference paper. I am very grateful to Hans Georg Niemeyer for his critical reading and detailed discussion of the archaeology of the Phoenician expansion in the west. This study owes a great deal to his recent work. I am also thankful to Christy Constantakopoulou and Katerina Panagopoulou for the splendid organization of a conference which opened up a wealth of thoughts and information of which I was barely aware.

Notes

[1] August. Epist. 13.

[2] Referring perhaps to the colour red and the artisanship of purple dyeing. In the Hebrew bible, Canaan is the son of Ham and the father of Sidon (Genesis 9:18; 10:15).

[3] Heliod. 10.323. For additional information on the author and his work Nesselrath, 'Heliodor'.

[4] Marcus 7:25–26.

[5] The notion of the Mediterranean as a common ground of the people living next to it, very present in Braudel, *Mediterranean*, has recently been revitalized and universalized by Horden and Purcell, *Corrupting Sea*. Cf. the instructive review by Shatzmann.

[6] On the Phoenician expansion in the Mediterranean in general, among others, see Whittaker, 'Western Phoenicians'; Sznycer, 'L'expansion'; Bunnens, *L'expansion;* Aubet, *Phoenicians*; Sommer, *Europas Ahnen*, 99–121; Sommer, *Phönizier*, 113–43.

[7] For the pottery, see Schreiber, *Cypro-Phoenician Pottery.*

[8] Niemeyer, 'Phoenicians in the Mediterranean'; Niemeyer, 'Phoenician or Greek', 45–46. Although the importance of agriculture in the Phoenico-Punic west has been re-emphasized in recent years (by among others Ameling, *Karthago*), the typological differences between the Greek and the Phoenician 'colonial' diasporas in this respect are quite straightforward. Even though Carthage relied heavily on agricultural production for its subsistence, the city was never dominated by an elite of landholders. For a survey of Carthage's economic and political patterns, see Sommer, *Europas Ahnen*, 221–22 and 248–49.

[9] See *pars pro toto* Boardman, *Greeks Overseas*, 8, where Phoenicians figure merely as 'powerful rivals' of the Greeks.

[10] For an excellent summary see Rauner, 'Ziemlich verknotet'. On the development of social network analysis as a sub-discipline of the social sciences, see Scott, *Social Network Analysis*, 7–38. For more detail on network theory, see Ian Rutherford's paper in this volume.

[11] Newman, 'Structure', 167: 'A network is a set of items, which we will call vertices or sometimes nodes, with connections between them, called edges. Systems taking the form of networks (also called 'graphs' in much of the mathematical literature) abound in the world. Examples include the Internet, the World Wide Web, social networks of acquaintance or other connections between individuals, organizational networks and networks of business relations between companies, neural networks, metabolic networks, food webs, distribution networks such as blood vessels or postal delivery routes, networks of citations between papers, and many others."

[12] The so-called Dunbar's number; see Dunbar, 'Social Network Size'.

[13] 'Community structure' is a property shared by most social networks: Vertices are not distributed equally and homogeneously within the network, but they form clusters ('groups', 'communities') which again are linked to each other. The links between individuals are usually more frequent than links between groups. Newman, 'Structure', 17–19.

[14] Powell, 'Neither Market', 300–305.

[15] On weak ties and the accelerated spread of ideas within open networks, Granovetter, 'Weak Ties'; Granovetter, 'Revisited', 214–20. The innovative forces of open networks is now best exemplified by the rapid distribution of open-source software via the internet.

[16] Scott, *Social Network Analysis*, 166–75.

[17] Beckmann, 'Von Thünen'; Mills, 'Efficiency'.

[18] Catling, *Knossos Area*; Coldstream, 'Greeks and Phoenicians'; Niemeyer, 'Die phönizischen Niederlassungen', 46–47.

[19] On the collapse, see Liverani, 'Collapse'; Zaccagnini, 'Ceremonial Exchange'; Sommer, 'Untergang'. On the resurgence of the Mediterranean trade in the Early Iron Age, see Liverani, *Antico Oriente*, 629–60; Sommer, *Europas Ahnen*, 97–99.

[20] Popham *et al.*, *Lefkandi I.*

[21] Gehrig, 'Phönizier', 27–31.

[22] Niemeyer, *Das frühe Karthago*, 23–25; Niemeyer, 'Phönizier am Mittelmeer'; Aubet, *Phoenicians*, 172–84; Sommer, *Phönizier*, 113–15.

[23] The Homeric epics are, of course, not historical writing in the proper sense; like their origin, their historical context is still the subject of fierce debate, but they can be read as reflections of basic social and mental patterns of the period when they came into being. Equally, Homer's Phoenicians are not historical figures, and even as fictitious protagonists they would make sense in their literary context only if they were recognizable to the contemporary audience (for a critical assessment see Latacz, 'Phönizier'; Winter, 'Homer's Phoenicians'; Sommer, 'Peripherie'). No less disputed is the historicity of the Hebrew bible. Modern scholarship tends towards a sceptical point of view which puts emphasis on the narrative's functioning as 'intentional' history. The bulk of the texts was compiled during, and in the aftermath of, the Babylonian Exile, i.e., centuries after the period in which the events described supposedly happened (most notably now, Liverani, *Oltre la bibbia*). This does, however, not basically rule out the inclusion of earlier texts into the corpus, such as the 'Lament over Tyre' (Ezekiel 27, see below).

[24] *Od.* 15.388–483.

[25] Counter to the opinion of Aubet, *Phoenicians*, 77–118, who believes that the model of centralized palace organization played a major role in the process of Phoenician expansion overseas. The details cannot be discussed here, but the author's point of view is exposed in Sommer, *Phönizier*, 97–105.

[26] A possible typological parallel is the Hanseatic League in northern-central Europe in its period of expansion. On the Hanseatic merchants and their operation, see Dollinger, *Hanse*, 17–24;

Schildhauer *et al.*, *Hanse*, 38–67; Friedland, *Hanse*, 36–71. For a detailed structural comparison, based on Weber, *Die Stadt*, see Sommer, *Europas Ahnen*, 156–72.

[27] *Il.* 23.741–52.

[28] This view is corroborated by a famous honorific inscription (367 BCE) for Straton, king of Sidon, with whom the *demos* of Athens agrees to exchange the *symbola*. Austin and Vidal-Naquet, *Gesellschaft und Wirtschaft*, 241–42.

[29] This stage of the Mediterranean long-distance trade network seems appropriately illustrated by the 'peer polity interaction' model, once popular in the archaeologies and used as a panacea to explain the internal forces of change within a given system of societies ('peer polities'). Renfrew, 'Introduction'.

[30] In more detail, Sommer, 'Peripherie als Zentrum', 233–35.

[31] For the account and commentary see Goedicke, *Report*.

[32] 2 Chr 9:21; 2 Chr 20:36–37; Ps 72:10. With reference to the Tarshish ships: 1 Kings 22:49; 2 Chr 9:21; Jes 2:16.

[33] 1 Kings 5–7.

[34] On Francavilla Marittima, see Maaskant-Kleibrink, 'Early Cults'. On the adaption of oriental pottery styles in Iron Age Italy, see Piergrossi *et al.*, 'Sviluppo'; Niemeyer, 'Die phönizischen Niederlassungen', 50–55; Niemeyer, *Das frühe Karthago*, 23.

[35] Archaeological evidence for use in dating the settlements is scarce. Niemeyer, *Das frühe Karthago*, 24–26; Docter *et al.*, 'Radiocarbon Dates', 568–70.

[36] Aubet, *Phoenicians*, 56–59.

[37] Frankenstein, 'Phoenicians'; Gittin, 'Neo-Assyrian Empire'; more skeptical are Oded, 'Phoenician Cities'; Garbini, 'Fenici in Occidente', 128–37. Contra, with strong arguments and with chronological considerations in mind, are Niemeyer, 'Phönizier am Mittelmeer', 188; Niemeyer, 'The Phoenicians and the Birth', 245–46.

[38] On the political institutions of the Phoenician city-states, see Sommer, *Phönizier*, 203–19. On the Levatine bronze age cities, see Liverani, *Antico Oriente*, 546–61.

[39] And possibly even direct control over Kition and its vicinity. For a detailed reconstruction of the political history, mainly based on the literary evidence, see Katzenstein, *History of Tyre*, 102–10. See also Aubet, *Phoenicians*, 27–35; Sommer, *Europas Ahnen*, 96–97; Sommer, *Phönizier*, 159–65 and the TAVO map by Röllig *et al.*

[40] Ciasca, 'Fenicia', 145–47.

[41] Pritchard, *Sarepta*, 111.

[42] Ezek 27:8–13.

[43] For a thorough analysis of the Ezekiel passage, see Liverani, 'Trade Network'.

[44] Or, in my opinion, less likely Tarsos in southern Anatolia. See Koch, *Tarschisch*.

[45] On Melqart as a focal point of urban identity throughout the Mediterranean Garbini, 'Continuità e innovazioni', 154–55; Bonnet, 'Culte'; Bonnet, *Melqart*; Bonnet, 'Héraclès'. On the cult of Melqart in Thasos, see Berchem, 'Sanctuaires'. The importance of Heracles-Melqart as a syncretistic mythic figure in the Mediterranean 'middle ground' is now highlighted by Malkin, 'Herakles'. On Delphi as a centre of communication and information in Archaic Greece, see Dunbabin, *Western Greeks*, 26–39.

[46] For an outline, see Huss, *Karthager*, 28–38; Moscati, *Chi furono*, 132–37; Garbini, 'Fenici in Occidente', 128–29, who speaks of 'occidentalizzazione' and a process of 'trapianto' towards the west. A truly comprehensive history of the Mediterranean in the period preceding the Roman expansion is still much desired.

[47] Though there may be traces of *tofetim* even in the east: for Tyre see Seden, 'Tophet', for Cyprus (Amathus) see Karageorghis, 'Cyprus'.

[48] See Karageorghis, 155–56; Moscati, *Adoratori*; Moscati, *Chi furono*, 63–64. For the archaeology of the tofet, see Moscati and Uberti, *Scavi al tofet*.

[49] Garbini, 'Fenici in Occidente', 136–37.
[50] Ibid., 136–37.
[51] Ibid., 135–36.
[52] SAA 2, §5.
[53] Katzenstein, *History of Tyre*, 322–34.
[54] Elayi, 'Cités phéniciennes'; *Économie*; *Sidon*; 'Phoenician Cities'.

References

Ameling, Walter. *Karthago Studien zu Militär, Staat und Gesellschaft*. Münche: C.H. Beck, 1993.

Aubet, Maria Eugenia. *The Phoenicians and the West. Politics, Colonies, and Trade*. Cambridge: Cambridge University Press, 1993.

Austin, Michel, and Pierre Vidal-Naquet. *Gesellschaft und Wirtschaft im alten Griechenland*. München: C.H. Beck, 1984.

Beckmann, Martin J. 'Von Thünen Revisited. A Neo-Classical Land Use Model'. *Swedish Journal of Economics* 74 (1972): 1–7.

Boardman, John. *The Greeks Overseas. Their Early Colonies and Trade*. London: Thames and Hudson,[4] 1999.

Bonnet, Corinne. 'Le culte de Melqart à Carthage. Un cas de conservatisme religieux'. In *Religio Phoenicia*, edited by Corinne Bonnet. Namur: Societé des études classicques, 1986.

———.'Héraclès en Orient. Interprétations et syncrétismes'. In *Héraclès. D'une rive à l'autre de la Méditerranée*, edited by Corinne Bonnet. Brussels: Institut Historique Belge de Rome, 1992.

———. *Melqart. Cultes et mythes de l'Héraclès Tyrien en Méditerranée*. Leuven: Peeters, 1988.

Braudel, Fernand. *The Mediterranean and the Mediterranean World in the Age of Philip II*. London: Harper Collins, 1992.

Bunnens, Guy. *L'expansion phénicienne en Méditerranée. Essai d'interprétation fondé sur une analyse des traditions littéraires*. Brussels and Rome: Institut Historique Belge de Rome, 1979.

Catling, Hector W. *The Knossos Area 1974–76*. Athens: British School at Athens, 1977.

Ciasca, Antonio. 'Fenicia'. In *I Fenici*, edited by S. Moscati. Milano: Bompiani, 1988.

Coldstream, John N. 'Greeks and Phoenicians in the Aegean'. In *Phönizier im Westen*, edited by H. G. Niemeyer. Mainz: Philipp von Zabern, 1982.

Docter, Roald F., Hans Georg Diemeyer, Albert Nijboer, and Hans van der Plicht. 'Radiocarbon Dates of Animal Bones in the Earliest Levels of Carthage'. *Mediterranea* 1 (2001): 557–77.

Dollinger, Philippe. *Die Hanse*. Stuttgart: Kröner, 1966.

Dunbabin, Thomas J. *The Western Greeks. The History of Sicily and South Italy from the Foundation of the Greek Colonies to 480 BC*. Oxford: Oxford University Press, 1948.

Dunbar, Robin. 'Social Network Size in Humans'. *Human Nature* 14 (2003): 53–72.

Elayi, Josette. 'Les cités phéniciennes entre liberté et sujetion'. *Dialogues d'Histoire Ancienne* 16, no. 2 (1990): 93–113.

———. *Économie des cités phéniciens sous l'Empire Perse*. Napoli: Istituto Universitare Orientale, 1990.

———.'The Phoenician Cities in the Persian Period'. *Journal of the Ancient Near Eastern Society of Columbia University* 11 (1980): 13–28.

———. Sidon. *Cité autonome de l'empire Perse*. Paris: Ed. Idéaphane, 1989.

Frankenstein, Susan. 'The Phoenicians in the Far West. A Function of Neo-Assyrian Imperialism'. In *Power and Propaganda. A Symposium on Ancient Empires*, edited by Mogens Trolle Larsen. Copenhagen: Akademisk Forlag, 1979.

Friedland, Klaus. *Die Hanse*. Stuttgart: Kohlhammer, 1991.

Garbini, Giovanni. 'Continuità e innovazioni nella religione fenicia'. In *I fenici. Storia e religione*, edited by Giovanni Garbini. Napoli: Istituto Universitario Orientale, 1980.

———.'I fenici in occidente'. In *I fenici. Storia e religione*, edited by Giovanni Garbini. Napoli: Istituto Universitario Orientale, 1980.

Gehrig, Ulrich 'Die Phönizier in Griechenland'. In *Die Phönizier im Zeitalter Homers*, edited by Ulrich Gehrig and Hans Georg Niemeyer. Mainz: Philipp von Zabern, 1990.

Gittin, Seymour. 'The Neo-Assyrian Empire and its Western Periphery. The Levant, with a Focus on Philistine Ekron'. In *Assyria 1995*, edited by Simo Parpola and Robert M. Whiting. Helsinki: Neo-Assyrian Text Corpus Project, 1997.

Goedicke, Hans. *The Report of Wenamun*. Baltimore and London: Johns Hopkins University Press, 1975.

Granovetter, Mark S. 'The Strength of Weak Ties'. *American Journal of Sociology* 76 (1973): 1360–80.

———.'The Strength of Weak Ties. A Network Theory Revisited'. *Sociological Theory* 1 (1983): 201–33.

Horden, Peregrine, and Nicholas Purcell. *The Corrupting Sea. A Study of Mediterranean History*. Oxford: Blackwell, 2000.

Huss, Werner. *Die Karthager*. 2nd edn. München: C.H. Beck, 1994.

Karageorghis, Vassos. 'Cyprus and the Phoenicians: Achievement and Perspectives'. In *I fenici. Ieri, oggi, domani. Ricerche, scoperte, progetti*. Roma: Istituto per la Civiltà Fenicia e Punica, 1995.

Katzenstein, H Jacob. *History of Tyre. From the Beginning of the Second Millennium B.C.E. Until the Fall of the Neo-Babylonian Empire in 538 B.C.E.* Jerusalem: The Schocken Institute for Jewish Research, 1973.

Koch, Michael. *Tarschisch und Hispanien. Historisch-geographische und namenkundliche Untersuchungen zur Kolonisation der iberischen Halbinsel*. Berlin: de Gruyter, 1984.

Latacz, Joachim. 'Die Phönizier bei Homer'. In *Die Phönizier im Zeitalter Homers*, edited by Ulrich Gehrig and Hans Georg Niemeyer. Mainz: Philipp von Zabern, 1990.

Liverani, Mario. *Antico Oriente. Storia–società–economia*. Bari: Laterza, 1988.

———.'The Collapse of the Near Eastern Regional System at the End of the Bronze Age: The Case of Syria'. In *Centre and Periphery in the Ancient World*, edited by Michael Rowlands, Mogens Trolle Larsen, and Kristian Kristiansen. Cambridge: Cambridge University Press, 1987.

———. *Oltre la bibbia. Storia antica di Israele*. Bari: Laterza, 2003.

———.'The Trade Network of Tyre According to Ezek. 27'. In *Ah, Assyria*... (Festschrift Hayyîm Tadmor), edited by Mordechai Cogan, and Israel Eph'al. Jerusalem: Hebrew University Magnes Press, 1991.

Maaskant-Kleibrink, Marianne. 'Early Cults at the Athenaion at Francavilla Marittima as Evidence for a Pre-Colonial Circulation of nostoi Stories'. In *Die Ägäis und das westliche Mittelmeer. Beziehungen – Wechselwirkungen, 8. bis 5. Jh. v. Chr.*, edited by Fritz Krinzinger. Wien: Verlag der Österreichischen Akademie der Wissenschaften, 2000.

Malkin, Irad. 'Herakles and Melqart. Greeks and Phoenicians in the Middle Ground'. In *Cultural Borrowings and Ethnic Appropriations in Antiquity*, edited by Erich Gruen. Stuttgart: Franz Steiner, 2005.

Mills, Edwin S. 'The Efficiency of Spatial Competition'. *Papers of the Regional Science Association* 25 (1970): 71–82.

Moscati, Sabatino. *Gli adoratori di Moloch. Indagine su un celebre rito cartaginese*. Milano: Jaca Book, 1991.

———. *Chi furono i fenici*. Torino: Società Editrice Internazionale, 1992.

Moscati, Sabatino, and Maria L. Uberti. *Scavi al tofet di Tharros. I monumenti lapidei*. Roma: Consiglio Nazionale delle Ricerche, 1985.

Nesselrath, Heinz-Günther. 'Heliodor'. In *Lexikon antiker Autoren*, edited by Oliver Schütze. Stuttgart and Weimer: J.B. Metzler, 1997.

Newman, Mark E. J. 'The Structure and Function of Complex Networks'. *Society for Industrial and Applied Mathematics Review* 45 (2003): 167–256.

Niemeyer, Hans G. *Das frühe Karthago und die phönizische Expansion im Mittelmeerraum*. Göttingen and Zürich: Vandenhoeck und Ruprecht, 1989.

———.'Phoenician or Greek? Is there a Reasonable Way Out of the Al Mina Debate?'. *Ancient West and East* 3 (2004): 38–50.

———.'The Phoenicians and the Birth of a Multinational Mediterranean Society'. In *Commerce and Monetary Systems in the Ancient World. Means of Transmission and Cultural Interaction*, edited by Robert Rollinger and Christoph Ulf. Stuttgart: Franz Steiner Verlag, 2004.

———.'The Phoenicians in the Mediterranean. A Non-Greek Model for Expansion and Settlement in Antiquity'. In *Greek Colonists and Native Populations. Proceedings of the First Australian Congress of Classical Archaeology, Sydney 9–14 July 1985*. Oxford: Oxford University Press, 1990.

———.'Die Phönizier am Mittelmeer. Neue Forschungen zur frühen Expansion'. In *Die nahöstlichen Kulturen und Griechenland an der Wende vom 2. zum 1. Jahrtausend v. Chr. Kontinuität und Wandel von Strukturen und Mechanismen*, edited by Eva Braun-Holzinger and Hartmut Matthäus. Möhnesee: Bibliopolis, 2002.

———.'Die phönizischen Niederlassungen im Mittelmeerraum'. In *Die Phönizier im Zeitalter Homers*, edited by Ulrich Gehrig and Hans Georg Niemeyer. Mainz: Philipp von Zabern, 1990.

Oded, Bustany. 'The Phoenician Cities and the Assyrian Empire in the Time of Tiglath-pileser III'. *Zeitschrift des Deutschen Palästinavereins* 90 (1974): 38–49.

Piergrossi, Alessandra, Silvia Kortenaar, and Valeria Acconcia. 'Lo sviluppo e la circolazione della ceramica di impasto rosso in Etruria meridionale e nel Lazio'. In *Metodi e approcci archeologici. L'industria e il commercio nell'Italia antica*, edited by Eric C. De Sena and Hélène Dessales. Oxford: Archaeopress, 2004.

Popham, M. R., L. H. Sackett, and P. G. Themelis. *Lefkandi*. Vol. I: *The Iron Age Settlement, the Cemeteries, Text (British School at Athens Supplementary Volume)*. London: Thames and Hudson, 1980.

Powell, Walter W. 'Neither Market nor Hierarchy. Network Forms of Organisation'. *Research in Organizational Behaviour* 12 (1990): 295–336.

Pritchard, James B. *Recovering Sarepta, a Phoenician City. Excavations at Sarafand, Lebanon, 1969–1974*. Princeton: Princeton University Press, 1978.

Rauner Max. 'Ziemlich verknotet.' *Die Zeit*, no. 10, 26 February 2004.

Renfrew, Colin. Introduction to *Peer Polity Interaction and Socio-Political Change*, edited by Colin Renfrew and John F. Cherry. Cambridge: Cambridge University Press, 1986.

SAA 2: Parpola, Simo, and Kazuko Watanabe. *Neo-Assyrian Treaties and Loyalty Oaths (State Archives of Assyria*. Vol. 2*)*. Helsinki: Helsinki University Press, 1988.

Schildhauer, Johannes, Konrad Fritze, and Walter Stark. *Die Hanse*. Berlin: VEB Deutscher Verlag der Wissenschaften, 1977.

Schreiber, Nicola. *The Cypro-Phoenician Pottery of the Iron Age*. Leiden: Brill, 2003.

Scott, John. *Social Network Analysis. A Handbook*. London: SAGE Publications, 1991.

Seden, Helga. 'A Tophet in Tyre?'. *Berytus* 39 (1991). Available from http://almashriq.hiof.no/ddc/projects/archaeology/berytus-back/berytus39/seeden-tophet/.

Shatzmann, Israel. Review of *The Corrupting Sea*, by Peregrine Horden and Nicholas Purcell. *Sripta Classica Israelica* 23 (2004): 279–83.

Sommer, Michael. *Europas Ahnen. Ursprünge des Politischen bei den Phönikern*. Darmstadt: Wissenschaftliche Buchgesellschaft, 2000.

———.'Die Peripherie als Zentrum. Die Phöniker und der interkontinentale Fernhandel der Eisenzeit'. In *Commerce and Monetary Systems in the Ancient World. Means of Transmission and*

Cultural Interaction, edited by Robert Rollinger, and Christoph Ulf. Stuttgart: Franz Steiner Verlag, 2004.

———. *Die Phönizier. Handelsherren zwischen Orient und Okzident*. Stuttgart: Alfred Kröner, 2005.

———.'Der Untergang des hethitischen Reiches. Anatolien und der östliche Mittelmeerraum um 1200 v. Chr.'. *Saeculum* 52 (2001): 157–76.

Sznycer, Maurice. 'L'expansion phénico-punique dans la Méditerranée occidentale. Problèmes et méthodes'. In *Actes du 2e Congrès international d'études des cultures de la Méditerranée occidentale*. Alger: 1976.

TAVO. *Tübinger Atlas des Vorderen Orients*, edited by Sonderforschungsbereich 19 Tübinger Atlas des Vorderen Orients der Universität, Tübingen.

Van Berchem, Denis. 'Sanctuaires à Hercule-Melquart. Contribution à l'étude de l'expansion phénicienne en Méditerranée'. *Syria* 44 (1967): 73–109, and 307–36.

Weber, Max. *Die Stadt (Wirtschaft und Gesellschaft. Die Wirtschaft und die gesellschaftlichen Ordnungen und Mächte,* Teilband 5), edited by Wilfried Nippel. Tübingen: Mohr-Siebeck, 1999.

Whittaker, Charles R. 'The Western Phoenicians. Colonisation and Assimilation'. *Proceedings of the Cambridge Philological Society* 200 (1974): 58–79.

Winter Irene J. 'Homer's Phoenicians. History, Ethnography or Literary Trope?'. In *The Ages of Homer: A Tribute to Emily Townsend Vermeule*, edited by Jane B. Carter and Sarah P. Morris. Austin: University of Texas Press, 1995.

Zaccagnini, Carlo. 'Aspects of Ceremonial Exchange in the Near East During the Late Second Millennium BC'. In *Centre and Periphery in the Ancient World*, edited by Michael Rowlands, Mogens Trolle Larsen, and Kristian Kristiansen. Cambridge: Cambridge University Press, 1987.

Networks of Rhodians in Karia

Riet van Bremen

Introduction

According to a now traditional designation introduced by Peter Fraser and George Bean in *The Rhodian Peraea and Islands*, Rhodian territorial possessions on the mainland fell into two parts: the 'incorporated' and the 'subject' Peraia. The former covered roughly the Loryma peninsula, bordering on the territory of Kallipolis in the north, its eastern limits approximately where the extensive territory of Kaunos began, its western boundary at the entrance to the Knidian peninsula (Figure 1).[1] This region was incorporated at an early stage, probably before the end of the fifth century. Its inhabitants were fully integrated into the Rhodian state: their localities became Rhodian demes and were distributed among the island's three tribes: *Ialysia*, *Kameiris*, and *Lindia*, themselves the island's former *poleis*, whose synoikism had, in 408/7 BCE, created the new Rhodian state.[2] Inside this state they were known by their demotic, while abroad they were *Rhodioi*, indistinguishable in the epigraphic record from Rhodians of the island.

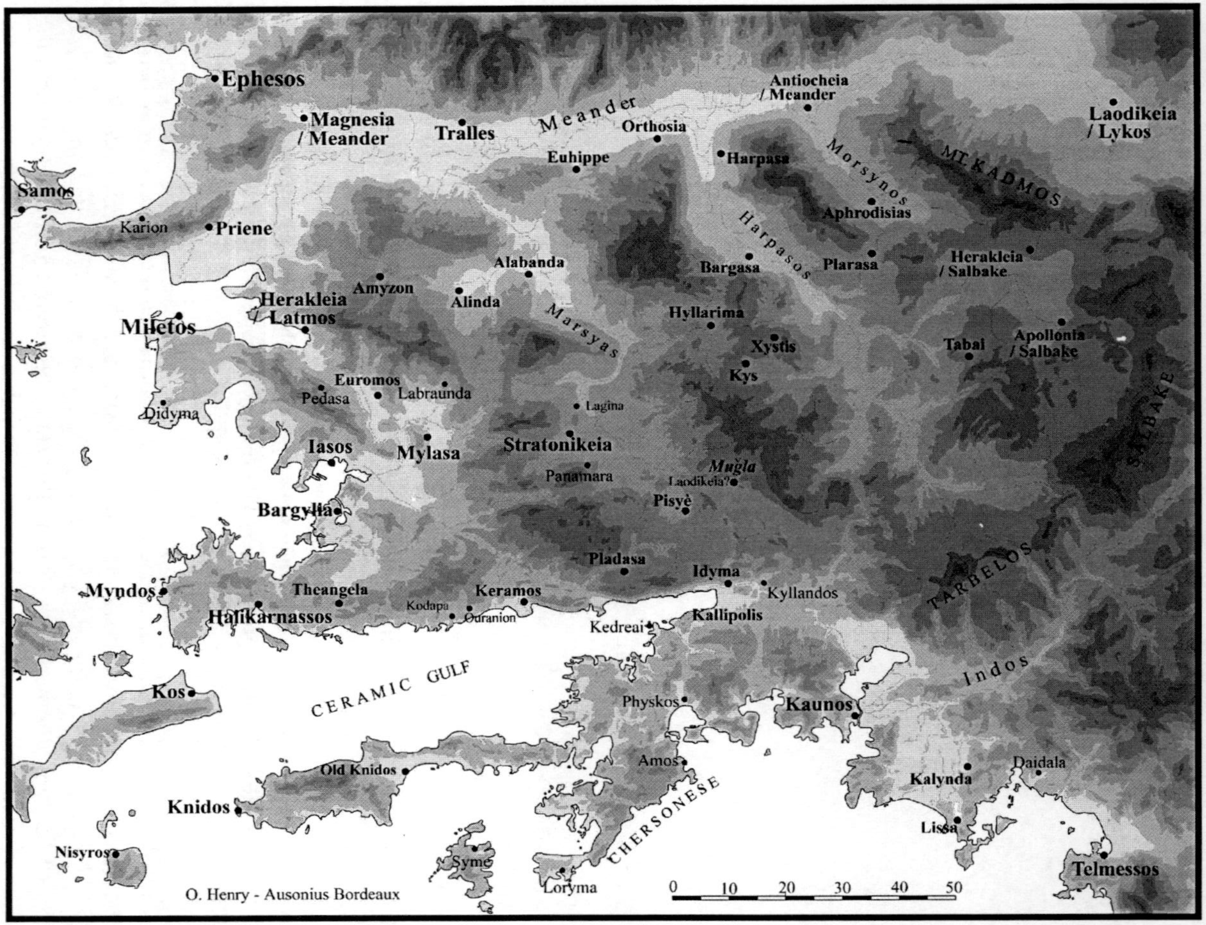

Figure 1 Karia with the 'subject' and incorporated *Peraiai* (Publications Ausonius, Bordeaux - O. Henry).

The core of the subject Peraia (a term not used in Antiquity)[3] consisted of the part of Karia that lies north of the Keramic Gulf: between Keramos and Idyma on the coast, and with Stratonikeia, Hyllarima, and Muğla on its northern, north-eastern, and eastern extremes (Figure 1). The extent and chronology of Rhodian domination in this region are slowly becoming better understood as more evidence comes to light, though much is still unclear.[4] Rhodes's acquisition of territorial possessions in Karia pre-dates by at least half a century Rome's grant, at the peace of Apameia in 188 BCE, of all territory south of the Maeander to the island state (minus a number of cities).[5] The process of expansion probably took place in the 260s and 250s, though the surviving evidence does not allow us to reconstruct precisely how this region fell into Rhodian hands.[6] Only in two cases are we told more: the city of Stratonikeia, a Seleukid foundation of the 260s or 250s, came to the Rhodians as a gift of the Seleukid kings, perhaps as early as the 240s. Kaunos, near the Lykian border, was bought from Ptolemy's generals for 200 talents, in 197 or soon after.[7] Stratonikeia was briefly lost to the Macedonian king Philip V in 201, but was recovered in 197. The Rhodians held on to both cities until 167 BCE.[8] All other possessions in this part of Karia, including the region around Muğla, were retained.

If Stratonikeia was given to Rhodes in the 240s, the region between the Keramic Gulf and that city must by that time already have been brought under Rhodian control. An inscription published in 2001 appears to confirm this. Found at Arslanlı, one of the central settlements of ancient Pisye (Figure 2), and dated to between 275–225 BCE, it lists financial contributions from members of local communities to the building of dockyards (*neoria*) on the coast at Akbük—in the territory of Pladasa, previously a *polis* (some of?) whose citizens had formed a *koinon* with the Pisyetai.[9] One reconstruction of the inscription's fragmentary first lines has the *demos* of the Rhodians as recipient of the promised contributions. If Rhodes was indeed behind the building of these dockyards, as I believe it was, then the location of the inscription itself, some thirty kilometres inland, is important. This text is in fact the earliest to refer to the Pisyetai and Pladaseis as a *koinon*, though precisely how we should imagine this new polity is unclear.[10] The concentration of epigraphical documents at Pisye's main settlements, Yeşilyurt, Arslanlı, and Aldıran Asarı, and the extensive archaeological remains there, show this to have been one of the major centres of the entire subject Peraia, while Pladasa as a community disappears almost completely from view: for the period after the *sympoliteia* we have only one brief funerary inscription (for a Rhodian) from its territory.[11] The editors suggest that a kind of overarching super-*koinon* was formed, stretching all the way from Pisye to the sea, and taking in all the smaller communities in between that are listed as contributors in the *neoria* inscription (see Figure 2 for its possible extent). But other scenarios are possible, such as a complete removal of (most of the?) Pladaseis to Pisye, giving the former *polis* site at and around Akbük over to Rhodian military use. If so, there is no real necessity for the smaller *koina* to have been incorporated into a new structure.[12]

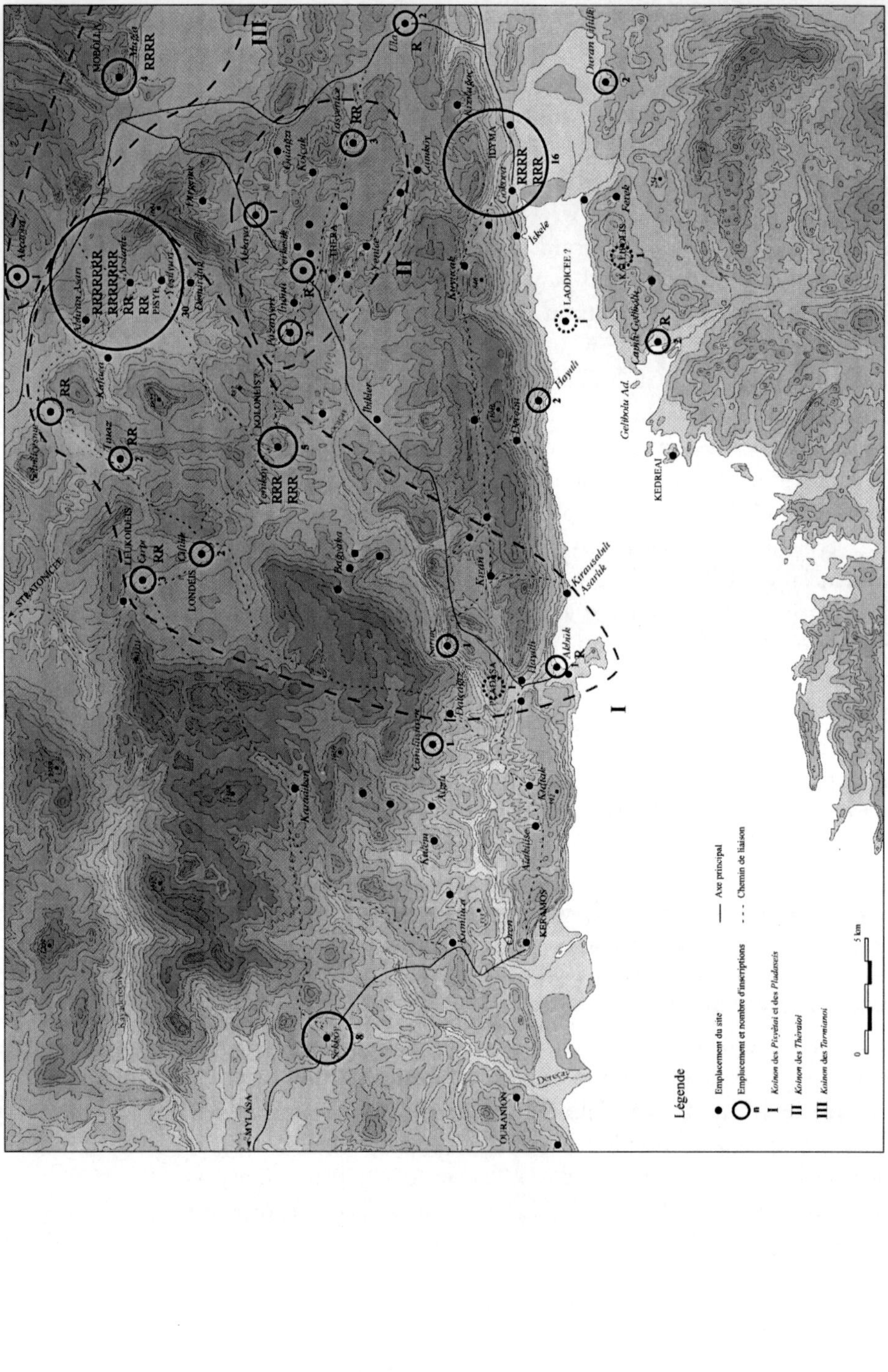

Figure 2 The Karian Uplands. Each 'R' represents one inscription in which Rhodioi feature (Publications Ausonius, Bordeaux).

This was certainly a region of *koina*: with the exception of Stratonikeia and Kallipolis,[13] no cities developed in the course of the Hellenistic period, even though the large plain of Muğla with its impressive acropolis, or the plain around Pisye itself, or that around Yerkesik, ancient Thera, were all capable of sustaining a sizeable city.[14] It has been said that the *koinon*—a loose federation of village settlements with a common political and religious structure—was a typical Karian phenomenon that somehow sprouted fully-formed from the region's physical geography. But village settlements are not unique to this part of the Mediterranean, and in many other parts of the Greek world similar groups of settlements did turn themselves into *poleis* and developed *polis* institutions (as indeed happened elsewhere in Karia). In fact, many communities in this part of Karia, too, had begun to acquire *polis* identity already in the fourth century.[15] The loss of that identity was the direct result of Rhodian domination: within the wider Rhodian sphere of influence the only true *polis* acknowledged was that of Rhodes itself.[16]

However we choose to describe the process—as a kind of arrested development, a freezing of a situation, or even a turning back of the clock—the result was a Rhodes-generated network of dependent *koina*.

Patterns of Rhodian Presence

This part of Karia has always been little known and under-explored. The steeply mountainous coast of the Keramic Gulf blocks access to its hinterland, and there are few routes through to the Maeander Valley. The recent publication of *Les hautes terres de Carie*—the fruit of many years of systematic exploration in the late 1980s and early 1990s, by a Franco-Turkish team—has done much to change this picture, and has succeeded in bringing out the relative density of settlement in these 'Karian uplands', as well as their complex infrastructure. Although many of the sites had been seen and described before, in particular by W. R. Paton and J. L. Myres, and then by P. M. Fraser and G. Bean, the Bordeaux team was able in almost all cases to add substantially to earlier descriptions and put a number of new sites on the map.[17] The epigraphic section of the book confirms this impression of a region coming at last into sharper focus: about half of the inscriptions are published here for the first time. What is immediately striking is that around half of the just under a hundred entries specifically mention Rhodians (*Rhodioi*).[18] If we discount the incomprehensible fragments (six), the Roman milestones (twelve), and the inscriptions pre-dating the third century BCE (five),[19] then the relative predominance of *Rhodioi* becomes greater still: about two-thirds. Where Fraser and Bean were aware of fourteen inscriptions from the entire region mentioning Rhodians, *HTC* has fifty: a virtual tripling of evidence. From the *koinon* of the Pisyetai and Pladaseis alone we now have a total of thirty inscriptions (including some of Roman date) of which seventeen either feature *Rhodioi* or have clear Rhodian associations.[20] Only two of these were known to Fraser and Bean.

These peculiar patterns raise questions about the impact and nature of Rhodian domination. Much has been written about the island state's formal controlling mechanisms over its subject territories: even if they cannot be reconstructed in any detail from the surviving evidence, some of the outlines are visible. The most recent discussion is by H.-U. Wiemer in *Krieg, Handel und Piraterie* (251–60), to which the reader is referred for further details.[21] Wiemer, rightly in my view, emphasizes the military nature of Rhodian control, before, during, and after Apameia, of which we catch glimpses through titles such as that of *strategos en toi Peran* and *hagemon epi Karias*. Rhodian garrisons are attested for Stratonikeia, Kaunos, Pisye, Kyllandos, and Idyma; *epistatai* are known from Idyma, Muğla, and Panamara; and a single *strategos* is attested at Idyma. *Epistatai* and *strategoi* may have been responsible also for non-military administration: the adjudicating of disputes and the upkeep of roads and bridges are both attested. They may also have collected tribute, though the evidence for this is virtually nonexistent.[22]

My aim in this brief article is not to offer a complete (re)interpretation of the nature, mechanisms, and impact of Rhodian control, although such a full treatment of the relations between Rhodes and its dependencies, in Karia, Lykia, and the islands—not forgetting Crete—is much needed.[23] My modest focus here is on those Rhodians of whose widespread presence we have for the first time a clearer view, those whose names are recorded in the region's many funerary and honorific inscriptions, but who were not, or at least not obviously, part of any administrative-military machine. Their presence, when plotted on a map (Figure 2), shows up interesting patterns of Rhodian individuals and groups of individuals across the entire region, with a dense concentration in the Pisye region. Who were these *Rhodioi*? Why do they dominate the epigraphy of this part of Karia to such a high degree, almost to the exclusion of locals? What interests did they have, and what role did they play vis-à-vis the indigenous communities?

In *HTC*, Alain Bresson proposes a tentative answer. The frequency with which Rhodian ethnics are encountered, he writes, is best explained by the gradual extension of Rhodian citizenship to the elites of local communities. His general idea—elaborated in the discussion of individual inscriptions—is that the *Rhodioi* whom we encounter in the epigraphic record of this region are not (or not only) Rhodians from the island and the incorporated Peraia who had settled in subject territory, but are members of 'indigenous' elites: the wealthiest, most important families of their communities, to whom had been granted Rhodian citizenship, *at a certain moment* and as a special mark of distinction, but without this citizenship being extended *en bloc* to their communities, and without those communities themselves being incorporated into the system of demes, as were those of the incorporated Peraia.[24] This is a striking suggestion, and an important one, but one that does not sit easily with what we know about the nature of Greek citizenship. It needs to be tested and its implications further explored. The economic implications, for instance, are not easy to think through: who, in these split communities, would have been liable to pay tribute to Rhodes? Presumably not the *Rhodioi*.

More recently, Bresson discussed the economic interests that individual, 'true' Rhodians may have had in subject territory on the mainland in the years between 188 and 167 BCE, when the island's possessions were at their most extensive: 'Le détail nous est encore trop mal connu, mais on peut considérer que les Rhodiens purent désormais librement acquérir des terres, jouer un rôle d'intermédiaire commercial et surtout financier en prenant la ferme de la collecte des impôts ou en accordant des prêts, sources de juteux profits.'[25] A passage in Polybios lends some substance to this assessment: 'The Rhodians, delivered from their difficult position, now breathed freely and sent Kleagoras on an embassy to Rome to beg that Kalynda might be ceded to them and to ask the Senate to allow those of their citizens who owned property in Lykia and Karia to hold possession of it as before' (Pol. 31.4.3). This appeal to the Senate (of 164 BCE) specifically concerns those territories over which the Rhodians had lost control in 167—and which they had thus held for only twenty years.[26] It seems however safe to assume that a similar situation prevailed in areas they had controlled since the mid-third century. Among the terms of the treaty of Apameia itself (Pol. 21.42.16) we find the following: 'All houses (οἰκίαι) that belonged to the Rhodians and their allies in the dominions of Antiochos shall remain their property as they were before he made war on them.' This must refer to possessions gained in the course of the third century BCE, which Antiochos had briefly claimed between 203 and 189. These passages leave no doubt as to the reality of Rhodian landed possessions in Karia, but they do not allow us to see how widespread such ownership was.[27] There is no direct evidence for Rhodian commercial and financial profiteering in the region, but given what is known of Rhodian activities elsewhere, this would not be surprising.[28]

Are Bresson's two models of Rhodian presence in Karia contradictory? Not necessarily: they could be argued to be different facets of the same historical situation, in which 'real' Rhodians going about their profitable business in subject territory mixed with members of local elites, who were gradually assimilated to the desirable status of *Rhodios*. But how likely is the second part of this proposition? The answer depends almost entirely on how we identify the numerous *Rhodioi* in our inscriptions: if we see them as Rhodians of the island or of the incorporated Peraia, then their presence would put some flesh on the bones of the 'land-grabbing and financial profiteering' model; if it can be shown that they were instead (or mostly) members of indigenous elites to whom full Rhodian citizenship-without-deme-membership had been extended, then a rather different picture of Rhodian relations with the 'subject' Peraia emerges, one which led Bresson to argue that the boundaries between the two Peraiai—and the institutional, social, and political divisions between its inhabitants—were more fluid than was previously thought.[29] We might in that case need to question the very term of 'subject' Peraia: for there is surely an inherent contradiction in the idea of subjecting one's fellow *Rhodioi*.

Rhodioi

Rhodioi occur in the epigraphic record from the late third century BCE onwards, and continue into the third century CE, with the majority of inscriptions dated to the first centuries BCE and CE. Many inscriptions are only approximately datable, and margins of about a hundred years are not uncommon. They are recorded for the following communities (see Figure 2): the *koinon* of the Pisyetai; Tınaz (site of a small, unnamed, *koinon*); the *koina* of the Leukoideis, the Londeis, the Koloneis; Pladasa/Akbük; the *koina* of the Theraioi(?) and Tarmianoi; Ula and Idyma. Nine *Rhodioi* are mentioned in a recently published inscription from Stratonikeia, possibly as property owners.[30] Three (?) Rhodians, one of whom was certainly an *epistates*, are honoured in inscriptions issued by the *koinon* of the *Panamareis* (early second century BCE?).[31]

The earliest of the private honours for *Rhodioi* date from between ca. 225 and 150 BCE (*HTC* 7 and 9; both from Pisye). In one, two brothers, *Kelimareis* (a small *koinon* whose location is not certain),[32] honour a Rhodian as their 'saviour and benefactor'; in the second, two *Laodikeis* honour a Rhodian and his wife as their 'benefactors and saviours'. Private funerary dedications for *Rhodioi* by *Rhodioi* cluster around 100 BCE (with margins of ca. fifty years on either side); the first public honours are of similar date. Many of the inscriptions are in fact of a funerary nature, whether private or public.[33] In their architectural form and in their manner of commemoration they are recognizably 'Rhodian': square bases for round funerary altars; a type that has been well described by P. Fraser in *Rhodian Funerary Monuments*.[34] The round marble shields with which the *Tarmianoi* honoured two Rhodians (*HTC* 62 and 63: see below) are identical to a type much in fashion on Rhodes itself, especially for military honours.[35]

The family groups of dedicators include both agnatic and cognatic relatives, extending over several generations: a mode of commemoration characteristic of Rhodes and its dependencies.[36] Once or twice, private commemoration and public honours by *koina* go hand in hand. A typical example comes from Yeniköy, the site of the *koinon* of the *Koloneis*.[37] Here, two funerary dedications for the same man survive: one private, one mixed public/private (second or first century BCE). The first text (*HTC* 41: a square base) is restored on the basis of the second (42: a cylindrical altar):[38]

[For]
[Dionysios, son of Herodes, *Rhodios*]
Dionysios and Herodes and Iason
4 sons of Dionysios, for their father, Panar-
ista d. of Pyrrhos, *Ladarmia*, for her husband,
Dionysios and Panarista children of Herodes, *Rhodioi*,
for their grandfather, Iason, son of Pyrrhos, by adoption of
8 Python, *Rhodios*, for the husband of his sister…

No. 42 is a dedication of several local *koina* to the same man, in association with (some) members of his family (for a reconstruction of the stemma, see *HTC*, 152):

For
Dionysios, son of Herodes, *Rhodios*,
honoured by the *koinon* of the *Londeis*
4 and the *Koloneis*, and by the *koinon*
of the *Pisyetai* and by the *koinon* of
the *Theraioi*, and crowned with gold
crowns, deemed worthy both of a public
8 funeral and of the erection of statues;
Herodes, son of Dionysios [for his father];
Panarista daughter of Herodes [for her grandfather],
Iason and Dionysios [for their //////]
12 [------] his/her brother MO[--]

Altogether, eleven members of this family are mentioned. Dionysios may have resided in the territory of the *koinon* of the *Londeis* and *Koloneis*, because, despite its small size, it is this *koinon* that comes first in the honouring queue. The *Pisyetai* and *Theraioi* were both much larger units, but adjoined the territories of the *Londeis* and *Koloneis*. To be thus honoured, with a public funeral, gold crowns and statues, Dionysios must have been a person of considerable influence and power beyond his chosen place of residence. Was he an administrator? A wealthy Rhodian who lent money to local communities and who served as their protagonist at the centre, thus earning their gratitude and their gold crowns? Or was he, instead, as Bresson suggests, a member of the local elite, a *Koloneus* by birth, who was able to exercise his influence on behalf of the indigenous communities by virtue of having gained access to Rhodian citizenship and thus to the Rhodian decision-making bodies?

One can see the two different models taking shape. In his commentary on 41 (p. 152), Bresson writes: 'Cette inscription et la suivante émanent du même groupe familial, qui devait constituer la principale famille de notables du lieu. Les Cariens hellénisés de la "Pérée sujette" ... n'étaient pas inclus dans le système des dèmes. Mais on voit qu'ils n'étaient nullement des Rhodiens de seconde zone.' This means, first, that he takes Dionysios and family to be of indigenous ('Hellenized Karian') descent, and, secondly, that their acquired status as *Rhodioi* gave them access to the full range of rights and duties that citizens belonging to the deme-system had. A further assumption is that by promoting men like Dionysios to the status of *Rhodios*, a two-tier system was created, in which some *Koloneis* had become Rhodians, while others remained *Koloneis*.[39] How can we test Bresson's hypothesis? One crucial element in the argument is the assumption that Dionysios, a 'naturalized' Rhodian of the Peraia, had been able to marry a woman from one of the old Rhodian demes: his wife Panarista came from the important (Lindian) deme of the *Ladarmioi*. The right of marriage after all—a marriage that would produce legitimate citizen offspring—was one of the central mechanisms determining and upholding Greek civic identity, and Rhodes, as is known, had particularly strict rules about the entitlement to citizen status by the offspring of mixed marriages.[40] I shall come back to this point, but want first to bring another case into the discussion, one that bears on the question of civic rights and duties. Did the (presumed) extension of Rhodian citizenship to Peraian elites bring with it such rights as membership of assembly and council, and unrestricted access to central magistracies, military commands, and priesthoods?

In no. 56, part of a larger monument from Yerkesik (Thera?), dated to between 87 and 70 BCE, Phanias Philippou, *Rhodios*, who had been *strategos kata polemon*, receives *post mortem* honours.[41] His father, Philippos Polycharmou, *Rhodios*, heads the commemoration, joined by six other men, all *Rhodioi*, all uncles (though with five different filiations).[42] A Philippos Polycharmou is recorded as *damiourgos* at Kameiros in 83 BCE (*TC* 3Δa l. 45): very likely the man in our inscription. Bresson, although keeping open the possibility that Philippos the *damiourgos* was a Rhodian by birth, opts for seeing him as a born 'Peraian' who owed his Kameiran office to his acquisition of Rhodian citizenship.[43] This allows him to interpret the damiurgy as a remarkable instance of the integration of a Peraian *Rhodios* into the very heart of Rhodian society. For if our man had been a Rhodian of the island, having 'acquired estates in subject territory, having married there and set up house', Bresson argues, would he not have '*signalé son démotique, comme la Panarista du no 42*?' (41 is meant; italics are added). It is true that the damiurgy, the office most closely linked to separate Kameiran identity within the Rhodian state, could in fact be held by non-Kameiran Rhodians, including men from the demes of the incorporated Peraia. In this it differs from the much more exclusively guarded priesthood of Athena at Lindos which was limited to men from the Lindian demes of the island.[44] But I am not sure if this relative openness warrants the leap of faith which we are asked to make. Should we not rather consider the holding of the damiurgy as a very strong indication of membership of the deme-system, and thus as evidence *against* Philippos's Peraian origin (an argument valid also in the case of his son Phanias's strategy)? A further, and practical, objection, as P. Fraser has shown, has to be that, like others on the island, this office was filled through a system of rotation based on deme and tribal membership. Our hypothetical Peraian *Rhodioi* were not so integrated, and it is hard to see how the system would have been adjusted to take these *novi homines* in without it showing up in the evidence.

Unfortunately, there is no way of telling how many of this family were resident in Theraian territory. The monument's location must count for something, and residency (of some, of all?) is an obvious explanation. But another might be that Phanias had fallen in action in Theraian territory.[45]

Different Statuses?

The case of Panarista's Rhodian demotic (*Ladarmia*) (above, no. 41) has turned out to be important in the discussion of no. 56. Let us therefore return to her. In the dedications 41 and 42, Panarista's status of *Ladarmia* stands out among that of her relatives, all of whom are *Rhodioi*. On Bresson's interpretation, in this family a *Rhodios* of the subject Peraia had married a woman with the true-Rhodian demotic of *Ladarmia*.[46] This 'fact' is then used as proof of the full equivalence of their different statuses. But can we be so sure? The principle that a Rhodian (or any Greek citizen) was known by his ethnic ('Rhodian', 'Athenian') *outside* his home city, but by his demotic ('Ladarmian, 'Acharnian') within the latter's boundaries, barely needs reiterating. It is this very principle that underlies the distinction scholars have made between the incorporated and the subject Peraia. Panarista's case appears to undermine it.[47] How far can we generalize from this one case? As an example of the use of a Rhodian demotic *outside* Rhodes, and among relatives who themselves are called *Rhodioi*, it is unique. Panarista's brother, Iason, whom we might have expected to sport the same demotic as his sister, is called *Rhodios*. Bresson explains this from his adoption by a certain Python 'qui devait être un Rhodios de la "Pérée sujette"'. On Bresson's interpretation Panarista would be the only real 'Rhodian of the island' in this entire company (and by extension in the entire epigraphical corpus of this region): all the rest are 'Rhodianized' Peraians.

Do onomastics help us to decide genuine 'Rhodian-ness'? In the case of the *damiourgos* Philippos Polycharmou and his son Phanias, the names are inconclusive: all three are well-attested on Rhodes, but they are by no means exclusive to the island. Interestingly, Panarista, the one genuine *Ladarmia*, has a name that is completely unattested on Rhodes (*LGPN* vol. I *s.v.*; the male name Panaristos is also unattested).[48] Iason's adoptive father, Python, on the other hand, supposedly of the subject Peraia, has a name that occurs widely on Rhodes (and elsewhere).[49] In some instances we can put more faith in the distinctiveness of a name, as in the case of a man honoured in a brief, private dedication (*HTC* 9, from Pisye, above p.119). The name Hypsikles, together with the consistent use of the Doric dialect (ὑπὲρ Ὑψικλεῦς Ἀργεάδα καθ' ὑοθεσίαν δὲ Μενεκλεῦς Ῥοδίον καὶ τᾶς γυναικὸς Φιλοκρατείας Νικάρχου Ῥοδίας εὐεργετῶν καὶ σωτήρων) goes some way to suggest native Rhodian-ness for the honorands.[50]

The subject of intermarriage between those of the island and members of indigenous Karian communities comes up in a different way in *HTC* no. 37: an inscription found at Yeşilyurt, seat of the *koinon* of the *Pisyetai* and *Pladaseis* (ca. 50 BCE–CE 50).[51] In it, the *koinon* grants a public burial to a *Rhodios*, Diokles, son of Aristarchos. The deceased's mother, Artemis, daughter of Menippos, *Leukoidis*, and Diokles's two sons, Aristarchos and Diokles, *Rhodioi*, join in. We do not have the ethnic of Artemis's husband. Bresson comments (ad locum):

> Tout d'abord, [ce texte] montre comment une femme de statut "indigène" (*Leukoidis*) pouvait avoir une descendence rhodienne. Il est bien dommage que l'on ne possède pas l'ethnique du père. *Néanmoins, il ne paraît pas déraisonnable d'admettre que, du côté paternel aussi, la famille était originaire du cru* et non pas des territoires anciennement rhodiens. [italics added]

However, 'pas déraisonnable' is a slippery term, when what is needed is proof. On Rhodes itself, the son of a Rhodian father and a non-Rhodian mother had the status of *matroxenos*, and, though inscribed in his father's deme, did not automatically acquire full citizenship.[52] The implication here is that Peraians were not *xenoi* in the true sense of the word but had already entered the outer circles of Rhodian-ness, enough to facilitate access to the status of *Rhodios* for their offspring. It is possible, and a suggestion worth considering, but I do not think that this example alone can show it. This is, of course, an interesting case whatever the father's status. If he was in fact a Rhodian of the island, as I think he may well have been, does that make the marriage a truly mixed one? I suspect Bresson would say it was not 'mixed' in the sense that the marriage between a Rhodian man and, say, a woman from Aspendos would have been. In the latter case, the *matroxenos* rule applied. But even if the Peraian wife were in a more favourable position, would that make her son a '*Rhodios* of the Peraia' (without membership of the deme-system) or would it rather facilitate his being properly naturalized at the centre and allocated to one of the Rhodian demes?[53] *HTC* no. 83, from Idyma (Hellenistic?), provides a parallel at least for the mixing: it is a brief epitaph of Anaxikratea, d. of Sopatros, *Idymia*, wife of Menekrates s.o. Menekrates, *Rhodios*.[54] Here, no children are evident and so a level of uncertainty remains, but there is no reason for assuming that Menekrates was a Peraian/Idymian raised to the status of *Rhodios*, rather than from the island or incorporated Peraia.[55]

In all the cases so far discussed, unless we make *a priori* assumptions about the identity of our Peraian *Rhodioi*, there is no way of proving that what we have are instances of naturalization of indigenous elites. The moment we take this assumption for granted, we are on dangerous ground. Take no. 63, an honorific inscription on a shield (84–54 BCE), in which a Rhodian commander (*hagemon*), Chrysippos, son of Apollonidas, who had served at Artouba and Parableia (in Kaunian territory, at that time once again under Rhodian control) and who had campaigned on kataphract ships (ll. 4–7), is honoured, for his *eunoia*, by the *koinon* of the *Tarmianoi*, based at Muğla. Why here, when nothing in the inscription specifically binds the man to the *koinon*? Bresson's answer (p. 189): 'Chrysippos a fort bien pu lui-même être un Rhodien de la *Pérée* et un Tarmianos en admettant que les élites des *Tarmianoi* avaient reçu à cette date la citoyenneté rhodienne.' Honoured here, then, because he was, underneath his Rhodian-ness, of Tarmianian descent? I venture an alternative: Chrysippos was stationed with the *Tarmianoi* and they had reason to be grateful to him. A similar, near-contemporary, shield-inscription (no. 62) dated by the Rhodian priest of Halios, and honouring another *Rhodios*, Sosikrates son of Sosinikos, *epistates*, 'for his *eunoia* and *dikaiosyne* (justice)', implies that this may be the more obvious interpretation. The *Tarmianoi* did have a reason for honouring Sosikrates: he was placed in a position of power over them and had used it with discretion.[56] Sosikrates is also attested on Rhodes, in the great *epidosis* list *IG* XII, 46, l. 454. He was very likely a Rhodian (and here Bresson does not disagree). Are we then justified in treating these two men as belonging to two different species of *Rhodioi* when their dates and contexts are so similar?

Integration, Transformation, Assimilation?

From the same *koinon* of the *Tarmianoi* comes a dedication by Nikolaos, son of Leon, *Rhodios*, who had served as ephebarch and gymnasiarch, 'to Hermes and Herakles, and to the *koinon* of the *Tarmianoi*' (*HTC* 64, second/first century BCE). This text was known to Fraser and Bean, who rightly pointed out the interest of a Rhodian holding a local magistracy in a subject community: 'valuable evidence for the penetration by a member of the ruling people into the local administration of the conquered'.[57] Even if the terminology reflects a view of Rhodian 'conquest' from which we need to move away, I see no need to disagree with their view of the man's origin. I would argue that his association with *ephebeia* and gymnasium (both of which, as I have suggested elsewhere, must go back to a pre-Rhodian situation during which there existed a *polis* on this site) can be understood in tandem with the two inscriptions just discussed. Together they suggest that the territory of the *Tarmianoi*, with its strategically useful acropolis at Muğla, was a nodal point in Rhodes's military presence in south-western Karia.[58] The gymnasium (the only one of its type attested in this region) may have served as a training centre for local men: why not under Rhodian supervision? Auxiliary troops from the Peraia had assisted the Rhodian general Pausistratos in 197 BCE in regaining territory taken by Philip V (Livy 33.18); among them were the *Tarmianoi* and *Pisyetai* (the two *koina* were also closely connected in honouring a Rhodian, Moschos, s.o. Antipatros, with a funeral at public expense: *HTC* 4, from Arslanlı). The Rhodian *epistates* and his men may themselves have trained here. From this perspective, Nikolaos's involvement in the supervision of the local youth does not seem so out of place.[59]

Two similar cases of integration of Rhodians come from the *koinon* of the *Leukoideis*. The first (*HTC* 36) is an honorific decree dated to between 107 and 80 BCE by the Rhodian priest of Halios, for Sopatros son of Theon, *Rhodios*, who had greatly benefited the *koinon* by having become their advocate (*ekdikos*), presumably representing their interests in Rhodes. Bresson's view is that Sopatros was most likely a local man, but I see no inherent reason why he should not be a Rhodian of the island: a patron, rather than a man risen from the ranks, someone long (διὰ προγόνων) settled in subject territory and integrated into the daily affairs of the *Leukoideis*, affairs that very likely affected his own estates.[60] Possibly as much as a century later, *HTC* no. 38 (50 BCE–CE50?) is a funerary commemoration by the same *Leukoideis*, for Euphranor (a good Rhodian name), son of Antimachos, who had been the *koinon*'s priest (*neokoros*), wine-treasurer (*oinotamias*), and *komarchos* and had 'sorted out all the *koinon*'s business'. The level of integration in this case is striking and says much about the assimilation of Rhodians settled for many generations among the local communities. Euphranor at least (at last?) got into the spirit of things, some two centuries after the Rhodians first took control of this region. But it does not prove that he was a born *Leukoides*.

In none of the evidence discussed is there proof that the *Rhodioi* whose monuments and presence are attested throughout the subject Peraia were in fact members of local elites to whom Rhodian citizenship of a certain kind had been extended. Though I have not dissected every single piece of evidence, I have taken into account all those that are problematic, or telling, or puzzling, or have been singled out in *HTC* as supporting the 'transformation' model. One can see the thinking behind this model: if at some point in the fifth century the Rhodians had made the Karians of the Loryma peninsula fully Rhodian, then why stop there?[61] It seems a fair point. The problem to my mind lies not with any cultural obstacle but rather with the two-tier model of citizenship that we are asked to accept, whose stages and implications are completely opaque and for whose principles there are no direct parallels in the Greek world even if some of its ingredients are familiar. New models may be created from old ingredients, if circumstances demanded it, but my—boring—point is that the evidence does not compel us to think along these lines.

Bresson's idea of the creation of elites of *Rhodioi* in the subject Peraia, is fundamentally a dynamic and enabling one: local men become Rhodians and have the option of participating in the decision-making processes at the centre, thus at the same time representing 'their' communities and serving the Rhodian state as military men and administrators. Though never explicitly mentioned, the example of Roman citizenship and the—near-contemporary—creation of an elite of Roman citizens in the provinces cannot have been far from Bresson's mind. Did the Rhodians look specifically to Rome for such an example? Or did the Romans, on the contrary, initially take their cue from Greek models? After all, the Rhodians' incorporation of their Peraia and the extension of Rhodian citizenship to the Karian Chersonnese had been completed by the end of the fifth century. After that early stage, we need not even postulate one-way influence, but may be able to think instead in terms of interactions of ideas and practices.

While it is tempting to develop such possible parallels and implications in a more general sense, I have here set out reasons why I do not think that we are justified in applying a model of deliberate creation of an elite of *Rhodioi* in all its institutional implications to the region north of the Keramic Gulf. Over the years, through intermarriage, and subsequent conferring of full citizenship (including deme-membership), some Peraians may have become Rhodians, but my contention has been that the normal naturalization mechanisms served well enough in these instances. I wish to emphasize once more the strikingly undynamic, almost atrophied, picture that emerges from the epigraphical evidence, which is unusually monochrome: a picture of endless funerary or honorific commemorations for Rhodians, Rhodians, Rhodians. The relationships that emerge and the language used to describe them speak of distance and inequality, even in the case of men like Euphranor, or Sopatros: they are relations of patronage rather than of a citizen honoured by his fellow citizens. We need, of course, to get away from the concept of 'conqueror-subject', and be aware that there may have been changes over time, but we cannot entirely escape from the notion of control and military, administrative (and fiscal?) presence. Honorific inscriptions offering crowns and public funerals are wont to emphasize the beneficial patronage exercised by *Rhodioi*; they do not tell us what pressures these men were capable of exercising, or, put slightly more favourably, what central demands they were instrumental in conveying or capable of mitigating.

As a contrast to the monochrome picture presented in the funerary and honorific inscriptions stands a small altar, found at Yeşilyurt, the place with the greatest concentration of *Rhodioi*.[62] On it is a dedication to Zeus Atabyrios: the great Zeus of Rhodes. Above the text an eagle is depicted in relief, and on the altar's upper moulding is a rose, symbol of Rhodes (Figure 3). The rose has, however, been mutilated, and it looks as if someone has also tried to erase the text. It is a comment of sorts on the integration of Rhodians into local life.[63] We cannot of course hang our interpretation of an entire region's acceptance or rejection of its powerful neighbour on a single mutilated rose, any more than on the single occurrence of a Rhodian demotic. And although it may seem that what I have done in this article is no better than what the defacers of the small altar did to the proud symbol of Rhodes, I have, in between

Figure 3 Altar dedicated to Zeus Atabyrios (from *Les hautes terres de Carie*, p.250. Photo: A. Bresson).

the lines, argued for a much more comprehensive study of the interaction of Rhodians and non-Rhodians in southern Karia, one that asks questions about the balance between civic (very little) and military (considerable) architecture, about the economic advantages and disadvantages of control; about the wider cultural impact of Rhodian customs (commemorative practices, onomastics, dialect); and the institutional and political atrophy of the local communities. Bresson's 'koinè du golfe Céramique' (above, n. 55) seems a very promising concept from which to start.

Acknowledgements

I wish to thank Publications Ausonius in Bordeaux, and O. Henry, for making available the maps in Figures 1 and 2, and the photograph in Figure 3, and Alain Bresson for much stimulating discussion.

Inscriptions and Abbreviations Used

IG	*Inscriptiones Graecae.* Berlin 1873 (1903)–
I.Lindos	Blinkenberg, Chr. *Lindos. Fouilles de l'Acropole 1902–1914, II, Inscriptions.* Vol. I–II. Copenhagen, 1941.
I.Peraia	Blümel, W. *Die Inschriften der rhodischen Peraia. Inschriften von Kleinasien* vol. 38: Bonn, 1991.
I.Pérée	Bresson, A. *Receuil des inscriptions de la Pérée rhodienne.* Bordeaux, 1991.
I. Stratonikeia	Şahin, M.Ç. *Die Inschriften von Stratonikeia. 3 Vols. Inschriften von Kleinasien 21, 22,1 and 22,2.* Bonn 1981–1990.
HTC	Debord, P., and E. Varinlioğlu. *Les hautes terres de Carie.* Bordeaux, 2001.
LGPN	*A Lexicon of Greek Personal Names.* Oxford, 1987–
NSER	Pugliese Caratelli, G. 'Nuovo supplemento epigrafico rodio'. *ASAA*, n.s. 17–18, 1955–56: 157–181.
SER	Pugliese Caratelli, G. 'Supplemento epigrafico rodio'. *ASAA*, n.s. 14–16, 1952–54: 247–316.
TC	Segre, M., and G. Pugliese Caratelli, 'Tituli Camirenses'. *ASAA*, n.s. 11–13, 1949–50: 139–318.
TCal	Segre, M. 'Tituli Calymnii'. ASAA n.s. 6–7 1944–5 (1952).

Notes

[1] Fraser and Bean, *Rhodian Peraea*, ch. II.

[2] Many questions remain about the organization, integration and even location, of these demes. See e.g. Gabrielsen, *Naval Aristocracy*, 28–31; Papachristodoulou, 'Rhodian Demes'; Rice, 'Relations'; Jones, *Public Organization*, 242–52.

[3] On the meaning and definition of 'subject Peraia': Fraser and Bean, *Rhodian Peraea*, 53, 70–78, 98–117; Reger, 'Rhodes and Caria', 78–81; and, critically, Gabrielsen, 'Rhodian Peraia', 129–131 (with n. 4 for previous bibliography).

[4] Cf. H.-U. Wiemer's cautious assessment: *Krieg*, 254. Cf. Reger, 'Rhodes and Caria'; Bresson, 'Les interets rhodiens'; and van Bremen, 'Laodikeia', 372–74.

[5] Pol. 21.24.7–8; Liv. 37.55.5–6. On the cities excluded see e.g. Reger, 'Rhodes and Caria', 89, with n. 47; cf. Bresson, 'Les interets rhodiens', 185–88, on Rhodes's relations with these 'free' cities.

[6] On the region see now *HTC* 11–18; part III for the individual sites, with van Bremen, 'Karian Uplands'. On the legal aspects of the process of territorial change from royal to non-royal status, see especially Bresson, 'Les interets rhodiens'.

[7] For both cities: Pol. 30.31.6. The most recent accounts of the different interpretations of Stratonikeia's acquisition are Bresson, 'Les interets rhodiens' 180–81, and Wiemer, *Krieg*, 182–84. For Kaunos, see Wiemer, *Krieg*, 238.

[8] An inscription from Hyllarima, dated by the Rhodian priest of Halios to 197 BCE, shows the furthest extent of Rhodian-controlled territory before Apameia: Adiego *et al.*, 'La stèle caro-grecque'; cf. Wiemer, *Krieg*, 258, on the political nature of the cult of the Rhodian *demos* at Hyllarima. It is not known whether Hyllarima was given its freedom in 167, but, considering its location, it seems likely.

[9] *HTC* 1 with extensive discussion (justification of the date on p. 103). Cf. van Bremen, 'Laodikeia', 373–74.

[10] The designation: 'the Pladaseis who are with the Pisyetai' suggests that there may have been 'other' Pladaseis remaining. So Bresson in *HTC ad locum*; Gabrielsen, 'Rhodian Peraia', 133–34, n. 15 gives parallels.

[11] Pladasa in the Athenian Tribute Lists: *ATL* i, p. 538; as a *polis* in the fourth century: *HTC* nos. 47 and 48, with Bresson's commentary on pp. 161–67.

[12] Wiemer, *Krieg*, 257, argues that Rhodes may in fact actively have *diminished* city territories; cf. also his remarks on Kaunos: ibid., 237. Fraser and Bean saw the site at Sarniç as 'almost purely military' (*Rhodian Peraea*, 76); description of the site: *HTC* 57–64.

[13] On the autonomous status of Kallipolis see Gabrielsen, 'Rhodian Peraia', 140–42, and *HTC* 210. But close contact with Rhodes must be assumed, cf. e.g. *HTC* 86: a *koinon* of *Haliastai Polemakleioi* (Hellenistic).

[14] Descriptions: *HTC*, s.v.; on the possibility of a Seleukid city-foundation at Muğla see van Bremen, 'Laodikeia', 375–91.

[15] See now Debord, 'Cité grecque', esp. 142–74, with all refs. Two mid-fourth-century inscriptions recently found at the modern town of Sekköy (*HTC* nos. 90 and 91) list a large number of local communities as *poleis*.

[16] Broadly speaking, this is valid also for the designations of *demos* and *plethos*: the latter is almost always used for the political male community of a *koinon*. On the complexities of the use of these two terms see Bresson's discussion *HTC* 101–10.

[17] For the history of the region's exploration, see *HTC* 21–22 (R. Descat).

[18] Also noted by Debord, 'Cité grecque', 174.

[19] Fragments: nos. 40, 51, 52, 60, 82, 87. Milestones: 93A, B and C, 94, 95A, B and C, 96, 97A, B, C and D. Pre-third cent.: 47, 48, 50, 90, 91.

[20] Nos. 1, 4, 5, 7, 9, 10, 11, 12(?) 13, 14, 15, 16, 17, 19(?) 20, 21(?), 26.

[21] See also Gabrielsen, 'Rhodian Peraia' with a different emphasis, and Bresson, 'Les interets rhodiens'.

[22] So, speculatively, Wiemer (*Krieg*, 255 and 335) with ref. to Pol. 30.31.7. See also Pol. 21.24.7–8 and 46.2–3; Liv. 37.56.

[23] Alain Bresson's eagerly awaited *La société rhodienne* will doubtless fill this gap.

[24] *HTC* 82. Debord, 'Cité grecque', 174, accepts Bresson's suggestion: 'indubitablement les notables locaux et/ou partisans de Rhodes.'

[25] Bresson, 'Les interets rhodiens', 188–89.

[26] On Kalynda, neighbour to Kaunos, see Hanssen and Nielsen, *Inventory*, s.v.

[27] Reger, 'Rhodes and Caria', 86–88 for the chronology. For similar ownership of land in the Samian and Chian Peraiai, see Carusi, *Isole*, 227–28.

[28] Gabrielsen, *Naval Aristocracy*, 81–84, on credit activities.

[29] So Bresson, *HTC* 82; Funke, 'Peraia', 68–71, and Carusi, *Isole*, ch. V, both point out for other islands and their Peraiai the almost total lack of evidence on matters of (shared) citizenship.

[30] Şahin, 'A Hellenistic List of Donors'. The date is almost certainly after 167, but the status quo referred to in the text may well be that of an earlier phase in the city's history. I intend to discuss this text elsewhere.

[31] *I.Stratonikeia* 5, 6 and 9. See photo of 9 in *MDAI* 25 (1975) pl. 60, 3: the letters could be late third or early second. In 9, the name of the *epistates* is erased: hostility to Rhodes? Or just to this particular *epistates*? The honorands in 5 and 6 are nameless and the stones are fragments.

[32] See *HTC* 28–29, 99–100, 107.

[33] Nos. 4, 5, 10, 11, 13, 16, 17, 19, 20, 31, 34, 37, 42, 44, 45, 67?, 78, 83.

[34] See also Berges, *Hellenistische Rundaltäre Kleinasiens*, and *Rundaltäre aus Kos und Rhodos*.

[35] See e.g. the series of shields from Kameiros: *Tit. Cam* 66, 70–78, with the discussion in *HTC* 185.

[36] Rice, 'Prosopographia rhodiaka', 220–21. Fraser (*Rhodian Funerary Monuments*, 58 and nn. 323–25) notes that this kind of meticulous recording of extended family groups 'is repeated at both Telos and Nisyros in the Rhodian period and is also found at Cos'. 'Such dedicatory monuments have no real parallel in Athens, where they are few in number and are limited to dedications by fathers, brothers, etc.'. *HTC* no. 56, discussed below, is part of a larger family monument, of a kind familiar from e.g. Lindos.

[37] On the site see *HTC* 45–50 (P. Brun).

[38] The letter forms allow for a date in the late second century, though the editors suggest first century BCE. They do not consider the possibility of the base and altar belonging together. Photos of 42: pp. 254–55.

[39] See also the comments on p. 142 (on the Leukoideis) and 189 (on the Tarmianoi and Pisyetai).

[40] Vérilhac and Vial, *Mariage grec*, ch. II; on Rhodes, see esp. pp. 65–68.

[41] The context is very likely that of the Mithridatic war. See further the commentary *ad locum*.

[42] They may be uncles on both father's and mother's side, and of two different generations. All refer to Phanias as their *anepsios*.

[43] Adducing the location of the text, the likely local family connections and the mixed dialect (*koine*/Doric), though with a cautious '?'.

[44] Fraser, 'Tribal Cycles', 40; Rice, 'Rhodes and Peraia', 50–51, with examples.

[45] A similar problem arises for no. 49: an epitaph from Akbük for Hekaton Sopatrou, *Rhodios* (ca. 100–50 BCE). A Sopatros, son of Hekaton, probably the man's son or his father, is attested in two inscriptions from the island. Does this make Hekaton a 'true' Rhodian settled in subject territory, or does it show instead the free movement into the island of naturalized Peraian Rhodians? *NSER*, 4, col. 2, l. 6 and *IG*, XII.1, 46, l. 463. For dating and further discussion: *HTC* 172.

[46] Bresson's comment (*HTC* p. 156, at no. 44) that what we have here is an 'union matrimoniale, cette fois indubitable, d'un Péréen prenant une épouse dans l'île de Rhodes' goes beyond the evidence. There is one near-parallel. A brief two-line epitaph from Idyma commemorates Panito, *Kedreatis*, i.e., of the deme of Kedreai on the coast of the incorporated Peraia: *HTC* 79. Bresson does not comment on the demotic in this case. One wonders if there is a particular reason why in both cases the bearer of the demotic is a woman. There are possible Rhodian demotics (*Pedieus*, *Pat*[*ureus*?]) in another inscription from Idyma (no. 75), which has led some to argue that Idyma was part of the incorporated Peraia. Bresson disagrees (*ad locum*), as does Gabrielsen, 'Rhodian Peraia', 143, n. 63.

[47] So, e.g., succinctly, Wiemer, 'die Grenze wird für uns dadurch erkennbar, daß Rhodier auf rhodischem Territorium das Demotikon, auf fremden Territorium aber das Ethnikon

(oder gar keine Herkunftsangabe) führen': *Krieg*, 254. Cf. Fraser and Bean, *Rhodian Peraea*, 53–54 with n. 2.

[48] E. Matthews kindly confirms this from the up-to-date *LGPN* database.

[49] *LGPN* vol. I, s.v.

[50] So Bresson *ad locum*. There is no space here to discuss the use of the Doric dialect vs. the *koine* in this region, but see on this now Bresson, 'Karien und die dorischen Kolonisation'.

[51] Although the inscription is, slightly misleadingly, listed under 'Leukoideis/Çırpı'.

[52] Vérilhac and Vial, *Le mariage grec*, 65–66, showing evidence of a son so naturalized ending up in a deme different from that of his father.

[53] Not necessarily that of his father: see previous note.

[54] Bresson does not comment on the husband's status in this case, other than to say that the name Menekrates was common on Rhodes.

[55] For the relatively late case of a Myndian, his Knidian wife and naturalized son, who has the demotic *Kedreates* (CE 100–200), see *I.Pérée* 11 (= *I.Peraia* 560). Bresson (*I. Pérée*, p.54) sees it as an example of the 'koinè du golfe Céramique'. For Myndians (relatively many) in the subject Peraia, see *HTC* 6, 14, 15, 32. The man honoured in 6 by a small local *koinon*, himself commemorates his Rhodian *euergetes* in 14: one could happily speculate about the ties that bound them.

[56] 'Gouverneur rhodien', so Robert and Robert, *La Carie*, 309; cf. Fraser and Bean, *Rhodian Peraea*, 73–74. For a different opinion: Gabrielsen, 'Rhodian Peraia', 136–37.

[57] *Rhodian Peraea*, 129–30. Differently: Gabrielsen, 'Rhodian Peraia', 138–39.

[58] For a Ptolemaic garrison at Muğla in the 270s, see *T.Cal.* no. 8.

[59] On the gymnasium and *ephebeia* at Muğla, see van Bremen, 'Laodikeia', 388–89. A second gymnasial dedication, to 'Helios, Hermes, Herakles and the *Tarmianoi*', was made by two *Kenendolabeis* (unknown community) who had each held the position of ephebarch. The fact that the inscription is on a shield reinforces my suggestion of a Rhodian (military) connection, as does the inclusion of Helios in the dedication; the date is early first cent. BCE, close enough to nos. 62, 63 and 64 to belong to the same overall context.

[60] The text, which is too long to give in its entirety, strongly emphasizes the social distance between honorand and *plethos* in a language frequently used for superior outsiders (καλὸς καὶ ἀγαθὸς ὑπάρχων διὰ προγόνων πολὰς (*sic*) καὶ μεγάλας χρήας παρέχεται τῶι πλήθει ἐν π[αν]τὶ καιρῶι πρόνοιαν ποιούμενος τῶν [σ]ụ[μφ]ερόντ[ων] τῶι κοινῶι τῶι Λευκοιδέων............ἀεί τινος ἀγαθοῦ παραίτιος γεῖνος (*sic*) τῶι πλήθει (11. 3–9).

[61] As Bresson has shown, a complex mythography existed to justify Rhodian claims on the Karian Chersonnese: Bresson, 'Grecs et Cariens'.

[62] *HTC* no. 26, with photo. Date: first/second century CE.

[63] It takes its place alongside the Amyzonian inscription in which the first list of *stephanephoroi* after 167 BCE begins 'after the Karians were liberated' (with the comments of L. Robert, *Amyzon*, 250), and the inscription from Panamara (*I. Stratonikeia* 9) in which the name of the Rhodian *epistates*, Polykratidas Dailochou, has been erased (though not, in this case, the rose above the text). Of course, in neither case do we know when the defacing took place.

References

Adiego, I.-J., P. Debord, and E. Varinlioğlu. 'La stèle caro-grecque d'Hyllarima (Carie)'. *REA* 107 (2005): 601–53.

Berges, Dietrich. *Hellenistische Rundaltäre Kleinasiens*. Freiburg: Kommisionsbetrieb Wasmuth, 1986.

———. *Rundaltäre aus Kos und Rhodos*. Berlin: Gebr. Mann, 1996.

Bresson, Alain. 'Rhodes and Lycia in Hellenistic Times'. In *Hellenistic Rhodes. Politics, Culture, and Society*, edited by V. Gabrielsen. Aarhus: Aarhus University Press, 1999.

———.'Grecs et Cariens dans la Chersonnèse de Rhodes'. In *Origines gentium*, edited by V. Fromentin, and S. Gotteland. Bordeaux: Editions Ausonius, 2001.

———.'Les intérêts rhodiens en Carie à l'époque hellénistique, jusqu'à 167 a.C'. In *L'Orient méditerranéen de la mort d'Alexandre aux campagnes de Pompée*, edited by F. Prost. Rennes, 2003.

———.'Karien und die dorische Kolonisation', In *Die Karer und die Andern*, edited by Frank Rumscheid. Forthcoming, 2007.

Carusi, Cristina. *Isole e peree in Asia minore. Contributi allo studio tra poleis insulari e territori continentali dipendenti*. Scuola normale Superiore, Pub. Cl. Lettere 28. Pisa, 2003.

Debord, P. 'Cité grecque—village carien. Des usages du mot *koinon*'. *Studi Ellenistici* 15 (2003): 115–80.

Fraser, Peter M. 'The Tribal Cycles of Eponymous Priests at Lindos and Kamiros'. *Eranos* 51 (1953): 23–47.

———. *Rhodian Funerary Monuments*. Oxford: Oxford University Press, 1977.

Fraser, Peter M and George E. Bean. *The Rhodian Peraea and Islands*. Oxford: Oxford University Press, 1954.

Funke, Peter. 'PERAIA: einige Überlegungen zum Festlandbesitz griechischer Inselstaaten'. In *Hellenistic Rhodes. Politics, Culture, and Society*, edited by V. Gabrielsen. Aarhus, 1999.

Gabrielsen, Vincent. *The Naval Aristocracy of Hellenistic Rhodes*. Aarhus: Aarhus University Press, 1999.

———.'The Rhodian Peraia in the Third and Second Centuries BC'. *Classica et Medievalia* 51 (2000): 129–83.

Hansen, Mogens H., and Thomas H. Nielsen. *An Inventory of Archaic and Classical poleis*. Oxford, 2004.

Jones, Nicholas F. *Public Organization in Ancient Greece*. Philadelphia: American Philosophical Society, 1987.

Meyer, Ernst. *Die Grenzen der hellenistischen Staaten in Kleinasien*. Zürich and Leipzig: Orell Füssli, 1925.

Papachristodoulou, Ioannis. 'The Rhodian Demes within the Framework of the Function of the Rhodian State'. In *Hellenistic Rhodes. Politics, Culture, and Society*, edited by V. Gabrielsen. Aarhus: Aarhus University Press, 1999.

Reger, Gary. 'The Relations between Rhodes and Caria from 246 to 167 BC'. In *Hellenistic Rhodes. Politics, Culture, and Society*, edited by V. Gabrielsen. Aarhus: Aarhus University Press, 1999.

Rice, Ellen E. 'Prosopographica Rhodiaka'. *ABSA* 81 (1986): 209–50.

———.'Relations between Rhodes and the Rhodian Peraia'. In *Hellenistic Rhodes. Politics, Culture, and Society*, edited by V. Gabrielsen. Aarhus: Aarhus University Press, 1999.

Robert, L., and J. Robert. *La Carie* II. Paris: de Boccard, 1954.

Robert, L., and J. Robert. *Fouilles d'Amyzon en Carie, I Exploratíon, histoire, monnaies et inscriptions*. Paris, de Boccard, 1983.

Şahin, Mehmet Ç. 'A. Hellenistic List of Donors from Stratonikeia'. *EpAnat* 38 (2005): 9–12.

Van Bremen, Riet. 'Laodikeia in Karia'. *Chiron* 34 (2004): 367–98.

———.'The Karian Uplands'. *Review of HTC. CR* (2005): 226–9.

Vérilhac, Anne-Marie, and Claud Vial. *Le Mariage grec, du VIe siècle av. J.–C. à l'époque d'Auguste*. *BCH* Supplément 32. Paris and Athens, 1998.

Wiemer, Hans-Ulrich. *Krieg, Handel und Piraterie. Untersuchungen zur Geschichte des hellenistischen Rhodos*. Berlin: Klio Beiheft. Neue Folge Bd. 6, 2002.

Libanius' Social Networks: Understanding the Social Structure of the Later Roman Empire

Isabella Sandwell

> [They] try to avoid imposing assumptions about the boundaries of aggregates. They do not assume that analysis can proceed on the basis of a few discrete categories such as proletariat and bourgeoisie or core and periphery. They do not assume that tightly bound groups are the fundamental building blocks of large-scale social systems... it is the network members' crosscutting membership in multiple social circles that weave together social systems.[1]

Introduction

Our understanding of the later Roman empire is often posed in terms of an opposition between the imperial centre and the periphery of Greek cities, between the imperial bureaucracy and civic elites. As one scholar has recently restated for us, the strengthening 'of the administration of central government', 'the consolidation and systematization of distinct areas of responsibility under the supervision of high ranking office holders associated with the emperor himself', and the constantly advertised 'benefits of close association with the imperial regime' could alienate the civic elites.[2] This tension between the power and needs of the imperial centre and imperial bureaucracy, and the power and needs of the civic elites, is seen as one of the key problems of the running of the empire. Re-establishing the links between these two groups, which had become increasingly opposed, is seen as the solution to this problem.[3] In this view, the question of the unity of the social structure of the empire is seen in terms of bounded aggregates, discrete categories, and tightly bound groups; on the one hand the imperial centre and imperial bureaucracy, and on the other hand the civic elites. This has normally been the case in studies of the Syrian city of Antioch, as it has for the study of the Roman empire in the fourth century more generally. Some recent works have challenged the normal picture of the subjugation of the city's local elites and civic freedom to the imperial authorities.[4] However, they still tend to see the relationship between the two as a problem that needs to be understood and still work with a model of these groups as discrete categories.[5] This view of the later Roman empire can be seen echoed in the writings of Libanius, the teacher and orator from Antioch. Libanius is often described as a typical representative of traditional values of Greek civic life and Greek culture, particularly those of his home city of Antioch. He was a member of the curial class and civic elite of Antioch, if exempt from serving on the council due to his position as a teacher. He was a teacher of rhetoric, a distinguished orator, and a lover of classical literature. In his orations he appears as a typical representative of these groups. He often defended the role and values of the civic council (*Orations* 48 and 49), the teaching profession and Greek language and literature (*Ors*: 31 and 42), and spoke out against the growing role of the imperial bureaucracy (*Or.*2.58–59 (F.I.258–59)) and of the professions of law (*Or.*1.154 (F.I.156) and *Epistle* F.1375), shorthand (*Or.*31.26–28 (F.III.137–38), *Or.*62.8–10 (F.IV.350–52) and *Or.*2.58 (F.I.258)) and the military (*Or.*47.13–16 (F.IV.410–12)).[6] He thus reinforces the opposition that we see to exist between the imperial centre and bureaucracy, on the one hand, and the civic elites, on the other hand, and so also to present a picture of the world divided into discrete categories and bounded aggregates. He thus makes it very easy for us to continue with our traditional kinds of social analysis of the later Roman empire.

The problem is that the situation as presented in Libanius' writings is not as simple as it might at first appear. Libanius' letters, in comparison to his orations, often show us Libanius acting against the values of the very groups of people he is supposed to represent. On numerous occasions he helps young men to escape civic duties and find positions for themselves in the expanding imperial bureaucracy of the fourth century.[7] At the same time, Libanius had connections with a whole range of people with whom we would not expect him to have allegiance. Bradbury has pointed out that Libanius' letters show him enmeshed in 'patronage networks' involving the 'interaction of the provincial elites with the imperial administration'.[8] These networks show Libanius' social connections extending way beyond the confines and concerns of his home city of Antioch. As Bradbury has recently stated, Libanius' letters suggest the importance of 'travel and the maintenance of social networks between cities and across whole provinces'.[9] It is thus quite reasonable to suggest that Libanius' networks were designed precisely to overcome the tension between the centre of power surrounding the emperor and his highest officials, and the civic elites.[10] In this article I would like to push this argument further. As the quotation at the head of this article suggests, network theorists approach social analysis without resort to the categorization of society into discrete social groups. 'Instead of trying to sort units into categories defined by their inner attributes (essences)', the networks approach focuses more on relations and ties between individuals.[11] I want to suggest that from the point of view of Libanius' personal relationships, civic elite and the imperial centre were not clearly defined, distinct and opposed entities in the first place. It is not so much that he used his personal relationships to bridge some huge gap between civic elite and imperial centre. Rather, from the point of view of his personal relationships, such a gap was far less of a reality than we might expect.

Throughout, my discussion of Libanius will be based mostly on letters written during the period AD 354–58. Such a snapshot will best allow us to see how his social networks were functioning at a particular time.

The Bases of Libanius' Social Networks

Libanius' social networks were primarily based on personal relationships. The people included in his web of correspondence were often connected to him by ties of kin, friendship or shared culture.[12] During the early years of his return to Antioch in 354, Libanius still had a strong family network that ensured him a good position throughout the reign of Constantius.[13] Libanius' cousin, Spectatus, was one of Libanius' most important connections during this time. The son of Libanius' uncle Panolbius, Spectatus had left Antioch to take up the position of *tribunus et notarius* in the court of Constantius in the early 350s. This made Spectatus a very useful contact for Libanius, and we see Libanius sending him numerous letters during the 350s making requests both for himself and others.[14] Spectatus also introduced Libanius to other individuals who were important in Constantius' regime. It was because of Spectatus' words of encouragement that Libanius had dared to start an epistolary relationship with two successive Masters of Offices, Musonius 1 (*magister officiorum* 356–57) and his successor Florentius 3 (*magister officiorum* 359–61) (*Ep.* B.32.2 (F.558) and *Ep.* B.36.3–4 (F.510)).[15] Another important contact that Libanius had made through family connections was Anatolius 3, a citizen of Berytus and the Praetorian Prefect of Illyricum 357–60. Anatolius had been a good friend of Libanius' uncle Phasganius and Libanius had inherited this friendship when the two men got to know each other, during Libanius' time as a teacher in Constantinople between

350 and 353.[16] Like Spectatus, Anatolius received numerous letters from Libanius, and proved to be very useful to him during the 350s.[17]

Libanius also wrote numerous letters on behalf of family members. This included letters for an unnamed cousin and for Thalassius 2, grandson of Libanius' uncle Panolbius.[18] He also wrote a series of letters of introduction for the young philosopher Iamblichus 2 of Apamea, with whom he had some unstated kinship connection (*Ep.* B.19.5 (F.574) and *Ep.*N.24.1 (F.571)). Iamblichus came to Antioch in the mid-350s when his father died, but was soon summoned to the emperor's court, then based at Rome. Libanius sent Iamblichus on his way with nine letters of introduction to people based in different cities along his route, often using Iamblichus' kinship connections as the basis for the introduction.[19] In one letter to his friend Aristaenetus about Iamblichus, he first describes Iamblichus' closest family relations; he is 'the son of Himerius, nephew of Sopater, namesake of Iamblichus' (*Ep.*N.24.1 (F.571)).[20] He then goes on to describe Iamblichus' relationship to himself; he is 'my kinsman and friend' (*Ep.*N.24.1 (F.571)).[21] Ties of kinship to Libanius were important persuasive techniques in getting Libanius' contacts to carry out the favours requested of them. So important was this factor that it could be introduced even when no actual kinship tie existed. On numerous occasions in his letters, Libanius makes fictive or metaphorical references to kinship ties with those he was recommending. His Antiochene friend Letoius 1, for example, was 'not outdone by my kinsfolk' in 'his attitude towards me' (*Ep.*N.22.10 (F.552)).[22]

Old school friends and colleagues also provided important points of contact for Libanius.[23] In one letter he writes to Eugnomonius, a fellow student in Athens during their youth, who had gone on to become a chancellery official in 357–58 (*Ep.* B.31 (F.559)). Libanius writes to him on behalf of his friend Letoius, and in so doing prefaces his request with a reminder of their time spent in Athens: 'Do you still recall ... all those times we had in Athens? Or have you grown proud towards me and have nothing to say of old times?' (*Ep.* B.31.1 (F.559)).[24] Libanius also wrote on behalf of those he had studied with during his youth. Eugnomonius received another letter from Libanius, this time on behalf of Olympius 6, who had studied with Libanius at Antioch when the two were young boys. In order to persuade Eugnomonius to carry out the requested favour, Libanius refers to the fact that Olympius 'is a fellow citizen of mine and was a schoolmate of mine in youth' (*Ep.* B.34.4 (F.382)).[25] It was the very fact that Libanius had been school friends with someone that meant favours should be granted to them. As Libanius says in one letter to Anatolius 3 about another school friend Pelagius, he is 'a school companion of mine from old, he's always been a friend. Let him learn how much that means to you' (*Ep.* B.59.6 (F.563)).

The third type of personal relationship that formed a basis for Libanius' network was that with ex-pupils. Libanius stayed in contact with many of his ex-pupils after they had left his school and they provided invaluable contacts for him.[26] As Cribiore shows, it is possible to speak of a 'circle of alumni' around Libanius and his school and Libanius himself describes 'chains' of students connected by their study under him.[27] One ex-pupil to whom Libanius wrote on a regular basis was Andronicus 3, a young man who had studied with Libanius in Constantinople, and who was probably the nephew of Strategius Musonianus, Praetorian Prefect of the east in 354–58. Andronicus received thirty-nine letters from Libanius between 355 and 365. Many of these letters were about matters personal to the two men.[28] Others, however, were letters of reference and addressed Andronicus as a man who had some influence in the capital (*Ep.* B.87 (F.560) and *Ep.* F.553). Libanius' relationships with his students also gave him access to their families. Libanius could use the fact that he taught a man's son, or other male relative, as a pretext to write and ask for their assistance. In one letter of 357 he writes to Barbatio, Master of the Infantry from 355 to 359, at court in Rome to ask for his help in being allowed to stay in Antioch instead of returning to his teaching post in Constantinople. In so doing he refers to the fact that 'your Gessius is my pupil' (*Ep.* B.29 (F.436)). The most interesting example of Libanius' connections with his pupils and ex-pupils comes from a letter on behalf of his pupil Honoratus 3. The young Honoratus had been enrolled among the notaries so that he would have a post there when he finished his studies, but in 358 the emperor ordered all members of the palatine bureaux, which included the notaries, to appear in court or be struck from the lists. Libanius wrote a number of letters to request Honoratus' exemption so that he could continue his studies, but not lose his post.[29] Two letters were addressed to Bassus 5, who was head of the notaries 357–58. In these Libanius makes reference not only to the fact that Honoratus was a pupil of his (*Ep.* B.35.4 (F.366)), but also adds extra persuasive force to his letter by referring to the fact that Bassus' son Calliopus 3 had likewise been his pupil,

> I both loved and assisted your son when he was present. ... You too seem to me, after discovering learning in him, to feel gratitude towards the one who imparted it ... [so present] me with a gift—none would be easier for you, none greater for me. (*Ep.* B.35.1–2 (F.366))[30]

Again, it is Libanius' personal relationship to Bassus' son that was the persuasive force for Libanius' request to be granted.

Libanius had numerous connections with those he called friends.[31] A striking example of this can be seen in Libanius' description of the time he spent with his friend Clematius 2, during a visit Clematius made to Antioch in the winter of 355/56. The two men were so inseparable, Libanius states, that they became something of a standing joke among others in Antioch: 'One time Clematius visited Strategius [Musonianus, Praetorian Prefect of the East] when I wasn't present, and he was stunned and asked "Where's your other half!"' (*Ep.* B.25.8 (F.435)). Often, though, friends could not be together but had to continue their relationship through letters. Libanius' correspondence with Aristeaenetus 1 of Nicomedia stands out as an example of this. Between 355 and 358 Libanius wrote thirty-three letters to Aristaenetus on a number of subjects including both more personal matters, such as condolence over the death of Aristaenetus' wife (*Ep.*N.8 (F.414) and *Ep.*N.11 (F.430)), debates over the appropriate length for letters (*Ep.* B.173 (F.561) and *Ep.*N.25 (F.580)), and requesting favours for others (*Ep.* B.178 (F.381) and *Ep.*N.34 (F.21)). He also introduced friends of his to Aristaenetus. In a letter carried by the young Apamean philosopher Iamblichus from Antioch to Aristaenetus in Nicomedia, Libanius states that Aristaenetus should 'regard' Iamblichus 'as your friend too at first meeting. Do not unite yourself with the lad by degrees, but after my letter display everything to him' (*Ep.*N.24 (F.571)).[32] In fact, the language of friendship permeated Libanius' letters. Libanius often referred to emotional connections between himself and those to whom he wrote—for example, to sadness at their absence or when they are facing hard times, or to the long-standing nature of his friendships with them or with mutual acquaintances.[33] We also see the characteristic features of a desire for frankness (*Ep.*N.47.1 (F.81) and *Ep.* B.4.7 (F.219)), teasing of one another and humour (*Ep.*N.4.7 (F.391)), and of outspoken advice (*Ep.*N.11.2 (F.430)). References to long-standing friendship could be used as justification for asking a favour of a friend who was also an official, or as the reason for recommending someone whom Libanius was trying to promote (*Ep.*N.46.2 (F.80) and *Ep.* B.25.1–3 (F.435)). This language of friendship was, of course, a conventional and expected feature of ancient letter writing, and we should not be surprised to find it here.[34] However, the very fact that claims to friendship could be used in order to get things done and to have favours granted tells us something about the nature of Libanius' social networks.[35]

The relationships of friendship and kinship that tied Libanius to others cut across the stereotypical division of imperial bureaucracy versus civic elite. In fact, it can be argued that they not only cut across them but also challenge the idea that they were such distinct and opposed entities in the first place. From Libanius' point of view, what he had was an interconnected web of personal relationships with people from a variety of different backgrounds, holding different amounts and kinds of power and different levels of proximity to himself and to the emperor. These personal relationships could contain within them the supposed contradiction between imperial centre and civic periphery. Spectatus, for example, was both an Antiochene from the curial elite and a notary with influence at court in Constantinople and cannot easily be classified as belonging to one category or the other. The fact that Spectatus was also Libanius' cousin and gave Libanius an easy route of access to imperial power meant that home city and imperial court were brought together for Libanius too. Libanius' relations with other friends and acquaintances also show this to be true. Libanius had personal relationships with the highest officials such as the Praetorian Prefects Anatolius (Praetorian Prefect of Illyricum) and Strategius Musonianus (Praetorian Prefect of the East), the Master of Offices Florentius and the senator and philosopher Themistius (*Ep.*N.16 (F.476) and *Ep.* B.82 (F.40)). He was also friendly with a number of lower officials. These included notaries; as well as Spectatus, we can mention Bassus (*primicerius notarium*) (*Ep.*N.28 (F.359) and *Ep.* B.35 (F.366)), his pupil Honoratus 3 (*Ep.* B.35 (F.366); *Ep.* F.358 and *Ep.* B.5 (F.365)), and Jovianus 1 (*Ep.* B.25 (F.435)). It also included doctors (Marcellus (*Ep.* B.64 (F.362); *Ep.* F.359 and *Ep.* F.378), lawyers and advocates (Silanus (*Ep.* B.162 (F.433) and Gaianus 6 (*Epp.*B.131–6 (F.336, 119, 799, 800, 1422, 1218)) and military men (Barbatio (*Epp.*B.29–30 (F.436 and 556)). The fact that he knew these men personally—that he had studied with them, that close friends or relatives had introduced him to them, or that he had taught them or their male relatives—cut across any differences in loyalties and interest that there might normally be between them and him.

In fact, Libanius' letters generally reveal a world in which values and loyalties did not play out in ways in which we might at first expect. One example shows that Libanius did not always use the category of civic elite as his prime way to think about his social interaction. On his death, Libanius' uncle Phasganius had divided his estate between Libanius and his other nephew Spectatus, who were both exempt from curial liturgies — Libanius as the city's sophist, Spectatus due to his position in the imperial service. The council in Antioch was challenging this will and trying to stake a claim to some of the land which had originally been subject to curial obligations. In Libanius' representation of this incident, however, this is not described as a public conflict involving the civic council as an official body. Rather, it is presented as a personal battle between himself and one scheming individual who had a personal grudge against him (*Ep.*N.56.3 (F.115)). It was the personal relationship with an individual that was important to Libanius, not the fact that such a person might represent a larger body such as the civic council. Rhetoric and classical culture are other factors that we might expect to see connecting Libanius with people.[36] This is the case on a number of occasions and so could reinforce the picture of Libanius belonging to a bonded group of those who value classical culture in opposition to those who have taken up the new studies of shorthand and law (*Ep.* B.32.3 (F.558) and *Ep.* B.25.5 (F.435)). Other references to classical culture in Libanius' letters, however, confound this simple classification, as we see it as a factor linking Libanius to advocates and notaries. In one letter to Bassus 5, head of the notaries, Libanius recommends Honoratus 3, the pupil of his enrolled among the notaries, by talking of the education in eloquence that he was receiving at Libanius' school. Surprisingly for someone who is supposed to see eloquence and shorthand as in direct opposition, Libanius says that Bassus will appreciate this education being given to Honoratus and that it will make him a better notary! (*Ep.* B.35.4 (F.366)).[37] A similar point can be made about Libanius' conception of what it was to be a good governor. Bradbury and others have shown that Libanius 'treats his friends'' service as governors as an extension of the normal activities of a prominent *principalis* who helped cities flourish and promoted Hellenic culture.[38] Libanius' letters to governors were thus full of references to civic values, so collapsing the normal distinction between imperial bureaucracy and civic elite.

From these examples it is clear that when we look at the individual ties between people as expressed in Libanius' letters, normal assumptions about the opposition of certain roles and groups of people simply do not work. Libanius and those he knew were not acting according to strictly laid down rules about what it was to be a representative of the civic elite or the imperial centre. Rather, they took a much more practical approach and used whatever connections were available to link themselves to one another in order to achieve the desired outcome at that particular moment.[39] In these connections, personal relationships rather than membership of publicly recognized categories were often what were emphasized.

The Flexibility and Ephemerality of Libanius' Social Relations

Libanius' relationships with others often did not have an existence outside the medium of letter writing.[40] Bradbury and Cribiore have shown that Libanius was immersed in a geographically wide-ranging network that extended right across the eastern empire.[41] This meant that the members of Libanius' network did not ever meet together as a group and that they probably did not even all know one another. This does not have to imply that these relationships were entirely private in nature: the public nature of ancient letter-writing has long been accepted and Libanius himself often refers to the way that letters could be shown or read aloud to others.[42] Despite this, however, Libanius clearly envisaged that some of the opinions he expressed in letters could potentially remain private as he chides certain correspondents for revealing the contents of his letters when he had not expected them to do so.[43] This means that Libanius' letters do have the potential to show us a more personal and less visible side to Libanius.[44] At the same time, the relationships established in letters could be more ephemeral than those established through regular social contact in person, because Libanius had never met many of his correspondents and only wrote to them very infrequently. There were points on Libanius' social network where denser connections built up because Libanius wrote numerous letters to certain individuals who also knew one another and who did actually meet together on some occasions, but this was the exception, not the rule. Very often Libanius' social relationships with others existed in the connection made by the letters themselves.[45] This meant that letters, or the bearers of a letter, often had to stand in for *actual* meetings between Libanius and others. In one letter, to a father of one of his pupils, Libanius suggests that the pupil acted as a replacement for direct association with the father. Now the father was recalling the pupil, the father's letter would have to serve this function instead (*Ep.*N.28.1–2 (F.359) and also *Ep.*N.31.1 (F.370)). Letters could also make connections with people Libanius had not yet met, or whom he had only met briefly, such as those to the two Master of Offices Musonius and Florentius (*Ep.* B.36 (F.510 and *Ep.* B.32 (F.558)). Such introductory letters were, however, fraught with difficulties.[46] They risked appearing mercenary in their intent: 'Now I'm hunting for friendship through a letter, not that I might reap a benefit from your influence, for that would be characteristic of a businessman and not a man eager for friendship' (*Ep.* B.36.1 (F.510)). Or they could be rejected outright; a fear that Libanius expresses in his letter to Musonius and that eventually turned out to be well-founded—Musonius did not ever reply to the letters Libanius sent him (*Ep.* B.32.4 (F.558)).[47]

Such anxieties about receiving a reply permeated much of Libanius' correspondences.[48] He often wrote to complain to correspondents that others had received letters from them, whereas he had not.[49] Not receiving a reply or a successful response to a letter could mean the loss of influence and the loss of contact with those in power. As Libanius says to his cousin Spectatus, Spectatus should grant the request being made for their friend Sabinus 5 so 'that he [Libanius himself] won't seem to be employing friends weaker than other people', a criticism that would be 'neither tolerable for me nor noble for you' (*Ep.* B.2.2–3 (F.545)). If someone did not reply to Libanius, it meant the ending of a social relationship, a weakening of Libanius' social position, a marginalization of his person and so weakening of his network. It meant that he had fewer resources with which to pursue his practical manipulation of social relationships for both his own benefit and that of those he knew. Maintaining his networks of connections thus became a goal in itself for Libanius. The occasions when he sent an individual on their way with numerous letters of recommendation could serve a two-fold purpose; as in the nine letters carried by Iamblichus on his route from Antioch to Rome in 357. These letters could help introduce Iamblichus to powerful people in cities along his route. They could also, however, serve the additional purpose of keeping open a particular branch of Libanius' network by reminding his contacts along this route of his own relationship with them.

One reason that network theorists favour the network approach to social analysis is that it allows us to study more ephemeral social relations that take place in private.[50] This can be useful to us too as it allows us to see the significance of Libanius' networks, which were much less visible, less cohesive, and less publicly recognizable than those we might see in groups that met together regularly and in person. It was these characteristics of Libanius' social relations that enabled him to interact with people who were not his obvious allies. These relationships could transcend normal social categorization without people noticing the contradictions and inconsistencies in Libanius' patterns of friendship. As such, his social network can be described as a set of cross-cutting ties linking people across 'bounded institutionalized groups and categories' and showing the 'the ephemerality and inconsistency of social relations'.[51]

One of the primary purposes of Libanius' network, as with other social networks, was to enable the flow of resources between network members by getting favours done for himself and those he knew.[52] This process can be divided into three stages. Firstly, Libanius knew, or developed relationships with, a number of important officials and courtiers. These people themselves had the power to grant Libanius' requests, either directly through their own office or indirectly through the influence that they themselves could exert on their superiors. The former was the case for Libanius' friends Anatolius 3, Praetorian Prefect of Illyricum and Florentius 1, the Master of Offices. As Libanius describes, it was these people who could really get things done: 'Well it's my role to pray, since I don't have power, but it has fallen to you to help with concrete actions' (*Ep.* B.4.1 (F.219)). The latter appears to be the case for Libanius' cousin Spectatus, whose ability to exert his influence was much more patchy and inconsistent because it was due simply to his position at court and the people he knew there: 'Or is it that your influence is insufficient for the favour we're requesting? And yet you mocked Quirinus for omitting nothing in his zeal and you insinuated that you could perform favours of *that* sort even in your sleep!' (*Ep.* B.5.2 (F.365)).

The second stage we can identify is that Libanius was able to call on the influence of these individuals because of his personal ties to them. His personal relationships with people were precisely the means that he used to gain influence and get things done. As we saw above, Libanius often used his personal ties of friendship, kinship, or teacher–student relationship to achieve his goals. As he saw it, his connection to people meant that they owed him access to their influence. In a letter to Eugnomonius, the chancellery official who had been his friend in Athens, Libanius writes: 'Now if you didn't have influence, I would pray that you did. But now, since you do, I'm perhaps not unjustified in wanting to enjoy it' (*Ep.* B.34.5 (F.382)). The expectation generally seems to have been that it was right and proper for powerful individuals to share their influence. As Libanius himself says to the lawyer Silanus, 'If I had some influence and you weren't benefiting from it, I'd be in the wrong' (*Ep.* B.162.3 (F.433)).[53] The third stage of this process is that Libanius could himself accrue power and influence due to the connections he had with powerful individuals. He became known as someone who had strong connections with those who had authority, and so himself was seen as a useful person to know: 'don't be surprised if Marcellus couples my letter with Strategius' letter for the rumour has it that I've very great influence with you' (*Ep.* B.64.8 (F.362)). It is only this background of a perception of Libanius' sway over others that makes sense of the occasions when he has to explain his *lack* of influence where others expected him to have it.[54] Despite living and working in Antioch from the 350s onwards, Libanius did not see the power associated with the imperial centre as something far off and inaccessible to him. Rather, he saw himself as having direct access to it via family members and close acquaintances. This idea is captured nicely in a playful letter to his cousin Spectatus. In this letter, Libanius amuses himself by imagining Spectatus at the side of the emperor: the emperor converses with Spectatus, praises him for his skills in oratory and mention is even made of Libanius (*Ep.* B.7.1–3 (F.352)). Libanius is clearly mocking Spectatus' claim to influence with the emperor in this description, and he goes on to tease his cousin for not having used this imagined opportunity to help him in turn. But, this example still shows how easily Libanius could include the emperor in his network, how few degrees of separation he could perceive there being between himself and the imperial centre.[55]

It has been argued that it was such influence given to civic elites back in their home towns, or the granting of new ranks, titles and honours to the civic elites, that made imperial power acceptable to them.[56] I want to argue the stronger point that in fact such influence, and its dispersal throughout a number of points on his network, entailed that Libanius did not see imperial power as centralized or as in opposition to his own goals and needs. Rather, to him it would have appeared that power and influence were shared out among a number of different officials and bodies, including the Praetorian Prefects of his own and other regions, the governors of his own and other provinces, the Master of Offices and palatine bureaucracy, the senate at Constantinople, court circles with direct access to the empire and a range of lower officials. This dispersal of power meant that Libanius had a number of sources of influence on which he could choose to call, and very often his network of connections in fact allowed him to call on more than one source. This can be seen in 354/55, when he was attempting to get support for his return to Antioch rather than be called back to Constantinople to return to his teaching post there. He was able to call on the Praetorian Prefect of the East in 354, Thalassius 1, who was married to the daughter of

Libanius' uncle Panolbius (*Ep.*N.2 (F.16), the influential notaries Datianus (*Ep.*N.7 (F.409) and *Ep.*N.13 (F.441) and Jovianus 1 (*Ep.* B.25.12 (F.435)), Anatolius 3, later to be Praetorian Prefect of Illyricum but at this stage in between posts but influential at court (*Ep.*N.4 (F.3910) and *Ep.* B.55 (F.438)), and the *magister peditum* Barbatio (*Epp.*B.2930 (F.436 and 556) and *Ep.* F.491). At times, Libanius could also play officials off against one another. In one letter he suggests that Spectatus will act on his promises and help a family friend, Miccalus, because he was frightened of Miccalus' brother and Libanius' friend Olympius 3, who was a senator in Constantinople (*Ep.* B.9.4 (F.98)). On another occasion, Libanius asks his friend Anatolius, Praetorian Prefect of Illyricum, to intervene with Musonius, Master of Offices, on behalf of the doctor Marcellus and his sons: 'it is said that a nod from you is law with Musonius, and the custom is an old one that the prefect be arbiter of actions taken by those occupying his present position (that of *magister officiorum*)' (*Ep.* B.64.7 (F.362)).

When Libanius surveyed his relationship to the imperial centre he would not have seen imperial power as some distant, unreachable or distinct source of influence but rather as a constellation of smaller sources of power. He did not just tolerate imperial power at the centre of the empire because it gave him influence at home; rather, his whole network of relationships was imbued with small distillations of power and influence on which he could draw through his personal connections to people.

Conclusion

Some of Libanius' orations put forward a view of the world divided into the clear-cut categories and bounded aggregates of civic elite versus imperial bureaucracy and champions of classical culture versus notaries, lawyers, and military men. His letters, in contrast, tend to reveal a world of more personal relationships and individual ties that cut across such distinctions and belie the importance of such bounded groups. How can we understand this seeming contradiction? Wellman has stated that analysis in terms of bounded aggregates looks at 'normative prescription', where we *expect* to see people interacting socially, while analysis of networks looks at 'actual ties'.[57] One way to apply this idea to Libanius' writings might be to suggest that there is a contrast between some of his publicly delivered orations that express normative ideals of social organization, on the one hand, and his letters that represent the messiness of actual practice of social relationships, on the other. This contrast between the orations and letters is a little over-simplified because some of Libanius' orations too show the sophist adjusting his opinions in a practical way to suit the occasion rather than making statements about desired ideals. However, it does help us to get a slightly different perspective from the way that Libanius' writings are usually used as evidence for the social structure of the fourth-century Roman empire. If we move away from the 'normative prescription' of thinking in terms of discrete and opposed bounded aggregates and instead see practical manipulations of personal relationships as the fundamental building blocks of fourth-century imperial society, then the opposition between imperial centre and civic elites disappears and becomes a non-problem.

Notes

[1] Wellman, 'Structural Analysis', 37.

[2] Kelly, 'Bureaucracy and Government', 183, 195.

[3] Kelly, 'Bureaucracy and Government', 195–200; Heather, 'Senators and Senates', 185–6.

[4] Cabouret, 'Pouvoir municipal', 140.

[5] Cabouret does at the very end of her article perhaps suggest a more radical model in which the two powers are not as distant from one another as we often think—'Le clivage entre une aristocratie régionale cantonnée aux affaires locales et d'influence restreinte et une nouvelle "noblesse" de function que santionne l'appartenance au sénat constantinopolitain est bien moins marquee que ne pourrait le suggérer une typologie hâtive' (Cabouret, 'Pouvoir municipal', 140). However, she does not push this idea as far she might.

[6] Liebeschuetz, *Antioch*, 242–55, 115.

[7] *Ep*.B.177 (F.374); *Ep*.B.9 (F.98); *Ep*.N.53 (F.97); and see Pack, '*Curiales*'; Liebeschuetz, *Antioch*, 179.

[8] Bradbury, *Selected Letters*, 2. See also Liebeschuetz, *Antioch*, 17; Kelly, 'Empire Building', 178. Bradbury's new translation of, and introductions to, 183 of Libanius' letters provides a crucial tool for the kind of analysis found in this present work and the current author is indebted to it for this. Bradbury's forthcoming accompanying work on Libanius' social context is also to be awaited with great anticipation. The arguments presented in this current article have developed out of the author's work on religion as a factor in Libanius' social organization (Sandwell, *Religious Identity*) and is simply a small detour from that.

[9] Bradbury, 'Libanius' Letters', 74.

[10] Bradbury, 'Libanius' Letters', 75. See also Kelly, 'Empire Building', 178.

[11] Wellman, 'Structural Analysis', 20.

[12] Cribiore also uses the model of the network to describe the relationships that we see in Libanius' letters. She too emphasizes the importance of ties of friendship and culture in these networks (Cribiore, *School of Libanius*, 107–10). Criobiore's work was published just before I made the final corrections to this article. She and I make some similar points about the role of educational connections in Libanius' network. I have tried to incorporate references to Cribiore's work where I can in the present article.

[13] Wintjes, *Das Leben des Libanius*, 43–62; also Liebeschuetz, *Antioch*, 17.

[14] *Ep*.B.2 (F. 545); *Ep*.B.5 (F.365); *Ep*.B.40 (F.64); *Ep*.B.8–9 (F.74, 98); *Ep*.F.449; *Ep*.F.525; *Ep*.F.538; *Ep*.F.358; *Ep*.F.116.

[15] I use Jones *et al.*'s *The Prosopography of the Later Roman Empire* numbering for the officials mentioned throughout this article. Where there are translations into English, Norman's and Bradbury's translations of Libanius' letters are marked with a N. and a B. respectively. F. marks Foerster's edition of the text.

[16] Liebeschuetz, *Antioch*, 20; Bradbury, 'A Sophistic Prefect', 177; also *Ep*.N.10.13–14.

[17] *Ep*.N.4.9 (F.391); *Ep*.B.55 (F.438); *Ep*.B.57 (F.535); *Ep*.B.59 (F.563); *Ep*.B.19 (F.574); *Epp*.B.62–4 (F.339, 348, 362); *Ep*.N.40 (F.19).

[18] For the cousin, see *Ep*.B.2 (F.545); *Ep*.B.4 (F.219). For Thalassius, see *Ep*.B.11 (F.330); *Epp*.B.13–14 (F.377, 1404).

[19] *Ep*.F.569; *Ep*.F.570; *Ep*.N.24 (F.571); *Ep*.F.572; *Epp*.F.573, *Ep*.B.17 (F.474); *Epp*.F.575–77.

[20] Iamblichus was the son of Himerius 3 and the grandson of Sopater 1. His grandfather had studied with the philosopher Iamblichus of Apamea and his family were one of the leading families of Apamea.

[21] See also *Ep*.B.19.5 (F.574); *Ep*.B.2.1 (F.545); *Ep*.B.4.2 (F.219); *Ep*.B.11.1 (F.330).

[22] See also *Ep*.B.5.4 (F.365); *Ep*.B.11.4 (F.330).

[23] Cribiore also makes this point but uses different examples (*School of Libanius*, 107 and 110).

[24] See also *Ep*.N.23.1 (F.557) on Mygdonius.

[25] See also *Ep*.B.33.4 (F.604).

[26] Petit, *Les étudiants* 154; Liebeschuetz, *Antioch*, 17; Kaster, *Guardians*, 205; Bradbury 'Libanius' Letters', 75–7.
[27] Cribiore, *School of Libanius*, 104–7, for the former point, and 108, for the latter point which refers to Libanius *Or.*1.51 (F.I.109).
[28] *Ep.B.*54 (F.506); *Ep.*N.21 (F.515); *Ep.*N.17 (F.477); *Ep.*B.86 (F.339).
[29] See Bradbury's introduction to *Ep.*B.35 (F.366) (Bradbury, *Selected Letters*, 64) and Liebeschuetz, *Antioch*, 178.
[30] Compare this to *Ep.*N.28 (F.359) also to Bassus on Homoratus.
[31] On friendship in Libanius' letter, see Cribiore, *School of Libanius*, 107.
[32] See also *Ep.*N.9 (F.427); *Ep.*B.11 (F.330).
[33] For the former, see *Ep.*N.4.12 (F.391); *Ep.*N.6 (F.405). For the latter, see *Ep.*N.9 (F.427); *Ep.*N.11.12 (F.430); *Ep.*N.22.10 (F.552); *Ep.*N.23 (F.557); *Ep.*N.43 (F.70); *Ep.*N.46.2 (F.80).
[34] Stowers, *Letter Writing*, 29–31; Trapp, *Greek and Latin Letters*, 40–42.
[35] See Matthews, 'Letters of Symmachus', 64.
[36] As Cribiore states in *School of Libanius*, 107, 109.
[37] See *Ep.*B.131 (F.336) to the advocate Gaianus.
[38] Bradbury, 'Libanius' Letters', 79; Leibeschuetz, *Antioch*, 132–36; Petit, *Libanius et la vie*, 291–3. See also Cabouret, 'Pouvoir municipal', 140.
[39] This idea is based on the work of Pierre Bourdieu on the way individuals are able to strategically manipulate social rules (Bourdieu, 'Marriage Strategies'; *Outline of a Theory*, 95). For a fuller application of Bourdieu's ideas to Libanius' writings, see Sandwell, *Religious Identity*.
[40] As Cribiore has shown of the educational networks around Libanius (Cribiore, *School of Libanius*, 104).
[41] Bradbury, 'Libanius' Letters'; Cribiore, *School of Libanius*, 95, 99.
[42] Trapp, *Greek and Latin*, 17; Liebeschuetz, *Antioch*, 20; Bradbury, *Selected Letters*, 19–20. For examples of people reading aloud, see Libanius *Ep.*N.12 (F.434); *Ep.*F.773.5; *Ep* B.107 (F.779).
[43] *Ep.*N.16 (F.476), *Ep.*N.17 (F.477); *Ep.*N.40.1 (F.19); *Ep.*N.133.5 (F.1264); *Ep.*N.168.4 (F.957).
[44] On the ambiguous private/public nature of letters, see Conybeare, *Paulinus Noster*, 12, 21.
[45] Trout, *Paulinus of Nola*, 199; Conybeare, *Paulinus Noster*, 35, 131.
[46] Liebeschuetz, *Antioch*, 21.
[47] See also *Ep.*N.151 (F. 867); *Ep.*N.186 (F.1057)).
[48] Such complaints were to some degree part of social and epistolary convention, *Ep.*N.133.1 (F.1264).
[49] *Ep.*N.20.1–3 (F.509); *Ep.*N.22.2 (F.552); *Ep.*N.86.1–2 (F.725).
[50] Wellman *et al.*, 'Networks as Personal', 147; Eriksen, *Ethnicity*, 44.
[51] Wellman, 'Structural Analysis', 21; Vincent, 'Structuring of Ethnicity', 376.
[52] Wellman and Berkowitz, 'Thinking Structurally', 16; Handelman, 'Organization', 269; Eriksen, *Ethnicity*, 42.
[53] See also *Ep.* B.53.2 (F.503).
[54] *Ep.* B.28 (F.529); *Ep.* B.64 (F.362); *Ep.* B.162 (F.433); *Ep.*N.21 (F.515); also *Epp.* B.8–9 (F.74, 98).
[55] On Libanius' relationships with the emperors Julian and Theodosius and their officials, see Swain, 'Sophists and Emperors'.
[56] Kelly, 'Bureaucracy and Government', 196.
[57] Wellman, 'Structural Analysis', 22.

References

Bourdieu, Pierre. 'Marriage Strategies as Strategies of Social Reproduction'. In *Family and Society: Selections from the Annales*, edited by R. Foster and O. Ranum. Baltimore: John Hopkins

University Press, 1977 ('Les strategies matrimonial dans le système des strategies de reproduction'. *Annales* 4–5 (1972): 1105–27).

———. *Outline of a Theory of Practice*. Cambridge: Cambridge University Press, 1977 (*Esquisse d'une théorie de la pratique, précédé de trois etudes d'ethnologie kabyle*, 1972).

Bradbury, Scott. 'Libanius' Letters as Evidence for Travel and Epistolary Networks among Greek Elites in the Fourth Century'. In *Travel, Communication and Geography in Late Antiquity: Sacred and Profane*, edited by L. Ellis, and F. Kidner. Burlington: Ashgate Publishing. 2004.

———.'A Sophistic Prefect: Anatolius of Berytus in the Letters of Libanius'. *Classical Philology* 95 (2000): 155–77.

Cabouret, Bernadette. 'Pouvoir municipal, pouvoir imperial au IVe siècle'. In *Antioche de Syrie Histoire, images et traces de la ville antique*. colloque organisé par B. Cabouret, P.-L. Gatier et C. Saliou, Lyon, Maison de l'Orient et de la Méditerranée, 4–6 octobre 2001/actes édités par Bernadette Cabouret, Pierre-Louis Gatier et Catherine Saliou. Topoi Orient Occident, Supplément 5. Paris: De Boccard, 2004.

Conybeare, Catherine. *Paulinus Noster: Self and Symbols in the Letters of Paulinus of Nola*. Oxford and New York: Oxford University Press, 2000.

Cribiore, Raffaella. *The School of Libanius in Late Antique Antioch*. Princeton and Oxford: Princeton University Press, 2007.

Eriksen, Thomas H. *Ethnicity and Nationalism: Anthropological Perspectives*. London: Pluto Press, 1993.

Handelman, Don. 'The Organization of Ethnicity'. *Ethnic Groups* 1 (1977): 187–200.

Heather, Peter. 'Senators and Senates'. In *Cambridge Ancient History*. Vol. 13. *The Late Empire, AD 337–425*. 2nd ed., edited by Av. Cameron and P. Garnsey. Cambridge: Cambridge University Press, 1998.

Jones, A. H. M., John R. Martindale, and J. Morris. *The Prosopography of the Later Roman Empire*. Vol. 1, *AD 260–395*. Cambridge: Cambridge University Press, 1971.

Kaster, Robert A. *The Guardians of Language: The Grammarian and Society in Late Antiquity*. Berkeley: University of California Press, 1988.

Kelly, Christopher. 'Bureaucracy and Government'. In *The Cambridge Companion to the Age of Constantine*, edited by N. Lenski. Cambridge and New York: Cambridge University Press, 2005.

———.'Empire Building'. In *Interpreting Late Antiquity: Essays on the Postclassical World*, edited by G. W. Bowersock, P. Brown, and O. Graber. Cambridge, MA: Belknap Press of Harvard University Press, 2001.

Libanius. *Selected Letters of Libanius from the Age of Constantius and Julian*. Translated with an introduction by Scott Bradbury. Translated Texts for Historians 41. Liverpool: Liverpool University Press, 2004.

———. *Autobiography and Selected Letters. 2* vols. Translated by A.F. Norman. Loeb Classical Library 478–79. Cambridge and London: Harvard University Press, 1992.

———. *Selected Works*. 2 vols. Translated by A.F. Norman. Loeb Classical Library 452. London: Heinemann and Cambridge: Harvard University Press, 1969–1977.

———. *Opera*. Edited by R. Foerster. Stuttgart: Lipsiae with B.G. Teubner. 1903–1927.

Liebeschuetz, John H. W. G. *Antioch. City and Imperial Administration in the Later Roman Empire*. Oxford: Oxford University Press, 1972.

Matthews, John. 'The Letters of Symmachus'. In *Latin Literature in the Fourth Century*, edited by J. W. Binns. London and Boston: Routledge and Kegan Paul, 1974.

Norman, Albert F., ed. *Antioch as a Center for Hellenic Culture*. Translated Texts for Historians 34. Liverpool: Liverpool University Press, 2000.

Pack, Roger. '*Curiales* in the Correspondence of Libanius'. *Transactions of the American Philological Association* 82 (1951): 176–92.

Petit, Paul. *Libanius et la vie municipale à Antioche au IVe siècle après J.C.* Paris: Libraire orientaliste Paul Geuthner, 1955.

———. *Les étudiants de Libanius.* Paris: Nouvelles Éditiones Latines, 1957.

———. *Les fonctionnaires dans l'oeuvre de Libanius: analyse prosopographique.* Paris: Université de Besançon, 1994.

Sandwell, Isabella. *Religious Identity in Late Antiquity: Greeks, Jews and Christians in Antioch.* Cambridge: Cambridge University Press, 2007.

Stowers, Stanley K. *Letter Writing in Graeco-Roman Antiquity.* Philadelphia: Westminster, 1986.

Swain, Simon. 'Sophists and Emperors'. In *Approaching Late Antiquity: The Transformation from Early to Late Empire,* edited by S. Swain and M. Edwards. Oxford: Oxford University Press, 2003.

Trapp, Michael B. *Greek and Latin Letters: An Anthology with Translation.* Cambridge and New York: Cambridge University Press, 2003.

Trout, Dennis E. *Paulinus of Nola: Life, Letters and Poems.* Berkeley and Los Angeles: University of California Press, 1999.

Vincent, Joan. 'The Structuring of Ethnicity'. *Human Organization* 33. no. 4 (1974): 375–9.

Wellman, Barry. 'Structural Analysis: From Method and Metaphor to Theory and Substance'. In *Social Structures: A Network Approach,* edited by B. Wellman and S. D. Berkowitz. Toronto: Canadian Scholars' Press, 2003.

Wellman, B., P. J. Carrington, and A. Hall. 'Networks as Personal Communities'. In *Social Structures: A Network Approach,* edited by B. Wellman and S. D. Berkowitz. 2nd ed. Toronto: Canadian Scholars' Press, 2003.

Wellman, B., and S. D. Berkowitz. 'Thinking Structurally'. In *Social Structures: A Network Approach,* edited by B. Wellman and S. D. Berkowitz. 2nd ed Toronto: Canadian Scholars' Press, 2003.

Wintjes, Jorit. *Das Leben des Libanius.* Rahden/Westf.: Verlag Marie Leidorf GmbH, 2005.

Network Theory and Religious Innovation

Anna Collar

Introduction

The approach put forward in this paper, that of using network theory as a tool to analyse the social processes of religious innovation and conversion, arises from the uniformity observed in the dedications to the monotheistic Theos Hypsistos.[1] Although intrinsic qualities can be argued to 'explain' a belief system's attractiveness and diffusion, this method is incomplete, because it does not explain why some high-quality ideas fail. Combining some understanding of the ideology and utility of a religion with an analysis of the influences of social networks and environment allows a more subtle appraisal of the evidence. Ideas and technology are transmitted along social interconnections, and cultural and religious change can be understood as emergent phenomena through analysis of the variability and dynamics of these interconnections.

Network theory at a basic level allows the historian to visualize a node's degree of centrality or remoteness on a network; at a more complex level, it can help conceptualize the multifarious routes along which information travels. Theorizing networks takes the links between nodes and makes them as important as the nodes themselves for understanding the properties of social change. Its purpose is to explain how human interactions affect cultural change and influence the success and failure of innovations and ideas. Although the spread of monotheism in the Roman Empire was a long process, it is an idea that arises from small-world network theory, that of 'information cascade', which provides a useful way of re-approaching how and why the monotheistic 'innovation' was so successful across such a wide area.

Network Theory

There is a vast body of literature on network theory, especially in physics and sociology. What follows here is an overview of the key ideas: 'phase transition'—the sudden dramatic shift from one network state to another; the small-world social network, made up of clusters interlocking with each other through weak and strong ties; the 'short path-length' property of the small-world network, which allows any two individuals to connect relatively easily; and the relationship of individual thresholds and network connectivity to the process of 'information cascade', as a result of these properties of the network.

Phase Transition

Mathematicians have long known that on a randomly connected network a 'phase transition' between disconnection and connection occurs—when the network is connected enough for most of the nodes to have joined up into clusters, with a few interconnecting links between the clusters—through the addition of a relatively small number of new links. The result is the joining of these isolated groups into one interconnected cluster, known as the 'giant component'.[2] The network undergoes a dramatic and sudden leap from one state to the other. Physical examples of phase transition include the magnetization of iron molecules, or the freezing of water. What the giant component does is allow communications across the whole network; i.e., when a network is not connected by the giant component, events are only felt locally. Absence of centrality is associated with phase transition and is vital to understanding the spread of information across networks. No single molecule causes a piece of iron to magnetize. It is a decentralized, emergent process.

Small events, behavioural switches, and individual choices percolate through the system and can lead to widespread change, and they could come from anywhere. Far from the centre determining the action on a network, the centre is created by that action. Watts states that 'the network centrality of individuals, or any centrality for that matter, would tell us little or nothing about the outcome, *because the centre emerges only as a consequence of the event itself*'.[3]

Small Worlds, Short Paths and Overlaps

But randomly connected graphs or networks are not representative of the real world. Social networks are by nature neither entirely regular nor entirely random: they are made up of close-knit (regular) clusters, formed by geography, religion, family and so on, intersected by long-distance (random) links, for example, somebody with whom nobody else in the cluster has any contact. This is the 'small-world' network.[4] The long-distance connections transgress the boundaries of a local cluster, becoming shortcuts to other clusters in the network. It is a global network phenomenon that is couched in local network terms: the 'small-world' has a 'short path–length'.

A short path in a social network can be either a 'weak' or 'strong' tie, where 'strength' and 'weakness' refer only to the fidelity of the connection. 'Weak' ties are low-fidelity bridges between different groups, representative of a long-distance or random contact. Converse to their name, Granovetter observed weak ties to be extremely powerful at facilitating diffusion across the network.[5] This is because they cross network distance (the number of nodes one has to pass through to reach the desired node), directly accessing totally separate clusters or individuals, making them particularly useful for *certain types* of diffusion in social networks (for example, disease), or passing on information about new jobs.[6]

However, most social ties are 'strong', reflective of fundamental facets of identity. Social identity can be defined (somewhat simplistically) by group membership. By belonging (or choosing not to belong) to certain groups, people acquire aspects of identity that drive the makeup of their social network. Most of these groups are clusters, 'strong tie' groups—trusted people, who may form 'closed triads' (the sociological term for a situation where all three people know each other)—or else have other markers of strength, such as frequency or length of contact.[7] Socially, it is these people that exert the most influence on an individual when it comes to 'complex' transmission of ideas or information.

Mathematically, however, diffusion using strong ties poses a problem. The path length is long, that is, the information takes time to travel from one side of the network to the other, as it must make lots of little hops through the clusters. However, empirical social network data has shown that strong ties can still have a relatively short path length. Shi *et al.* found that removing the weak ties in a test case did not disconnect the network; rather, 'the network sheds some nodes and shrinks modestly'.[8] The result is that a high-fidelity strong-tie social network, made up of overlapping clusters, spreads information at almost the same efficiency as the network linked together with long-distance weak ties. A combination of the two is the most accurate representation of the real world, in which both overlaps and weak ties connect the network, like the giant component does in physics, and make possible an 'information cascade'—equivalent to a phase transition.

Networks and Diffusion

Power Laws and Restriction of Power Laws

An important development in understanding spread on networks resulted from the discovery of Barabási and Albert,[9] who found that instead of following a 'normal distribution'[10] when plotted graphically, many real-world networks are highly skewed, the majority of nodes are poorly connected, a few have massive connectivity. This demonstrates a 'power law', where many small nodes coexist with a few very large ones. Power-law networks have 'hubs', nodes that are disproportionately well connected. Hubs exist partly because of preferential attachment, i.e., the probability of a new node joining to an existing node is proportional to the number of links that node already has. A node attracts new links largely on the basis of already having links. Seniority within a network is one of the main reasons a node will become a hub, but other aspects of identity are also factors. These allow for the 'lock in' of fluctuations, amplifying differences over time.

Power laws have been found to occur frequently in many real-world situations (for example, distribution of wealth, popularity of websites). However, human interactions are different and more subtle than all other kinds of networks. The anthropologist Schnegg[11] observed that reciprocity and the recollection of beneficial past acts restrict the development of hubs and power laws in social networks. This leads to the conclusion that in social networks, maximum utility is not as important to most members as reciprocal altruism and being fair—as supported by biological data.[12] 'Blending reciprocity and memory into transactions reproduces networks much better correlated to the social world'.[13] This is not to say that hubs do not exist in social networks. Rather, they exist, but their 'hub' status does not automatically mean they are universally powerful in attracting links, as opposed to known nodes within a social setting, with which there has been positive historical interaction. Hubs play an important role in transmitting information over a network. In order to understand how hubs function in different network situations, it is necessary first to discuss the processes by which information diffuses through a network.

Using the Network: Transmitting Information

Effective diffusion requires the nodes vulnerable to the new idea, technology and so forth, to be found. Network distance—the number of steps between two nodes—is more important in analysing information transmission than physical distance.[14] Distance is not measured in just one way however, there are many types—geographical, social, professional—every aspect of identity can measure distance. People use knowledge of the identities of their local network to assess the best route, the shortest path to spread information across, known as 'directed search'. When the vulnerable nodes in a network can be linked together through this process of search, they form what is known as a 'percolating vulnerable cluster'.

The percolating vulnerable cluster allows information to cascade across it. This can be rapid, as with an out-of-nowhere success story, or slow, as with the change of societal norms. The spread of monotheism may be an example of the latter. Information cascades, like phase transition in physics, are demonstrated by the network as a whole displaying emergent behaviour: individuals stop acting like individuals and behave as though part of an organized group. They are self-perpetuating—picking up 'new adherents largely on the strength of having attracted previous ones. An initial shock can propagate throughout a very large system, even when the shock itself is small'.[15] But all kinds of shocks happen all the time, yet the system generally remains unaffected.

Vulnerability to Innovation: Thresholds and Neighbours

Shocks that trigger cascade do not necessarily represent superior stimuli. Their success is rather an indication of the network structure in which they happen to land, whether the local environment is vulnerable or stable with regard to innovation. There are two ways of assessing a node's vulnerability, either 'it has a low threshold (thus, a predisposition for change); or because it possesses only a very few neighbours, each of which thereby exerts significant influence'.[16]

People have a 'critical threshold' whereby an opinion about something can jump abruptly from one alternative to another. Sometimes, the threshold is very high, when complete unanimity of solicited opinion is required to cause the individual to switch (even then they may not do so). Often the threshold is much lower, with fewer obvious differences between alternatives.[17] At this level, neighbourly influence is vital to an individual's decision to switch. If an individual observes enough endorsement of a particular choice, they are likely to switch, especially if those people are part of socially powerful relationships, regarded as being of higher status, or known to have made 'good' decisions in the past.

The 'threshold rule'—the position of an individual's threshold—will be subject to change depending on the nature of the innovation. Some are willing to adopt something about which they have little information; others are more conservative. In the 1960s Rogers,[18] in a discussion of the adoption of technological and organizational innovation, identified five groups in the population: *innovators*, venturesome, intelligent, technologically aware, often with substantial financial resources; *early adopters*, socially respected opinion leaders; the *early majority*, deliberators central to legitimizing the innovation; the *late majority*, sceptical about the new who adopt out of necessity; and *laggards*, the traditionalists. This is more like a continuous scale of awareness to adoption, and the details of these categories may need modification to apply to innovations in ideologies and beliefs. Many factors affect an individual's position, including age, education, environment, class, gender, religion, and financial and social status. People also have varying degrees of trust in different sources of information.

Layton[19] observed that it is 'those individuals who are the first to adopt innovations who are of prime importance. They act both as a bridge and a buffer between their fellow villagers and the outside world by bringing detailed knowledge of innovations into that community'.[20] The innovators must be connected to early adopters if the innovation is to spread. The success or failure of an innovation may be defined by the criteria used to judge it: is success marked by the time taken for an innovation to be accepted, general depth of understanding, or legitimization by a social leader? Additionally, it must also be asked how the new information is converted into knowledge that is trusted and applied, not simply possessed.

Mathematically, nodes are more stable the more links they have, because they are less likely to be influenced by any one link. Watts uses the term 'critical upper degree' to describe the 'number of neighbours a node can have and still be activated by one neighbour'—a node's boundary between retaining the status quo and adoption of the innovation.[21] A node will be stable if it has more neighbours than this degree. By this logic, hubs will be the most stable nodes of all.

Boundaries and Cascade

Mathematically, cascades will not occur for three reasons. First, no node's threshold is low enough: the innovation does not persuade. Second, the network is not well connected enough: a disconnected network is vulnerable, but does not percolate fully; it is therefore unable to spread influence. Third, it is too well connected: the majority of nodes are stable because of neighbourly influence and so, even though they percolate fully, cannot be part of a vulnerable cluster. This creates two boundaries within which information cascades can occur: the lower boundary, below which the nodes have too few neighbours, and the upper boundary, above which the nodes become too densely connected.[22]

Full network cascades only happen if the innovation activates the vulnerable cluster of highly connected nodes near the upper boundary. The percolating vulnerable cluster is tightly integrated with the rest of the network here, and so initially stable nodes will be exposed to multiple early adopters, causing them to activate. The prerequisites are difficult to achieve, making the upper boundary cascade very rare, but when an innovation does activate the vulnerable cluster here, the high connectivity means the entire network will follow and global cascade occurs. This is known as 'crossing the chasm'.[23]

Because nodes at the upper boundary are so well connected, it follows that 'cascades are almost as likely to be triggered by an individual with an average number of neighbours as someone to whom many people pay attention; ... being simply well-connected is less important than being connected to individuals who can be influenced easily'.[24] Conversely, at the lower boundary, because nodes are vulnerable to neighbourly influence, 'highly connected individuals are disproportionately effective in propagating social contagion'.[25] This fits better with diffusion of innovation theory, which classes early adopters–opinion–leaders with both long-distance links and high social integration—as the most effective promoters of new ideas or technology.

Problems and Ways Forward

This leaves a number of issues to be addressed. First of all, why are highly connected individuals not effective at the upper boundary too? There are two types of cascade: lower boundary, where hubs play a central role in diffusion, and upper boundary, where hubs are too inherently stable to be effective transmitters. This may make sense mathematically, and perhaps works in some network examples, but it is socially paradoxical. Early adopters, by definition respected opinion leaders, are likely to be well-connected nodes, possibly even a local hub. In real terms, these individuals are the most effective at spreading information—having lots of neighbours is the fastest way of transferring innovations. It is difficult to see how information can spread on a social network without these nodes being involved, regardless of their *mathematical* stability.

These problems arise partly because identity has been left out of the mathematics for simplicity's sake. But social connections run in different directions: the node can be doing the connecting, or be being connected to. Sometimes this will be mutual, but there are divisions between active and passive links that need to be distinguished in the developed model of cascades. The frequency of the interaction, and its fidelity, must be considered in the theorization of diffusion in social networks: there is connective directionality and asymmetry as well as variance in values of different group participations.

Another issue is the role of the missing third link in the chain, the early majority. Mathematically, the network does not explain the legitimization process — it ignores it by making the stable nodes all identical, simply switching under the collective pressure of being connected to many early adopters. Socially, this is not the case. Understanding legitimization is central to understanding how innovations spread from being small pockets of adoption to being cascades of information.

Networks and Religious Innovation

Information spreads along both horizontal and vertical axes. Horizontal transmission refers to peer similarity, represented, for example, by copying without understanding, whereas vertical transmission requires a depth of knowledge and experience, for instance through conditioned learning.[26] Religious information is transmitted both ways. One can horizontally copy or superficially appreciate aspects of a new cult, for example, by acknowledging a deity that is represented in a familiar way, but full understanding will be a vertical process, of learning and adapting a particular ritual format, liturgy, or set of physical movements.

Monotheism as a Religious 'Innovation'

Although the philosophical argument for one god was well known amongst the intellectual elite, and a 'monotheistic trend' has been observed during the Imperial period, monotheism can be called a religious innovation within the milieu of Imperial polytheism. The future application of these theories in this investigation seeks to understand the reasons why monotheism was so successful *at this juncture*. What circumstances had changed in order for monotheism to take hold? To whom was monotheism attractive, why did they convert, how did it favour them individually, and how are they defined? Does 'vulnerability' to a new faith involve having disposable time or income, a particular occupation, is it to do with age, status, gender, or ethnic identity? Did transmission follow set routes, or random connections? Anthropologically,[27] in general it is young men who travel and trade, and they that access and transmit new ideas. However, socially, the elder generation possess the experience to be effective religious leaders. Can this conflict be reconciled?

One way to set about answering these questions is to combine the theory of innovation with that of religious conversion. Social interactions are dynamic and transitory, with a transactional nature: the value must be equal to, or greater than, the cost. Directionality (i.e., give and take) of transactions means that over time asymmetries in that give and take become 'locked in', as has been observed above in the emergence of network hubs. Power relationships and inequalities result from social interchange dependent on, for example, an individual's status, resources, talent, etc. These aspects are both self-determined as well as predetermined.[28]

Conversion and Evolution

Some sociologists of religion have sought to explain religious conversion (or 'recruitment') through an analysis of social interactions. Stark has shown that status, background, and environment are important, but that social contacts are crucial to an individual's receptiveness to a new cult.[29]

> The basis for successful conversionist movements is growth through social networks, through a *structure of direct and intimate interpersonal contacts*. Most new religious movements fail because they quickly become closed, or semiclosed networks. ... Successful movements discover techniques for remaining open networks, able to reach out and into new adjacent social networks.[30]

His analyses of the spread of early Christianity argued that the combination of Christian ideology/self-representation within historical environments—from the disease epidemics of the second and third centuries, to Christian (as opposed to Roman, not Jewish) treatment of women—reconfigured the ratios of pagan to Christian in social networks in a way that ensured the 'inevitable' triumph of Christianity. Instead of taking the quality of religious stimulus as given, conversion should be understood as a social process, conditioned by people's networks driving the religious change. This approach does not ignore transcendence, or judge the truth of the ideology; it instead engages with the social aspects of why humans believe. The ideology of a religion will always be part of the explanation of conversion. However, regardless of the persuasiveness of an ideology, or the transcendence of a vision, conversion does not happen in a social vacuum.

It is relevant at this point to consider religious innovation within an evolutionary framework. Simplistically, biological innovations are successful if they favour the individual gene's chances of survival—in a sense, the genes are motivated by self-interest, and direct the organism in which they are contained to maximize chances of survival, either individually through reproduction, or through kin-selection. The term 'meme' has been coined for ideas that propagate via human consciousness, just as genes propagate through organisms over generations.[31] This form of argument is familiar to archaeologists; Shennan examines cultural development from an evolutionary perspective.[32] He cites work by Boyd and Richerson suggesting that 'culture can usefully be seen as a system of inheritance' and which is subject, as a consequence, to random fluctuations and errors of the type seen in genetic mutation.[33] However, this refers to technological innovation. Ideological innovation is much more difficult to study, but traces can be found in records of new practices or rituals.[34] The utility of a technological innovation means that this kind of innovation can be classed as 'evolutionary'; it follows a (generally) strict, unidirectional linear pathway. Some biologists have criticized the network approach to understanding religious transmission, and argue instead that monotheism's success should be understood in terms of individual self-interestedness, that a religious innovation should possess inherent utility, and its adoption and diffusion involve a process akin to natural selection, propagating beneficial elements and discarding useless or harmful ones. Therefore, critics argue that it is the value of the innovation rather than connectivity of the network that determines success or failure.[35]

Responses

There is not enough space here to deal with these challenges in full; however, evolutionary and network approaches are not mutually exclusive. It is a given that any innovation possesses inherent value within the context in which it was developed. The network structure simply determines how profoundly it will spread. Random fluctuations, the 'genetic mutations' of biology, become set over time, initially (although not always) because they are better suited to the environment. This is particularly clear in the physical characteristics of networks, for example in the geographical location of nodes, communication routes, etc., but eventually the network will change if there is a more effective and efficient alternative. Can religion be claimed to follow a similar developmental trajectory? Ideas do not impose themselves, but depend on the stance that an individual takes in regard to them. It is perhaps more helpful to term religious innovation as a dialectic of ideas rather than strictly as an evolution.

However, the utility of religion can be identified in the social environment into which the convert is admitted, as well as in its intrinsic properties and ideology. Beliefs have utility for those who hold them, as they *believe* in their utility. An individual will only convert if it is in their interest, and this self-interest is likely to drive conversion only if there are social benefits to be gained. In certain periods, and in particular environments, the religious status quo will undergo sudden upheaval as a response to social factors, and the emergence of a religion that has more utility for its believers. Parts of the system will mutate, while others remain the same. The question is then: will these traditional forms, which have not adapted to the environment and needs of the believers, eventually become extinct?

Some parts of the system have more freedom to mutate than others. Mathematically, the looser structure of networks at the 'peripheries' (near the lower boundary) is more conducive to *short-lived* change than the upper boundary 'core'. Is the social situation similar? As Stark writes, 'typically, the fate of new religious movements is largely beyond their control, depending greatly on features of the environment in which they appear'.[36] Peripheral religious innovation makes sense; orthodoxy is generally imposed from the static, traditional centre, controlling standards through official shared rituals. For example, in the Roman Empire during the third and fourth centuries, Christian 'orthodoxy' ultimately spread despite competition from various 'heresies'. However, at this point, it was not really the orthodoxy; it was a version of Christianity that happened to have reinforcement from Imperial legislation, and thus provided the 'best fit' for most people's social environment—the expenditure of least effort and energy. It became central as a result of the action on the network. Elsewhere, other 'orthodoxies' prevailed.

Network Approaches to the Past and Future Directions for Theos Hypsistos

Ancient networks concerning trade, technology, and communication have already been studied, built on centralization/diversification and complexity models, or cultural transmission theories. Recently, archaeologists have used networks in a more theoretical way. Isaksen used GIS and network analysis software to demonstrate the correlation between the physical and network locations of key political sites in Roman Baetica.[37] Bentley and Maschner explored the potential of mathematical complexity theory in archaeology,[38] and Knappett has collaborated with physicists to simulate Minoan networks across the Mediterranean.[39] Broodbank conducted a Proximal Point Analysis (PPA) to simulate and predict Early Bronze Age Cyclades settlement patterns, drawing contact lines between a site and its three closest neighbours.[40] A useful result is the transparent demonstration of site centrality or remoteness, and although these values can be intuited correctly, the method highlights issues or anomalies missed by common sense.

Malkin uses networks and network language to explore Archaic Greek colonization: 'Awareness of "sameness" occurs not when people are close to each other ... but when they are far apart. It is distance that creates the virtual centre. The more the connecting cables are stretched, the stronger they become'.[41] This idea that distance helps to create identity may be relevant to the study of widely distributed religious communities, although a 'virtual centre' is not necessarily also created. Mitchell's analysis of the cult of Hypsistos argued for its distinction from, but close connection to, Jewish Diaspora communities: many of the believers were 'Judaizers'.[42] The networks of 'Hypsistarians' and Jews are geographically fairly close, but what about ideologically and socially, in their own self-perception, and as perceived by the societies of which they were part? Questions that will arise in my application of network theory to historical data include: is the Hypsistos cult an example of a heterarchical network, a religious network without a centre, fundamentally different from an institutionalized, centralized, sanctuary-focused hierarchical church? In other words, is pagan monotheism characterized by its network? Or is this kind of network a characterization of early-stage cults, before they ossify into the full-network norm, and become the 'traditional'?

These approaches should lead to developments in understanding of the cognitive or physical space for new cults in the ancient world. Comparison of networks will allow the observation of crossovers or interchanges and the domination of certain cult distributions. For example, perhaps once a cult is settled there is no 'room' for another similar one, and the evidence shows examples of unsuccessful, partial cascades. How would Christianity have used those half-activated vulnerable clusters? Mitchell concludes that 'without them [the Hypsistos worshippers], the transformation of ancient patterns of belief from pagan polytheism to the predominantly monotheistic systems of Judaism, Christianity, and Islam would not only have been far less tidy and unidirectional than it was, it might not have occurred at all'.[43]

Problems and Potential

Assessing network structure rather than the 'stimuli' of historical events leads to a different way of understanding the past. Instead of viewing historical success as a measure of inherent merit, using networks means the observed outcomes of historical situations need not be 'superior'. They are survivors. Biological survival depends on 'best fit'. However, the theory of cascades suggests that inputs and outcomes are not necessarily proportional, yet the biologists' view is the opposite — the input generally requires enough utility for it to be selected for transmission. Are social and biological selection processes representative of different phenomena with different rules, or can evolution be fully incorporated into an understanding of social emergence? This paper has attempted to go some way to reconciling these views, but this must be developed further. For example, does the fundamental difference in scale, the complexity and the active agency involved, render 'natural selection' too blunt a tool to apply to historical situations?

There are obvious problems with theoretical approaches to the ancient world, particularly the uneven patterning of evidence and the application of modern interpretive frameworks. Networks, through their simulative as well as data-responsive functions, may help to fill some gaps. Ancient history and archaeology can provide time depth in which to test network theory, and to investigate similarities and differences between information transmission in pre-modern and modern society.

However, modelling cascade on networks is not a finished answer to these problems. At present, the threshold rule is highly idealized, and further development of this concept is necessary. Certain assumptions have also been made in innovation theory that need modification for the ancient world—for example the financial independence of innovators in antiquity, or the likelihood of non-geographically defined friendships. Instead, inter-community contact, patronage systems, or technological innovation as prestige-indicator might be more useful terms to employ. Equally, although occasional individual early adopters have been identified (for example, the apostle Paul[44]) it is nearly impossible to distinguish such people from the information generally available. Identifying early adopting groups may be easier.

This paper has intended to demonstrate that network theory holds great potential for approaching ancient data. There is more to be thought through and tested, and many aspects of the theory need further development, but with additional refinement, the theory of networks and the diffusion of information could become as fundamental a tool for understanding social and religious change and communication in the past as it is proved for understanding these aspects of society in the present.

Acknowledgements

This paper summarizes the theoretical framework of my Ph.D. research as part of Stephen Mitchell's AHRC project at the University of Exeter: 'Pagan Monotheism in Its Intellectual Context'.

Notes

[1] See Mitchell, 'The Cult of Theos Hypsistos'.
[2] Watts, *Six Degrees*, 45–46.
[3] Ibid., 53.
[4] Watts and Strogatz, 'Collective Dynamics of "Small World" Networks'.
[5] Granovetter, 'The Strength of Weak Ties'.
[6] See Ibid.; Watts, *Six Degrees*, 49.
[7] Shi et al., 'Networks of Strong Ties', 1.
[8] Ibid., 2.
[9] Albert and Barabási, 'Statistical Mechanics of Complex Networks'.
[10] The line creates a curve indicative of the rule of the average — i.e., very few nodes, if any, have no connections, and very few nodes, if any, have an excessive number of connections.
[11] Schnegg, 'Reciprocity and the Emergence of Power Laws in Social Networks'.
[12] See Dawkins, *Selfish Gene*, 183–88.
[13] Schnegg, 'Reciprocity and the Emergence of Power Laws in Social Networks', 8.
[14] However, in the ancient world, physical distance constricts network closeness in a different way to modern networks.
[15] Watts, *Six Degrees*, 206.

[16] Ibid., 233. This distinction may be particularly pertinent in modelling religious change.
[17] Watts, *Six Degrees*, 224.
[18] Rogers, *Diffusion of Innovations*, 169–71.
[19] Layton, 'Pellaport'.
[20] Ibid., 50–51.
[21] Watts, *Six Degrees*, 234.
[22] Ibid., 237–38.
[23] Ibid., 242.
[24] Ibid., 243. This could aid understanding of how cults propagate, through deliberate 'search' for individuals with low thresholds or few friends to keep them 'stable'.
[25] Ibid., 240.
[26] See Knappett, *Thinking Through Material Culture.*
[27] Philip Jenkins, Penn State University, in response to paper by Bas van Os, University of Groningen, at the SBL/AAR AGM in Washington, DC, November 2006.
[28] Boissevain, *Friends of Friends*, 25–26.
[29] Stark, *The Rise of Christianity.*
[30] Ibid., 20.
[31] Dawkins, *Selfish Gene*, 189–201.
[32] Shennan, 'Cultural Transmission and Cultural Change', 330–46.
[33] Ibid., 334.
[34] Not to preclude simultaneity—technological innovation bringing with it an ideology or associated religious belief, for those who use it, or who carry it.
[35] In conversation with N. Collar, BirdLife International/Department of Zoology, University of Cambridge.
[36] Stark, Rise of Christianity, 191.
[37] Isaksen, *Network Analysis of Transport Vectors in Roman Baetica.*
[38] Bentley and Maschner, *Complex Systems and Archaeology.*
[39] Evans *et al.*, *Thinking Through Material Culture.*
[40] Broodbank, *An Island Archaeology of the Early Cyclades.*
[41] Malkin, 'Networks and the Emergence of Greek Identity', 59.
[42] Mitchell, 'The Cult of Theos Hypsistos'.
[43] Ibid., 128.
[44] Barabási used Paul as an example of an ancient hub in his introduction to *Linked*.

References

Albert, R., and A.-L. Barabási. 'Statistical Mechanics of Complex Networks'. *Reviews of Modern Physics* 74. (2002): 47–97.

Barabási, A.-L. *Linked—The New Science of Networks*. Cambridge, MA: Perseus, 2002.

Bentley, R. Alexander, and D. G. Maschner Herbert, eds. *Complex Systems and Archaeology.* Salt Lake City, UT: University of Utah Press, 2003.

Boissevain, Jeremy. *Friends of Friends.* Oxford: Basil Blackwell, 1974.

Boyd and Richerson. *Culture and the Evolutionary Process.* Chicago: Chicago University Press, 1985.

Broodbank, Cyprian. *An Island Archaeology of the Early Cyclades.* Cambridge: Cambridge University Press, 2000.

Dawkins, Richard. *The Selfish Gene.* Oxford: Oxford University Press, 1976.

Evans, Tim, Carl Knappett, and Ray Rivers. 'Using Statistical Physics to Understand Relational Space: A Case Study from Mediterranean Prehistory'. In *Complexity Perspectives on Innovation and*

Social Change, edited by D. Lane, D. Pumain, S. van der Leeuw, and G. West. New York: Forthcoming Springer Methodos series, 2007.

Granovetter, Mark. 'The Strength of Weak Ties'. *American Journal of Sociology* 6., vol. 78 (1973): 1360–80.

Isaksen, Leif. *Network Analysis of Transport Vectors in Roman Baetica*. Southampton: University of Southampton, 2005.

Knappett, Carl. *Thinking Through Material Culture*. Philadelphia: University of Pennsylvania Press, 2005.

Layton, Robert. 'Pellaport'. In *What's New? A Closer Look at the Process of Innovation*, edited by S. Van der Leeuw, and R. Torrence. London: Unwin Hyman, 1989.

Malkin, Irad. 'Networks and the Emergence of Greek Identity'. *Mediterranean Historical Review* 18. (2003): 56–75.

Mitchell, Stephen. 'The Cult of Theos Hypsistos'. In *Pagan Monotheism in Late Antiquity*, edited by P. Athanassiadi and M. Frede. Oxford: Oxford University Press, 1998.

Newman, Mark. 'The Structure and Function of Complex Networks'. (2003). Available at http://arxiv.org/abs/cond-mat/0303516.

Rogers, Everett. *Diffusion of Innovations*. New York: Free Press, 1962.

Michael Schnegg. 'Reciprocity and the Emergence of Power Laws in Social Networks.' (2006) Available at http://arxiv.org/abs/physics/0603005. Website accessed 22 May 2006.

Shennan, S. 'Cultural Transmission and Cultural Change'. In *What's New? A Closer Look at the Process of Innovation*, edited by S. Van der Leeuw, and R. Torrence. London: Unwin Hyman, 1989.

Shi, Xiaolin, Lada, A. Adamic, and Martin J. Strauss. 'Networks of Strong Ties'. (2006) Available at http://arxiv.org/abs/cond-mat/0605279. Website accessed 16 May 2006.

Stark, Rodney. *The Rise of Christianity*. Princeton, NJ: Princeton University Press, 1996.

Watts, Duncan. *Six Degrees*. New York: Norton, 2003.

Watts, Duncan, and Steven H. Strogatz. 'Collective Dynamics of "Small World" Networks'. *Nature* 393. (1998): 440–442.

Commercial Networks in the Mediterranean and the Diffusion of Early Attic Red-figure Pottery (525–490 BCE)

Dimitris Paleothodoros

In this article, I wish to provide an answer to a series of questions often asked in recent scholarship, but seldom answered in a satisfactory way, if at all.[1] Why did Athenian painters invent a new drawing technique, the red-figure technique, which was far more laborious and difficult to achieve than the earlier and long-established black-figure? What prompted customers abroad to embrace a major technical innovation, such as the introduction of the red-figure technique, to the detriment of the black-figure vases to which they had been accustomed for centuries? Through which channels were the newly produced vases exported? Are we allowed to speak of trade networks, interdependent markets, or other meaningful patterns of export of vases from the Athenian metropolis to the receiving sites, throughout the Mediterranean?

The basic thesis of this study is that the pattern of distribution of attic vases in Greece and the Mediterranean[2] is not random, or biased by the excavation record, but that, at least to a certain degree, it depended on the existence of networks, meeting the clients' expectations and demands. In that respect, the role of middlemen is of paramount importance, not only in trade, but also in the transmission of information about the newest developments in offer and demand.

The red-figure technique appeared in Athens around 525 BCE. The figures were drawn in outline, with line for all interior detail of anatomy and dress, and the background was painted black. At first there was a very little incision, for details of hair and the like, but it was soon abandoned, as was any added colour detail on dress. 'The brush had taken over from the graver, and superficially the result looked like a reversal of the black-figure'.[3] Indeed, the painter was achieving his red figures by painting the background and leaving the figures reserved in the natural red color of the attic clay. Inner details of body musculature were marked by a thinner slip, the so-called dilute glaze line. The red-figure technique is more laborious and riskier than the earlier, black-figure technique. It was required from the painter to know exactly what he wanted to draw, before putting in the background in black gloss, since mistaken lines might be difficult to erase. A notoriously difficult detail was the so-called relief line, a three-dimensional line achieved by using a thicker mixture of slip than usual, applied with a very thin brush.[4] In terms of *energy expenditure*, the red-figure technique was an expensive experiment.

Scholars place the invention of attic red-figure into the context of the fervent atmosphere of experimentation that dominated in the Athenian Kerameikos during the period between 530 and 520, and led to the invention of innovative techniques such as the white ground, the coral-red ground, the outline on white ground and the so-called Six-techniques.[5] However, the red-figure technique was not just another one among a series of innovations; from the first moment of its invention, it was a very successful experiment, which brought dramatic changes to the organization of the ceramic production in Athens.

The *communis opinio* considers the invention of the red-figure technique as a necessary innovation in order to revitalize the earlier, black-figure technique.[6] Robin Osborne sees this revolution 'in the context of this [i.e., the painters'] desire to forge a more intimate relationship between the viewer and the scene viewed. The use of black background kept the stark contrast between figure and background that rendered gesture so readily legible'.[7] A more attractive view was expressed recently by Richard Neer: the red-figure technique gained favour because it made possible an especially rich, ornate pictorial style, comparable to the '*poikilon ethos*' of sympotic poetry.[8]

However, one is tempted to object that the aforementioned achievements are visible not to the work of the earlier painters, such as the Andokides Painter, Psiax, Oltos and Epiktetos, who did not 'use the possibilities of the new technique to the full', but to the work of the following generation of painters, the so-called Pioneers.[9] It is rather unlikely that the inventors of the new technique were visionaries who, although unable to explore all the potential of the new technique, had in mind future developments. Consequently, the aesthetic motives should not be overemphasized.

I will argue, instead, that the success of the new technique lay in the fact that it met the expectations of foreign, overseas customers.[10] Five arguments may be adduced:

1. The wide distribution of attic red-figure pottery is inextricably linked to its ultimate success in supplanting all other figured styles within a period of a generation. Eretrian, East Greek, Laconian, 'Etrusco-Ionian' and Chalcidian black-figure were extinct by 500 or a little later,[11] and only the Etruscan, Campanian and Boeotian black-figure workshops survived, albeit in a marginalized form.[12]
2. The monopoly that the painters of the red-figure vases enjoyed for a period of almost three quarters of a century is certainly an important supplementary factor for the enormous success of the technique in overseas markets. To copy an attic red-figured vase was not an easy task. The earliest imitations come from Etruria, during the first quarter of the fifth century, and are aesthetically unpleasing: the contours are confused, the glaze is of poor quality and the vases are misfired.[13] The span between the invention of the technique in 525 and the appearance of the first competent non-Athenian red-figure is remarkably long: truly satisfactory copies were produced in Boeotia and South Italy, only when Athenian-trained potters and painters were involved in the effort, shortly before the middle of the fifth century.[14]
3. Attic red-figure vases were enthusiastically received in Italy, especially in Etruria and Campania. Customers in mainland and insular Greece, even at the home market in Attica, were less willing to abandon the customarily black-figure vases in order to embrace the unusual innovation of the red-figure. This situation became evident very soon after the invention of the technique: since the earliest phase of the development of the red-figure, the vast majority of early red-figure vases were exported to Etruria. This diffusion of early red-figure to the West continued until approximately the second quarter of the fifth century BCE, when the market in Attica, as well as in other markets to the West and to the East, grew in importance.

   Since the situation in the West has often been the subject of detailed examination, I will present only a very broad outline of the finds.[15] Most scholars are willing to assign the use of vases found in Etruria and Campania to the local élites, not least in the light of iconographic parallels with funerary tomb paintings from Tarquinia, on which, incidentally, painted pots are represented in feasts of the members of Etruscan aristocracy.[16] However, the recent and thorough study on the presence of Attic vases in Etruria has demonstrated, beyond reasonable doubt, that Athenian vases are omnipresent in Etruria, not just in coastal, wealthy sites, in major sanctuaries and in aristocratic chamber-tombs, but virtually at all social levels.[17] Even if it is impossible to provide a trustworthy sample for an accurate statistic evaluation of the presence of early Attic red-figure in Etruria and generally in central Italy, we may easily conclude that the vast majority of the material had been deposited in Etruscan tombs.[18] As a rule, lesser, inland and remote sites were not rich in finds; late black-figure and local imitations took the lead.

Scholars have not yet felt the necessity to study closely the distribution of Athenian pottery within the Greek mainland.[19] Consequently, a detailed and accurate picture of the distribution of Attic vases in continental and insular Greece is lacking. The mechanisms through which Athenian vases were diffused in the Greek mainland and in the Aegean are not properly understood. One of the reasons for this was undoubtedly the lack of interest by modern scholars in measuring the impact of Athenian pottery upon Greek sites. That new and old excavation finds are not always published adequately, if at all, is an additional feature blurring our picture of the distribution of early Attic red-figure pottery in Greece. This does not necessarily mean, however, as is sometimes argued, that a general evaluation cannot be achieved.[20]

When examining the range of Athenian figured pots from excavation sites throughout Greece, time and again we encounter a uniform pattern: black-figure is modestly represented before 500 BCE, it becomes prolific during the first half of the fifth century BCE and is directly replaced by Classical red-figure vases of base quality after the middle of the century.[21] Greek customers were generally less willing to purchase the vases painted in the new, red-figure technique. The number of sites where early red-figure pottery is well represented is limited (Corinth, Eretria, Aegina, Samos, Thasos and Rhodes). These sites were either close to Athens and/or important commercial centres maintaining strong ties with both eastern and western markets. This fact underlines the importance of maritime trade in the transport of early red-figure vases. In other Greek-speaking areas—namely, Asia Minor—a paradox occurs: one encounters fewer red-figure vases in Greek cities like Miletus or Ephesos than in Persian Daskyleion, in Hellespontine Phrygia, or in Lycian Xanthos. We should single out the finds from Clazomenae, which are dated to the period before the abandonment of the city in 494 BCE.[22] In other parts of the Achaemenid empire, one might single out the massive presence of vases by the Pithos Painter in the Levant, and less so in Cyprus and Carthage, but otherwise, the penetration of early red-figure material was very limited.[23] In the Black Sea Region, early red-figure vases, mainly cups, were quite popular.[24]

4. The study of shapes shows that early red-figure vases were in the main oriented towards exports. Indeed, the early period of the new technique displays a rather limited range of shapes. More than eighty per cent of the vases known today are cups. This extraordinary one-sidedness is, at least partially, due to the fact that in the later sixth century, storage, mixing and pouring vessels are still produced in the old, black-figure technique. The second, and to my mind, more important factor is the wide range of activities associated with cups: the kylix is not simply a drinking vessel used at the symposion. It is also widely utilized for libations, or in cultic banquets inside sanctuaries, while it was popular as a grave offering. Finds of cups from Greece, Thasos, and the Black Sea area were mostly confined to sanctuaries. On the contrary, in Italy, cups were more popular in tombs. The sanctuary at Gravisca was apparently an exception, since it was usually frequented, at least in the sixth century BCE, by Greeks, as is plainly shown by the inscriptions scratched on dedicated pots.[25]

More uniform is the distribution of eye-cups, the most popular form of drinking vessel in the later sixth century BCE: whether bilingual or red-figure, eye-cups are rare in Greece, in Asia Minor, in the Levant, in Northern Africa, in Sicily and in the West.[26] Finds are prolific in Etruria, in Campania and in Latium; there are also seven specimens from coastal Black Sea sites.[27] Cups in the coral-red technique from the Kachrylion workshop were more widely distributed: cups from the Black Sea, Thasos and Samos are matched with finds from Athens, Corinth and Hermione, even if the vast majority of finds come from Etruscan and South Italian sites.[28]

The distribution of shapes through mainland Greece offers a different picture. Approximately one hundred early red-figure mugs survive, of the so-called oinochoe 8A type, and these are mostly painted by a single painter, the Painter of Berlin 2268.[29] This shape enjoyed an extraordinary diffusion (finds occur in mainland Greek sanctuaries—at Olympia, at Perachora, at Isthmia, at the Cabirion at Thebes—as well as in North African sites—at Naucratis and at Cyrene).[30] Mugs were rare in Etruria (except from Gravisca, where four specimens have been found in the Greek sanctuary and one in the new sacred deposit, which was excavated recently),[31] and are absent from the Black Sea Region and from Thasos. Alabastra, aryballoi, and lekythoi are found only occasionally in Etruria, Thasos, and the Black Sea Area, but are present in sanctuaries and tombs all over mainland and insular Greece.[32] Cups were very popular (there are almost a hundred from Greek sites), more often dedicated at sanctuaries than buried in tombs. Closed vessels and kraters in the new technique were rare, until approximately the late sixth century BCE. Kraters, amphorae, pelikai, stamnoi and hydriae are usually found in Etruscan tombs (with hydriae more often found at Vulci, because of the peculiar custom of being used there as mixing bowls),[33] while stamnoi and neck-amphorae, especially of the earlier fifth century BCE, are especially popular in Campanian sites.[34] These shapes are not unknown Greek sites, but they were rare prior to 500 BC.

5. Robin Osborne investigated in a recent study whether the organization of Athenian pot production by workshops affected the nature of the pottery output or its marketing and distribution.[35] He was led by his case study, on the Penthesilean workshop, to a negative answer. However, one might be cautious in generalizing from so limited a range of examples. It may, in fact, be argued that the situation was quite different during the earliest phases of the red-figure technique. It is a commonplace, for instance, that the earliest red-figure vases, painted by the Andokides Painter, are wholly absent from Athens and Attica, but were massively exported to Etruria.[36] Red-figure vases from the Nicosthenic workshop are attested in three regions: in Etruria, in Thasos and in the Black Sea.[37] By contrast, all black-figure vases signed by Nikosthenes have been exported in Etruria, the vast majority being the so-called Nicosthenic amphorae, which were especially designed for export.[38] The work of Psiax, a junior contemporary of the Andokides Painter, is not represented in Thasos, but some of the few red-figure vases decorated by him have been directed to the Black Sea Region.[39]

The question of the 'consumption' of vases and their imagery has been deliberately left aside. The general consensus is that vases assumed different functions in Etruria from those they held in Greece and in the Black Sea Area. It is tempting to conclude that images also had different meanings. Attic ceramic images were adaptable to the value systems of foreign consumers.[40] However, the uniform diffusion of vases of given shapes, sometimes produced in a limited number of workshops, might be taken as an indication of a certain degree of interdependence.[41]

Middlemen might have played a crucial role in the diffusion of products of certain pottery workshops. There is no question of commercial agents acting on behalf of the owners of the workshops, nor of members of local élites who arrived at Athens in order to buy vases on behalf of their compatriots. Literary and epigraphical sources, such as the graffiti engraved beneath the bottom of Athenian vases, indicate that Aeginetan and Ionian sailors played a crucial role in the commercialization of Athenian pottery before 490 BCE, in both Italy and the North.[42] The abundant material from Gravisca, with more than 600 fragments of early Attic red-figure vases, mostly cups, which had been dedicated by the late sixth/early fifth century BCE both by Greek (Aeginetan and Ionian) traders and by Etruscan customers, shows the relative importance of pottery trade.[43]

Notes

[1] I would like to thank the organizers of the conference for their invitation and Professors Martin Bentz, Bettina Kreuzer and Jean-Jacques Maffre for invaluable information on unpublished finds from Olympia, Samos and Thasos respectively. Thanks are also due to the anonymous reader, who provided a detailed and critical commentary of my views. All dates are BC, unless otherwise stated. *Add*^{2} = Carpenter, Thomas H., *Beazley Addenda*^{2}. *ARV*^{2} = Beazley, *Attic Red-Figure Vase-Painters*^{2}. *CVA* = *Corpus Vasorum Antiquorum*. *Para* = Beazley, *Paralipomena*.

[2] Stissi, 'Production, Circulation and Consumption', *passim*. See also Johnston, 'Greek Vases'. For a general picture of the distribution of attic vases, see Boardman, 'The Athenian Pottery Trade'.

[3] Boardman, *History of Greek Vases*, 79. For the dating of the invention in 525, see Cohen, *Attic Bilingual Vases*, 113, and Pécasse, 'Le trésor de Siphnos', 309–14. For revisions of the established chronological system (which depends on Langlotz, *Zeitbestimmung der strengrotfigurigen Vasenmalerei und der gleichzeitigen Plastik*), see below, note 8.

[4] On the technical requirements of the red-figure style, see Noble, *Techniques of Attic Painted Pottery*, 116–21, and the recent account of Cuomo di Caprio, *Ceramica in Archeologia 2*, 462–65.

[5] Williams, 'The Invention of the Red-Figured Technique', 103, Robertson, *Art of Vase-Painting*, 8–9, and Sparkes, *The Red and the Black*, 17. Experimental and special techniques: Cohen, *The Colors of Clay*.

[6] Smith, *Catalogue of Vases in the British Museum*, III, 1, Cook, *Greek Painted Pottery*, 155, Boardman, *Athenian Red Figure Vases*, 11. On the other hand, Furtwängler, in his review of Hartwig's *Griechische Meisterschalen*, in *Berliner Philologische Wochenschrift* (1894), col. 112 and Lane, *Greek Pottery*, 45, traced the origin of the style to the influence of contemporary wall- and panel-painting. Klein, *Euphronios*, 31, assumed that the technique evolved from the Gorgoneia of black-figure cups.

[7] Osborne, *Archaic and Classical Greek Art*, 136–37, echoing Villard, *Les vases grecs*, 10–11.

[8] Neer, *Style and Politics*, 14–23. In this important and innovative study, Neer (186–215) proposed a synchronism between the beginning of activity of the so-called Pioneers and the birth of democracy, resulting from the downdating of the beginning of the career of Euphronios

from 520 to 510 BCE. This was an ingenious proposition, based on the study of the early red-figured vases from the Agora wells. But the radical revision of the Pioneer Group is not necessary: Paleothodoros, Review of Richard. T. Neer, 405–11. Other scholars have attempted to revise the established chronology previously: Tölle-Kastenbein, 'Bemerkungen zur absoluten Chronologie'; Canciani, 'Considerazioni'; and ultimately Rotroff, 'Early Red-Figure in Context'. Rotroff observed the absence of red-figure vases in the deposits of the Athenian Agora that might be dated, according to the traditional chronological scheme, before 515 BCE. On this basis, she downdated the invention of the red-figure technique by about fifteen years. The argument, based on the 'visibility' principle, ignores the stylistic development of the early red-figure technique, since a few very early red-figure vases are indeed found in the Agora: Moore, 'Attic Red-Figure Painters', 469; see also below, note 36. The overall situation of the finds from Agora—which are comparable to the finds from the Acropolis, since in both sites earlier vases are rare and red-figure vases are less common than black-figure ones—lend further support to our thesis that the earlier red-figure pottery was much less successful in the home market than abroad. See Hannestad, 'The Athenian Potter', 224–27.

[9] The quotation is from Beazley, *Development of Attic Black-Figure*, 76. See also Villard, *Les vases grecs*, 70; Sparkes, *Greek Pottery*, 98.

[10] Cook, *Greek Painted Pottery*, 155, argues that the technique is unrelated to the needs of the market, since the production and export of black-figure vases did not stop after the invention of the red-figure. *Contra*, Boardman, *History of Greek Vases*, 79: 'The motivation was mainly aesthetic, no doubt, but it also represented a new line for the growing export market, and the response was more immediate in Athens and Italy (Greek and Etruscan) than in the rest of Greece.'

[11] Cook and Dupont, *East Greek Pottery*, 89, 94, 101–07 and 108–13. The end of the East Greek vase-painting styles is linked to the troubled events of the Ionian Revolt (499–4), but the decline is apparent already before 500 BCE. Local painted vases have been supplanted by Athenian imports, although the art of painting on clay survived in the decoration of the Clazomenian sarcophagi of the second quarter of the fifth century BCE. Northampton amphorae, Caeretan hydriae and Campana dinoi were painted until about 510–500 BCE. Note that Hemelrijk, 'Four New Campana Dinoi', 377–78, now regards the Campana dinoi and the Northampton group as East Greek products. Laconian and Euboan in Greece, as well as Chalcidian black-figure in South Italy, last until about 500: Boardman, *Early Greek Vase Painting*, 271.

[12] The production of Etruscan black-figure diminished in the early fifth century BCE and died out before the middle of that century. Rizzo, 'La ceramica a figure nere', 39. Campanian black-figure derives from later Etruscan: Parise Badoni, *Ceramica campana*, 133–39. Later Boeotian black-figured vases (other than the highly idiosyncratic cabirian) were crude imitations of contemporary Attic: Maffre, 'La collection Paul Cannellopoulos VIII', 496.

[13] The earliest copy is an amphora from Orvieto: Cristofani, 'La ceramica a figure rosse', 43. Soon, the painters realised that the technique of added red colour was easier to achieve. The inventor of the technique was a native of Greek origin, Arnth Praxias: Szilágyi, 'Zur Praxias-Gruppe'. The signature of Praxias is in lakonian alphabet betraying attic influence. He might be a Tarentine, according to Wachter, *Non-Attic Greek Vase Inscriptions*, 194–95. A single find, an Attic red-figure fragment of the 440s with an Etruscan painted inscription, implies the presence of Athenian painters in Etruria, albeit in a very limited number: Gill, 'METRU MENEKE'.

[14] Boeotian red-figure first appears after 480 BCE. See the pelike Munich 2347 (Sparkes, 'The Taste of a Boeotian Pig', 123); Lullies, 'Zur böotisch rotfigurigen Vasenmalerei'. The establishment of the Lucanian workshop around the middle of the fifth century BCE is attributed to the activity of the Pisticci Painter, who was trained in Athens: Denoyelle, 'Attic or non-Attic?', with full bibliography. An independent atticizing school appeared by the end of the second quarter of the fifth century in Campania. One need not infer continuity from the earlier, Campanian black-figure, as was advocated by Hadzisteliou-Price, 'Αμφορέας τύπου Nola ἐν Σικάγωι'.

On the emigration of Athenian potters, see McDonald, 'The Emigration of Potters', who overemphasizes the effects of the Peloponnesian War upon the Athenian pottery industry.

[15] I do not intend to present a systematic review of earlier scholarship. For studies until 2001, see Paleothodoros, 'Pourquoi les Étrusques'. Later publications include the major study by Reusser, *Vasen für Etrurien*, and two important studies by R. Osborne: 'The Anatomy of a Mobile Culture', which is an expanded form of his 'Why did Greek Pots Appeal to the Etruscans?', and 'Workshops'. See also the papers in Giudice and Panvini, *Il Greco*, vols II and III, Bentz and Reusser, *Attische Vasen*, de La Genière, *Les clients de la céramique grecque*, and Avramidou, 'Attic Vases in Etruria'.

[16] Arafat and Morgan, 'Athens, Etruria and Heuneburg', 117. Vases depicted on tombs: see recently Reusser, *Vasen für Etrurien*, I, 191–202 and II, 101–18. Vases found in painted tombs: Wiel-Martin, 'Vasi reali'. Note that Small, 'Scholars, Etruscans, and Attic Painted Vases', 38, has suggested that the painted vases depicted on Etruscan tombs are of Etruscan origin.

[17] Reusser, *Vasen für Etrurien*: no less than 80 domestic sites and 40 sanctuaries, as well as the vast majority of tombs from the sixth and fifth century, have yielded amounts of Athenian pottery. For domestic and sacred contexts, see the papers in Bentz and Reusser, *Attische Vasen* and the publication of the finds from Adria by Wiel-Martin, *La ceramica attica*.

[18] The available studies dealing with the statistical aspect of the distribution of attic pottery are not trustworthy. The number of vases from illegal and uncontrolled excavations available in the market, but in reality found in Etruria, is very important. Earlier finds are also poorly recorded. On sampling, see Stissi, 'Modern Finds and Ancient Distribution'; Paleothodoros, 'Pourquoi les Étrusques', 140–41; and Morgan, *Attic Fine Pottery*, 21–23.

[19] Exceptions should be noted: Hannestad, 'The Athenian Potter' and 'Athenian Pottery at Corinth'; Villanueva-Puig, 'Les vases attiques'.

[20] We should also keep in mind that the situation in Italy, Cyprus and Turkey is not necessarily better than in Greece, while in the countries surrounding the Black Sea, things are certainly worse.

[21] See Appendix.

[22] Tuna-Nörling, 'Archaische attische Keramik', 97 (a cup from Old Smyrna), 100 (a cup from Phokaia), 101 (eye-cups, several late-sixth-century cups by Epiktetos and the Wider Circle of the Nikosthenes Painter and an amphora of the earlier fifth century from Clazomenae), 102 (an eye-cup from Miletus, as well as a later cup and an askos postdating the destruction of the city in 494), 102 (cup by the Pithos Painter from the Ephesian Artemision). Daskyleion: Tuna-Nörling, *Daskyleion I*, nos. 379–80 and 'Attic Pottery from Dascylium', 119, fig. 10 and 122, fig. 22. Xanthos: Metzger, *Xanthos IV*, pl. 74-75, no. 333. Sardis: Snyder Schaeffer, *Corinthian, Attic and Lakonian Pottery*, no. Att 126, pl. 42. Gordion: De Vries, 'Attic Pottery from Gordion', 448, fig. 4.

[23] Paleothodoros, 'The Pithos Painter', with addenda in Paleothodoros, 'Η διάδοση της πρώιμης ερυθρόμορφης', n. 18.

[24] See now Paleothodoros, 'Η διάδοση της πρώιμης ερυθρόμορφης', with a list of 80 early red-figured vases, superseding earlier statistics based on Beazley's indexes and listing a very limited number of vases. For a detailed picture of the imports of Athenian vases in a single area of the Black Sea Region, the Taman Peninsula, see Morgan, *Attic Fine Pottery*, 149–235.

[25] Johnston and Pandolfini, *Le iscrizioni*.

[26] Athens: *ARV*^{2} 1625 and Agora P 9414; Moore, *Attic Red-Figured and White Ground Pottery*, 316, pl. 129, no. 1400. Argive Heraion: *ARV*^{2} 71.5. Selinus: *ARV*^{2} 47.146. Miletus and Clazomenae: Tuna-Nörling, 'Archaische attische Keramik', 100 and 101. Sardis: Snyder Schaeffer, *Corinthian, Attic and Lakonian Pottery*, no. Att 126, pl. 42. Al Mina: *ARV*^{2} 43.66, 47.147 and 148. Naukratis: *ARV*^{2} 42.53, 43.69, 47.144; Möller, *Naukratis*, 236–7. Marseille: Villard, *La céramique grecque de Marseille*, 28, pl. 15.1–2.

[27] See *ARV*^{2} 37-52; Cohen, *Attic Bilingual Vases*, 240–522. Black Sea: Paleothodoros, 'Η Διάδοση της πρώιμης ερυθρόμορφης', 53–75; Hupe, *Der Achilles-Kukt*, pl. 11.2.

[28] Vases by Cachrylion: *ARV* ^{2} 107–08. There are two cups from the Black Sea Region: *ARV* ^{2} 17.20 and *ARV* ^{2} 108.25. Thasos: Maffre, 'Cachrylion, Euphronios', 380, fig. 1 and 382–83, fig. 2–4. There is one from the Artemision of Samos, but without the signature of the potter Chachrylion, who, however, signs a red-figure one from the same site (Bettina Kreutzer, pers. comm.). Corinth: McPhee, 'Attic Red-Figure from the Forum', 298–9. Athens: Serbeti, 'Ερυθρόμορφη κύλικα από την οδό Λέκκα'; Moore, *Attic Red-Figured and White Ground Pottery*, 316, pl. 129, no. 1403; McCamp, 'Excavations', 231, pl. 76, no. 36.

[29] A thorough study of all figured mugs is under preparation by the present author. A total of 230 vases have been collected, mostly dating from the period 510–450 BC.

[30] M. Bentz, in a personal communication, generously provided a list of nine mugs from Olympia: Four belong to the period under discussion here. Perachora, Isthmia and the Theban Cabirion, see below, note 55. Naucratis: *ARV* ^{2} 157.80. Cyrene: McPhee, *Cyrene* VI, pl. 33, nos. 78–80.

[31] Huber, *Le ceramiche a figure rosse*, 140, nos. 777–79 and 152–53, no. 883; Fortunelli, 'Anathemata', pl. II c–d. Mugs are rare in Etruria: *ARV* ^{2} 77.97 (Orvieto), 156.51 (Bologna), 157.73–4 and 78 (Tarquinia). Finds from Campania (*ARV* ^{2} 156.53, 156.55, 157.77, 158.3, 983.10 [Capua], 156.67 [Cumae], 156.62 and 158.1 [Czartoryski collection], 157.75 [Suessula], S. Italy and Sicily (i.e. *ARV* ^{2} 152.1, 156.59-60, 63, 67, 157.79bis, 1629.67bis, *Add* ^{2} 405) are far more numerous.

[32] Alabastra: Badinou, *La Laine et le parfum*, 155–218. The shape is second only to the cup in popularity in Greece: *ARV* ^{2} 7.4, 8.13, 98.2, 99.3, 99.4, 99.9, 99.10, 100.15, 100.16, 100.23, 100.24, 100.26, 157.87, 157.88. There is a single find from South Russia: *ARV* ^{2} 7.5. Red-figure lekythoi and aryballoi are very rare before circa 480 BCE outside Attica. Note a lekythos from Thasos (Maffre, 'Chachrylion, Euphronios', 386, Fig. 8) and an aryballos from Corinth (Boulter, 'Fifth Century Attic Red-Figure', pl. 77a–b).

[33] Reusser, 'La céramique attique', 173.

[34] Stamnoi: Rendeli, 'Rituali e immagini'. Neck-amphorae: Martelli, 'Arete ed Eusebeia'.

[35] Osborne, 'Workshops', 78.

[36] Athens: a single vase from the Agora is placed in the neighbourhood of the workshop of the Andokides Painter: Moore, *Attic Red-Figured and White Ground Pottery*, 139, no. 24 ('recalls the Andokides Painter more or less vividly'); two more (nos. 164 and 1684) are dated to 520. Some slight pieces, attributed by Beazley to the 'the manner of the Andokides Painter', have been removed by Cohen, *Attic Bilingual Vases*, 511. Indeed, they are primitive but not early, and date to the end of the sixth century. Etruria: *ARV* ^{2} 3.1, 2, 3, 5; 4.7, 8, 9, 10, 11, 12, 13; 5.14. A single piece has been found at Locri in South Italy: *ARV* ^{2} 3.6.

[37] Among ten red-figure vases signed by Nikosthenes, a kantharos, also signed by Epiktetos as painter, was found in the sanctuary of Achilles on the island of Leuke, off Berezan (*ARV* ^{2} 77.87), a skyphos by the Nikosthenes Painter was found at the Artemision of Thasos (*ARV* ^{2} 1627.25bis), and the remaining kantharoi, round pyxides and cups all come from Etruria (*ARV* ^{2} 123).

[38] Tosto, *The Black-Figured Vases*.

[39] An alabastron (ARV^2 7.5) and a cup (ARV^2 8.10).

[40] Ruystedt, 'Athens in Etruria', 505.

[41] The anonymous reader has objected that trademarks are found almost exclusively on pots exported to Etruria, while they are totally absent on vases found in the Black Sea Area. However, Johnston, *Trademarks on Greek Vases*, 18, notes similarities between trademarks from the Black Sea Area (where they are rare) and Vulci.

[42] Graffiti: Johnston, *Trademarks on Greek Vases*. Other epigraphic data: Johnston, 'The Rehabilitation of Sostratos' and Cristofani, 'Sostratos e dintorni'. Aeginetan merchantmen in the Black Sea Region: Hind, 'Traders and Ports-of-Trade'.

[43] Osborne, 'Pots, Trade and the Archaic Greek Economy', against Gill, 'Pots and Trade'.

[44] Finds from the Acropolis and the Agora are numerous. Agora: Moore, *Attic Red-Figure and White Ground Pottery*; Mc Camp, 'Excavations', 245–52. Acropolis: Langlotz, *Die Antike Vasen*; see also Hannestad, 'The Athenian Potter', 224–26. Kerameikos: Knigge, *Kerameikos IX*, pl. 28.1 (cup fragment by Smikros); *Archäologische Anzeiger* (1987), 498, fig. 24 (skyphos fragment by Epiktetos); *Éros grec, Amours des Dieux*, nos. 37, 38 and 47 (two alabastra and a lekythos); add also the unpublished finds from tomb SW 30 (a mug by the Painter of Berlin 2268 and a psykter of early fifth century date) in the Kerameikos Museum. A single vase from the of the Athenian Metro excavations has been published: Serbeti, ''Εργο του αγγειογράφου'. Note also a child tomb from Ilioupoli with an early red-figure alabastron: Pologiorgi, 'Παιδική Ταφή στην Ηλιούπολη'; other finds: Tzachou-Alexandri, 'Αλάβαστρο του Ζωγράφου', 85–98.

[45] Papaspyridi, 'Ελευσινιακά Αγγεία', 3–11, figs. 1–12 (ten early red-figure plates, skyphoi, cups and alabastra); Kokkou-Byridi, *Ελευσίς. Πρώιμες πυρές θυσιών*, pl. 36 (three vases from the late sixth century); Tiverios, 'Αγγεία-Αναθήματα'.

[46] Kahil, 'Quelques vases'. Early red-figure is lacking from Piraeus, Sounion, Thorikos, Marathon and Acharnai and in the minor cemeteries of the Attic countryside regularly published through the pages of the *Archaeologicon Deltion*. Stray finds are reported from Velanideza (*ARV*^{2} 110.5), cape Kolias (*ARV*^{2} 20), Koropi (*ARV*^{2} 157.86). Unspecified finds from Attica: *ARV*^{2} 22.4 and 99.7. An Attic provenance is also the most likely for two cups, one belonging to the Wider Circle of the Nikosthenes Painter and a second one having been attributed to the Pithos Painter (Paleothodoros, 'The Pithos Painter', 72, no. 102; Sabetai, *CVA Athens*, pl. 53–54).

[47] Eretria: *ARV*^{2} 98.15, 99.5; Gex, *Eretria IX*, pl. 56, nos. S178 (psykter) and S179 (column-krater), 59, nos. S194–5 (calyx-kraters), 88, nos. S378–85 (cups) and 99, nos. S450–52 (fragments of pots). Another important find is a deposit including an early red-figure amphora of type A, perhaps from the Andokides workshop: Serbeti, 'Attic Pottery', 494, fig. 5. No finds are reported from Chalkis and Karystos.

[48] Aegina should be singled out as a very important site in Greece where early red-figure pottery has been unearthed: a cup by Epiktetos picturing Athena (*ARV*^{2} 74.51) and a white ground showing Europa on the bull are the earliest noticed attic vases being found in a controlled excavation in Greece. They were excavated by a Bavarian team in 1811 and are cited for the first time in a letter of Wolff to Eduard Gerhard published in the *Bollettino di Corrispondenza Archeologica* of 1829 (Gerhard, 'Riposta di Prof. Gerhard', 118-9, n. 2). Later on, in 1830, the consul of France in Naples, Count Beugnot, settled on the island and excavated many tombs, whose contents enriched his personal collection: see Gerhard, 'Vasi dipinti dalla Grecia', 196-7. The cup by Epiktetos, and a fragment by the same painter, are the only red-figure vases from Aphaia dating before 500 BCE: Williams, 'Aegina, Aphaia-Tempel XI', 630, nos. A1 and A2. Temple of Apollo in Kolona: Felten, 'Attische schwarzfigurige', 48–49, pl. 21, nos. 271–79. Finds from cemeteries: *ARV*^{2} 36.1-2, 223.6.

[49] Isthmia: *Hesperia* 24 (1955) pl. 52a, 19 (a mug). Nemea: *Hesperia* 47 (1978) pl. 20b (cup fragment). Perachora: Dunbabin, *Perachora* ii, pl. 146, nos. 3831, 3834–6. Acrocorinth: Pemberton, *Corinth XVIII, Part 1*, pl. 38, no. 334, pl. 43, no. 368. Corinth: Boulter, 'Fifth Century Attic Red-Figure'; McPhee, 'Red-Figured Pottery from Corinth' and 'Attic Red-Figure from the Forum in Ancient Corinth' (approximately twenty early red-figure vases). Note also cups in Athens (*ARV*^{2} 95.122 and 140.36), Boston (*ARV*^{2} 179.1), the Louvre (*ARV*^{2} 176.1) and Berlin (*ARV*^{2} 176.2 and 177.1), said to be from Corinth. A cup by the Pithos Painter in Nantes is from Hexamilia: Paleothodoros, 'The Pithos Painter', 69, no. 16.

[50] The red-figure vases from Olympia are now studied by Prof. Martin Bentz. For an overview, see Bentz, 'Attic Red-Figure Pottery from Olympia' (there are only seven early red-figure vases). Argolid: note two cups from Hermione (*ARV*^{2} 1543) and an eye-cup from the Heraion (*ARV*^{2} 71.5).

[51] The red-figure vases from the Heraion of Samos will be published by Prof. Bettina Kreuzer in a following volume of the series *Samos*. Thanks to her generosity, I present here an overview: there are 192 pieces (three of them now in Berlin), mostly dating to the second quarter of the fifth and the first quarter of the fourth century. The earlier finds include a krater by Euphronios, one by Euthymides and four others; a stamnos; ten cups, among which one might note a cup signed by Chacrylion, one in white ground and another one with coral red; there is also an early red-figure plate. A single vase from the Artemision is published: *Αρχαιολογικά Ανάλεκτα εξ Αθηνών* 13.2 (1980): 314, fig. 9. Finds from tombs are somewhat later in date, and may be safely placed to the second quarter of the fifth century: Deltion 43 B 2 (1988): pl. 289b-291.

[52] Delos and Rheneia: *ARV*[2] 133.20, 141.57, 143.19 and 172.3.

[53] *ARV*[2] 80.11, 86b; 139.8, 139.9, 139.23, 140.26, 157.71 (Kameiros).

[54] *ARV*[2] 81.3 and 177.1.

[55] Tanagra: *ARV*[2] 25.1, 83.9, 83.12, 176.1; *Corpus Vasorum Antiquorum Athens* 1, pl. 2.1, 3, 5; 3.1-3. Akraiphia: *ARV*[2] 130.31 and Andreiomenou, 'Το εργαστήριον χαλκοτεχνίας της Ακραιφίας', 487, fig. 15, 488–89, n.48 (two cups tentatively attributed to Epiktetos); Rhitsona: *ARV*[2] 140.29 and 302.10. Thebes: *ARV*[2] 85.2 and Thebes 23425 (Sabetai, *CVA Thebes*, 1, pl. 73). Kabirion: *ARV*[2] 156.65.

[56] The absence of early red-figure and of painted pottery in general from Delphi is a well-known fact. Note a white ground alabastron from a tomb of the early fifth century (*ARV*[2] 98.2).

[57] A cup in Boston (MFA 01.8074) is said to be from Locris (*ARV*[2] 76.74; Paleothodoros, *Epictétos*, 168, no. 150). There is an unprovenanced cup by the Pithos Painter in the Lamia Museum (Paleothodoros, 'The Pithos Painter', 69, no. 18). Atalanti: *ARV*[2] 100.26 and 178.3.

[58] In Thessaly, red-figure vases of the later sixth and the early fifth century are very rare: a cup by the Euergides Painter in Volos (*Add*[2] 395). Later finds include a fragmentary krater from Krannon by the Syleus Painter (*ARV*[2] 1640), a fragment of a cup of the early fifth century from the sanctuary of Athena Itonia (Katarrachias *et al.*, *Αρχαιολογικά Ευρήματα Φίλιας*, 29), three vases of the second quarter of the fifth century from Palaiokastron (*Archaeologikon Deltion* 28 [1973] B2, pl. 284a) and unpublished sherds from a house in Pharsala. A pair of early white-ground lekythoi, dating from about 470–60, have been excavated by Maria Lakaki in Meliboia (Hagiokampos) and are now in the Archaeological Museum of Larissa. I wish to thank the excavator for valuable information on this important find which will be published soon.

[59] See Tzanakaki, 'Εισαγωγές αττικών ερυθρόμορφων': the earliest finds reported date from the second quarter of the fifth century BCE.

[60] From Corfu, Kanoni, one may note a cup which may be attributed to the Euergides Painter (*Archaeologikon Deltion* 29 [1973–4] B3, pl. 457[e]). Unpublished finds from the Mon Repos include red-figure pottery from the second quarter of the fifth century on. Note also two early red-figure cup fragments from Lefkas: *Archaeologikon Deltion* 48 B1 (1993), pl. 95ζ and 27, B2 (1972), pl. 418a.

[61] As far as published or exhibited pottery in museums is concerned, finds from Paros, Naxos, Amorgos, Thera, Mytilene, Lemnos, and Nisyros are very rare. On Paros and Naxos, see Bikakis, 'Archaic and Classical Imported Pottery'. Amorgos: Marangou, 'Céramique attique à Amorgos'. Red-figure vases are very rare in Kythnos, as Prof. Alexandros Mazarakis informs me, and none is earlier than the second quarter of the fifth century BCE. Kimolos: cup by a painter of the later sixth century (*Archaeologikon Deltion* 21 [1966], B2, pl. 410c). Psara: a mug with satyrs kneeling and holding wineskins that may be provisionally attributed to the Painter of Berlin 2268 (Vlachopoulos, *Αρχαιολογία του Αιγαίου*, 139, fig. 183). Chios: cup by the Pithos Painter (*ARV*[2] 141.70).

[62] A cup that might be provisionally attributed to Epiktetos or to a painter following him was found in Torone (presented by the excavator, Manthos Besios, in *Contacts et échanges technologiques entre Grecs et indigenes à la frontière des territories des colonies grecques (VIIIe–IIe*

s. av. J.-C.), École Française d'Athènes 15–17 mars 2007). In Akanthos, Sindos, Mikra Karaburun and Hagia Paraskevi the earliest red-figure vases date from the end of the first quarter of the fifth century BCE. In Chalkidiki, early red-figure vases occur in small numbers: *ARV*[2] 140.28 and 56, 141.67 (cups by the Pithos Painter from Vrastina Kalyvia and Olynthos). Argillos: Giroux, 'La céramique attique à figures rouges', 183–87 (a cup by the Pithos Painter, a late-sixth-century alabastron, a head vase and a fragment by Myson).

[63] Kahil, *La céramique grecque de Thasos*, pl. XLIV; Maffre, 'Chachrylion, Euphronios' and 'Coupes attiques': around 100 red-figure fragments have been found in the French excavations in Thasos. Most finds date to the later part of the sixth century and the beginning of the fifth century. Besides cups (around 20 fragments), there are alabastra, plates and lekythoi. Thrace: Para 336.88bis, alabastron from Galepsos). Bakalakis, *Ανασκαφή Στρύμης*, pl. 28, no. 7 (a late-sixth-century sherd from Stryme); Tüna-Nörling, 'Polyxena bei Hector Lösung', 44–46 (early-fifth-century krater from Rhaidestos). No early red-figure has been found at Samothrace: Dusenberry, *Samothrace 11*, 513.

References

Andreiomenou, A. 'Το εργαστήριον χαλκοτεχνίας της Ακραιφίας'. In *Καλλίστευμα. Μελέτες προς τιμήν της Όλγας Τζάχου-Αλεξανδρή*, edited by A. Alexandri and I. Leventi, 469–526. Athens: 2001.

Arafat, Karim, and Catherine Morgan. 'Athens, Etruria and Heuneberg: Mutual Misconceptions in the Study of Greek-Barbarian Relations'. In *Classical Greece. Ancient Histories and Modern Archaeologies*, edited by Ian Morris. Cambridge: Cambridge University Press, 1994.

Avramidou, Amalia. 'Attic Vases in Etruria: Another View on the Divine Banquet Cup by the Codrus Painter'. *American Journal of Archaeology* 110 (2006): 565–79.

Badinou. *La Laine et le Parfum*. Epinetra et alabastres. Forme, Iconographie et Fonction. Leuven & Leiden, 2003.

Bakalakis, G. *Ανασκαφή Στρύμης*. Thessaloniki: Peeters, 1967.

Beazley, John Davidson. *The Development of Attic Black-Figure*. 1st ed. Berkeley and LA: University of California Press, 1951.

———. *Attic Red-Figure Vase-Painters*. 2nd ed. Oxford: Clarendon Press, 1963.

———. *Paralipomena. Additions to Attic Black-Figure Vase-Painters and to Attic Red-Figure Vase-Painters*[2]. Oxford: Clarendon Press, 1971.

Bentz, Martin. 'Attic Red-Figure Pottery from Olympia'. In *Athenian Potters and Painters* II, edited by J. Oakley and O. Palagia. Forthcoming.

Bentz, Martin, and Christoph Reusser, eds. *Attische Vasen in etruskischen Kontext: Funde aus Häusern und Heiligtümern, CVA* Beiheft 2. Munich: C. H. Beck, 2004.

Bikakis, M. H. 'Archaic and Classical Imported Pottery in the Museums of Paros and Naxos'. Diss., University of Cincinnati, 1986.

Boardman, John. *Athenian Red Figure Vases. The Archaic Period. A Handbook*. London: Thames & Hudson, 1975.

———.'The Athenian Pottery Trade. The Classical Period'. *Expedition* 21 (1979): 33–9.

———. *Early Greek Vase Painting*. London: Thames & Hudson, 1998.

———. *The History of Greek Vases*. London: Thames & Hudson, 2004.

Boulter, Cedric. 'Fifth Century Attic Red-Figure at Corinth'. *Hesperia* 49 (1980): 295–308, pl. 77–90.

Canciani, F. 'Considerazioni sulla cronologia della ceramica attica dal tardo arcaismo allo stilo severo'. In *Studien zu Mythologie und Vasenmalerei. Festschrift Konrad Schauenburg*, edited by E. Böhr and W. Martini, 59–64. Mainz: Verlag Philipp von Zabern, 1986.

Carpenter, Thomas H. *Beazley Addenda*[2]. Oxford: Oxford University Press, 1989.

Cohen, Beth. 'Attic Bilingual Vases and their Painters'. Dissertation, University of New York, 1978.

Cohen, Beth, ed. *The Colors of Clay. Special Techniques in Athenian Vases.* Los Angeles: Getty Museum Press, 2006.

Cook, R. M. *Greek Painted Pottery.* 3rd ed. London and New York: Routledge, 1997.

Cook R.M., and Pierre Dupont. *East Greek Pottery.* London and New York: Routledge, 1998.

Cristofani, Mauro. 'La ceramica a figure rosse'. In *La ceramica degli Etruschi. La pittura vascolare*, edited by M. Martelli, 43–52. Navarra: Istituto Geografico di Agostino, 1987.

———. 'Sostratos e dintorni'. In *Céramique et peinture grecques. Mode d'emploi. Actes du colloque international. École du Louvre, 26–28 Avril 1995*, edited by M.-C. Villaneuva-Puig, F. Lissarrague, P. Rouillard, and A. Rouveret. Paris: La documentation française, 1999.

Cuomo di Caprio, Ninina. *Ceramica in Archeologia 2. Antiche tecniche di lavorazione e moderni metodi di indagine.* Rome: L'Erma di Bretschneider, 2007.

de La Genière, J., ed. *Les clients de la céramique grecque. Actes du colloque de l'Académie d'Inscriptions et Belles Lettres, Paris, 30–31 janvier 2004*, Cahiers du *Corpus Vasorum*. Paris: De Boccard, 2006.

Denoyelle, Martin. 'Attic or non-Attic? The Case of the Pisticci Painter'. In *Athenian Potters and Painters. The Conference Proceedings*, edited by J. H. Oakley, W. D. E. Coulson, and O. Palagia, 395–405. Oxford: Oxbow Books, 1997.

De Vries, Keith. 'The Attic Pottery from Gordion'. In *Athenian Potters and Painters. The Conference Proceedings*, edited by J. H. Oakley, W. D. E. Coulson, and O. Palagia, 447–55. Oxford: Oxbow Books, 1997.

Dunbabin, T. J. Perachora. *The Sanctuaries of Hera Akraia and Hera Limenia. Excavations of the British School of Archaeology at Athens, 1930–1933. II. Pottery, Ivories, Scarabs and Other Objects from the Votive Deposit of Hera Limenia.* Oxford: Clarendon Press, 1962.

Dusenberry, Elsbeth B. *Samothrace 11. The Nekropoleis. Catalogue of Objects by Categories.* Princeton, NJ: Princeton University Press, 1998.

Éros grec. Amours des dieux et des hommes. Athens and Paris: Greek Ministry of Culture, 1990.

Felten, Wassiliki. 'Attische schwarzfigurige und rotfigurige Keramik'. In *Ält-Ägina II, 1*, edited by H. Walter. Mainz: Verlag Philipp von Zabern, 1982.

Fortunelli, S. 'Anathemata: ceramici attici del nuovo deposito votivo di Gravisca'. In *Il greco, il barbaro e la ceramica attica, Attid el Convegno di Catania, 14-10 maggio 2001*, I, edited by F. Giudice and R. Panvini, 55–63, pl. I-II. Roma: L'Erma di Bretschneider, 2006.

Furtwängler, Adolph. Review of P. Hartwig, *Die griechischen Meisterschalen der Blütezeit des strengen rotfigurigen Stils*, Stuttgart and Berlin, 1893. In *Berliner Philologischer Wochenschrift* (1894): col. 105–14 and 141–47.

Gerhard, Eduard. 'Riposta di Prof. Gerhard all'antecedente lettera du S. E. il Principe di Canino'. *Bullettino di Corrispondenza Archeologica* 1 (1829): 116–29.

———. 'Vasi dipinti dalla Grecia'. *Bullettino di Corrispondenza Archeologica* 2 (1830): 196–7.

Gex, Kristine. *Eretria IX. Rotfigurige und weissgrundige Keramik.* Lausanne: Payot, 1993.

Gill, David W. J. 'METRU MENEKE. An Etruscan painted inscription on a mid-5th century BC red-figure cup from Populonia'. *Antiquity* 61 (1987): 82–7.

———. 'Pots and Trade: Spacefillers or Objets d'Art?' *Journal of Hellenic Studies* 111 (1991): 29–47.

Giroux, Hubert. 'La céramique attique à figures rouges'. In *Les clients de la céramique grecque. Actes du colloque de l'Académie d'Inscriptions et Belles Lettres, Paris, 30–31 janvier 2004,* Cahiers du *Corpus Vasorum*, edited by J. de La Genière, 53–57, 183–87. Paris: La documentation française, 2006.

Giudice, Filippo, and Rosalba Panvini, eds. *Il Greco, il Barbaro e la ceramica attica. Immaginario del diverso, processi di scambio e autorappresentazione degli indigeni. Atti del Convegno Internazionale di Studi, 14–19 maggio 2001, Catania, Caltanissetta, Gela, Camarina, Vittoria, Siracusa.* Vol. II: Rome, 2003; Vol. III: Rome: L'Erma di Bretschneider, 2006.

Hadzisteliou-Price, T. 'Αμφορεύς τύπου Nola εν Σικάγω και Επανεξέτασις του Owl-Pillar Group'. *Archeologike Ephemeris* (1974): 168–92 Pl. 44–57.

Hannestad, Lise. 'The Athenian Potter and the Home Market'. In *Proceedings of the 3rd Symposium on Greek and Related Pottery, Copenhagen August 31–September 4 1987*, edited by Jette Christiansen and Torben Melander, 222–30. Copenhagen: National Museum, Ny Carlsberg Glypotek, Thorvaldsens Museum, 1988.

———. 'Athenian Pottery at Corinth c. 600–470 BC'. *Acta Archaeologica* 61 (1992): 151–63.

Hemelrijk, Jan M. 'Four New Campana Dinoi, a New Painter, Old Questions'. *Bulletin Antieke Beschaven* 82 (2007): 365–421.

Hind, John. 'Traders and Ports-of-Trade (Emporoi and Emporia) in the Black Sea in Antiquity'. *Il Mare Nero* 2 (1995/1996): 113–26.

Huber, Kalinka. *Gravisca. Scavi nel santuario greco 6. Le ceramiche a figure rosse*. Bari: Edipuglia, 1999.

Hupe, Joachim, ed. *Der Achilles-Kult im nördlichen Shwarzmerraum vom beginn der griechische Kolonisation bis in die römische Kaiserzeit. Beiträge zur Akkulturationforschung. Archäologie*, 94. Münster: Scriptorium, 2006.

Johnston, Alan. 'The Rehabilitation of Sostratos'. *La Parola del Passato* 27 (1972): 416–23.

———. *Trademarks on Greek Vases*. Westminster: Thames & Hudson, 1978.

———. 'Greek Vases in the Marketplace'. In *Looking at Greek Vases*, edited by N. Spivey and T. Rassmussen, 203–31. Cambridge: Cambridge University Press, 1989.

Johnston, Alan, and Maristella Pandolfini. *Gravisca. Scavi nel santuario greco, 15. Le iscrizioni*. Bari: Edipuglia, 2000.

Kahil, Lilly G. *La céramique grecque de Thasos*. Paris: De Boccard, 1951.

———. 'Quelques vases du sanctuaire d'Artémis à Brauron'. In *Neue Ausgrabungen in Griechenland, Antike Kunst*, Beiheft 1, 1–29, pl. 1–16. Bern: 1963.

Katarrachias, K., and N. Karafyllis. *Αρχαιολογικά Ευρήματα Φίλιας και Άρνης-Κιερίου*. Karditsa: 1992.

Klein, W. *Euphronios. Eine Studie zur Geschichte der Griechischen Malerei*. 2nd ed. Wien: C. Gerlod's Sohn, 1886.

Knigge, Ursula. *Kerameikos IX, Die Südhügel*. Berlin: 1976.

Kokkou-Byridi, K. *Ελευσίς. Πρώιμες πυρές θυσιών στο Τελεστήρω της Ελευσίνος*. Athens: Greek Ministry of Culture, 1999.

Lane, A. *Greek Pottery*. London: Faber & Faber, 1948.

Langlotz, Ernst. *Zeitbestimmung der strengrotifigurigen Vasenmalerei und der gleichzeitigen Plastik*. Leipzig: E. A. Seamann, 1920.

———. *Die antike Vasen von der Akropolis zu Athen*. Berlin: De Gruyter, 1925.

Lullies, R. 'Zur bootisch rotfigurigen Vasenmalerei'. *Mitteilungen des deutschen archäologischen Instituts. Athenische Abteilung* 65 (1940): 1–27.

Maffre, Jean-Jacques. 'La collection Paul Cannellopoulos VIII. Vases béotiens'. *Bulletin de Correspondance Hellénique* 99 (1975): 428–520.

———. 'Chachrylion, Euphronios et quelque uns de leurs contemporains à Thasos'. In *Proceedings of the Third International Symposion on Greek and Related Pottery*, edited by J. Christiansen and T. Melander, 379–89. Copenhagen: National Museum, Ny Carlsberg Glypotek, Thorvaldsens Museum, 1988.

———.'Coupes attiques à figures rouges trouvées à Thasos'. In *Athenian Potters and Painters* II, edited by J. Oakley and O. Palagia, Forthcoming.

Marangou, L. 'Céramique attique à Amorgos: données archéologiques et tradition littéraire'. In *Les clients de la céramique grecque. Actes du colloque de l'Académie d'Inscriptions et Belles Lettres, Paris, 30–31 janvier 2004*, Cahiers du *Corpus Vasorum*, edited by J. de La Genière, 69–74. Paris: De Boccard, 2006.

Martelli, Marina. 'Arete ed Eusebeia: le anfore attiche nelle necropolis dell'Etruria campana'. In *Il Greco, il Barbaro e la ceramica attica. Immaginario del diverso, processi di scambio e autorappresentazione degli indigeni. Atti del Convegno Internazionale di Studi, 14–19 maggio 2001, Catania, Caltanissetta, Gela, Camarina, Vittoria, Siracusa*, edited by F. Giudice and R. Panvini. Vol. II 7–37, pl. I-IV. Rome: 2006.

McDonald, Brian R. 'The Emigration of Potters from Athens in the Late Fifth Century BC and its Effect on the Attic Pottery Industry'. *American Journal of Archaeology* 85 (1981): 159–68.

McCamp, John. 'Excavations in the Athenian Agora: 1994–1995'. *Hesperia* 65 (1996): 231–61, pl. 65–76.

McPhee, Ian. 'Red-Figured Pottery from Corinth. Sacred Spring and Elsewhere'. *Hesperia* 50 (1981): 264–85, pl. 63–74.

———. 'Attic Red-Figure from the Forum in Ancient Corinth'. *Hesperia* 56 (1987): 275–302, pl. 47–62.

———. *The Extramural Sanctuary of Demeter and Persephone at Cyrene, Libya*. Vol. VI Part II: *Attic Pottery*. Philadelphia: Pennsylvania University Museum, 1997.

Metzger, Herni. *Fouilles de Xanthos IV. Les céramiques archaïques et classiques de l'acropole lycienne*. Paris: De Boccard, 1972.

Möller, A. *Naukratis. Trade in Archaic Greece*. Oxford: Oxford University Press, 2000.

Moore, Mary B. *Attic Red-Figured and White Ground Pottery*. Excavations in the Athenian Agora 30 Princeton: American School of Classical Studies, 1997.

———. 'Attic Red-Figure Painters and the Athenian Agora'. In *Athenian Potters and Painters. The Conference Proceedings*, edited by J. H. Oakley, W. D. E. Coulson, and O. Palagia, 469–72. Oxford: Oxbow Books, 1997.

Morgan, Catherine. *Attic Fine Pottery of the Archaic to Hellenistic Periods in Phanagoria*, Colloquia Pontica 10, Leiden and Boston: Brill, 2004.

Neer, Richard. T. *Style and Politics in Athenian Vase-Painting. The Craft of Democracy, ca. 530-460 BCE*. Cambridge: Cambridge University Press, 2002.

Noble, J. V. *The Techniques of Attic Painted Pottery*. London: Thames & Hudson, 1988.

Osborne, Robin. 'Pots, Trade and the Archaic Greek Economy'. *Antiquity* 70 (1996): 31–44.

———. *Archaic and Classical Greek Art*. Oxford: Oxford University Press, 1998.

———. 'Why Did Greek Pots Appeal to the Etruscans?' *World Archaeology* 33 (2001): 277–95.

———. 'Workshops and the Iconography and Distribution of Athenian Red-figure Pottery: A Case Study'. In *Studies in Honour of Brian A. Sparkes*, edited by X. Keay and S. Moser, 78–94. Oxford: Oxbow Books, 2004.

———.'The Anatomy of a Mobile Culture: The Greeks, their Pots and their Myths in Etruria'. In *Mobility and Travel in the Mediterranean from Antiquity to the Middle Ages*, edited by R. Schlesier and U. Zellmann, 23–36. Münster: Lit Verlag, 2004.

Paleothodoros, Dimitris. 'Pourquoi les Étrusques achetaient-ils des vases attiques?' *Les Études Classiques* 70, nos. 1–2 (2002): 139–60.

———. 'The Pithos Painter'. *Eulimene* 4 (2003): 61–76.

———. *Epictétos*. Collection d'Études Classiques 18, Louvain–Namur: Peeters, 2004.

———. Review of Richard. T. Neer, *Style and Politics in Athenian Vase-Painting. The Craft of Democracy, ca. 530–460 BCE*. Cambridge: Cambridge University Press, 2002. In *Les Études Classiques* 73, no. 4 (2005): 405–11.

———. 'Η Διάδοση της πρώιμης ερυθρόμορφης κεραμικής στη Μαύρη Θάλασσα (525-480 π.Χ.)'. *Eulimene* 6 (2005–2006): 53–75.

Papaspyridi, Semni. 'Ελευσινιακά Αγγεία. Α. Ερυθρόμορφα και Λευκά'. *Archaeologikon Deltion* 9 (1924–1925): 1–52.

Parise Badoni, Franca. *Ceramica campana a figure nere*. Florence: Sansoni Editore, 1969.

Pécasse, Mariane. 'Le trésor de Siphnos et le Peintre d'Andocidès: rapprochement stylistique ou chronologique?'. In *Céramique et peinture grecques. Modes d'emploi. Actes du colloque international. École du Louvre, 26, 27, 28 avril 1995*, edited by M.-C. Villaneuva-Puig, F. Lissarrague, P. Rouillard, and A. Rouveret, 309–14. Paris: La documentation française, 1999.

Pemberton, Elizabeth. *Corinth XVIII, Part 1. The Sanctuary of Demeter and Kore. The Greek Pottery.* Princeton, NJ: American School of Classical Studies, 1989.

Pologiorgi, Melpomeni. 'Παιδική Ταφή στην Ηλιούπολη'. *Archaiologiki Ephemeris* 134 (1995): 231–45, pl. 63–5.

Rendeli, Marco. 'Rituali e immagini: gli stamnoi attici di Capua'. *Prospettiva* 72 (1993): 2–16.

Reusser, Christoph. *Vasen für Etrurien. Verbreitung und Funktionen attischer Keramik im Etrurien des 6. und 5. Jahrhunderts vor Christus.* Vols. I–II. Zürich: Akanthus, 2002.

———.'La céramique attique dans les tombes étrusques'. In *Le vase grec et ses destins*, edited by P. Rouillard and A. Verbanck-Piérard, 167–78. Munich: Biering & Bringmann, 2004.

Richter, Gisela. *Attic Red Figure Vases. A Survey.* New Haven, Conn. 1946.

Rizzo, Maria Antoanetta. 'La ceramica a figure nere'. In *La ceramica degli Etruschi. La pittura vascolare*, edited by M. Martelli, 31–41. Navarra: Istituto Geografico di Agostino, 1987.

Robertson, Martin. *The Art of Vase-Painting in Classical Athens.* Cambridge: Cambridge University Press, 1992.

Rotroff, Susan. 'Early Red-Figure in Context'. In *Athenian Potters and Painters* II, edited by J. Oakley and O. Palagia. Forthcoming.

Ruystedt, Eva. 'Athens in Etruria. A note on Panathenaic amphorae and Attic ceramic imagery in Etruria'. In *Across Frontiers. Etruscans, Greeks, Phoenicians and Cypriots. Studies in Honour of David Ridgway and Francesca Serra Ridgway*, edited by E. Herring, 497–506. London: Accordia, 2006.

Sabetai, Victoria. *CVA Thebes, Archaeological Museum.* Athens: Academy of Athens, 2001.

———. *CVA Athens, Benaki Museum.* Athens: Academy of Athens, 2006.

Serbeti, Eleftheria. 'Ερυθρόμορφη κύλικα από την οδό Λέκκα'. In *Στήλη. Τόμος εις Μνήμιν Νικολάου Κοντολέοντος*, 321–7. pl. 146–47. Athens: Society of Friends of N. Kontoleon, 1980.

———. 'Attic Pottery from a Deposit in Eretria'. In *Athenian Potters and Painters. The Conference Proceedings*, edited by J. H. Oakley, W. D. E. Coulson, and O. Palagia, 491–99. Oxford: Oxbow Books, 1997.

———.'Εργο του αγγειογράφου Επικτήτου στη συλλογή του Πανεπιστημίου Αθηνών'. *Archaiologikon Deltion* 56 A (2001): 153–62.

Small, Joselyn Penny. 'Scholars, Etruscans, and Attic Painted Vases'. *Journal of Roman Studies* 7 (1994): 34–58.

Smith, Cecil. *Catalogue of the Greek and Etruscan Vases in the British Museum*, III. London: British Museum Press, 1893.

Snyder Schaeffer, J., ed. *The Corinthian, Attic and Lakonian Pottery from Sardis*, Archaeological Exploration of Sardis 10. Cambridge, Mass. and London: Harvard University Press, 1997.

Sparkes, Brian A. 'The Taste of a Boeotian Pig'. *Journal of Hellenic Studies* 87 (1967): 116–30, pl. XII–XXI.

———. *Greek Pottery. An Introduction.* Manchester and New York: Manchester University Press, 1990.

———. *The Red and the Black. Studies in Greek Pottery.* London and New York: Routledge, 1996.

Stissi, Vladimir. 'Production, Circulation and Consumption of Archaic Greek Pottery—Sixth and Early Fifth Centuries BC)'. In *The Complex Past of Pottery. Production, Circulation and Consumption of Mycenaean and Greek Pottery (Sixteenth to Early Fifth Centuries BC). Proceedings of the ARCHON International Conference held in Amsterdam, 8–9 November 1996*, edited by J. P. Crieelard, V. Stissi, and G. J. van Wijngaarden, 83–113. Amsterdam: J. C. Gieben Publisher, 1999.

———. 'Modern Finds and Ancient Distribution'. In *Céramique et peinture grecques. Modes d'emploi. Actes du colloque international. École du Louvre, 26, 27, 28 avril 1995*, edited by

M.-C. Villaneuva-Puig, F. Lissarrague, P. Rouillard, and A. Rouveret, 351–55. Paris: La documentation française, 1999.

Szilágyi, J. G. 'Zur Praxias-Gruppe'. In *Miscellanea Casimiro Majewski Oblata, Archeologia Polona 15*, 95–114. Warsaw: Institute of Archaeology and Ethnology, Polish Academy of Sciences, 1973.

Tiverios, Michalis. 'Αγγεία-Αναθήματα στο Μεγάλο Ιερό της Ελευσίνας'. In *Athenian Potters and Painters* II, edited by J. Oakley and O. Palagia. Forthcoming.

Tölle-Kastenbein, R. 'Bemerkungen zur absoluten Chronologie spätarchaischer und frühklassischer Denkmäler Athens'. *Archäologische Anzeiger* (1983): 573–84.

Tosto, Vincent. *The Black-Figured Vases Signed NIKOSTHENES EPOIESEN*. Amsterdam: Allard Piersom Museum, 1999.

Tuna-Nörling, Y. *Daskyleion I. Die Attische Keramik. Arkeoloji Dergisi* 6. Izmir: 1999.

———. 'Polyxena bei Hektor Lösung. Zu einem attisch-rotfigurigen Krater aus Tekirdag (Bisanthe/Rhaidestos)'. *Archäologische Anzeiger* (2001): 44–127.

———. 'Attic Pottery from Dascylium'. In *Achaemenid Anatolia. Proceedings of the First International Symposium on Anatolia in the Achaemenid Period, Bandirma, 15–18 August 1997*, edited by T. Bakir. Leiden: Nederlands Institut voor het Nabije Osten, 2001.

———. 'Archaische attische Keramik in Ionien'. In *Greek Settlements in the Eastern Mediterranean and the Black Sea*, edited by G. R. Tsetskhladze and A. M. Snodgrass, 97–129. Oxford: Archaeopress, 2002.

Tzachou-Alexandri, O. 'Αλάβαστρο του Ζωγράφου του Ευεργίδου'. *Archaeologiko Deltion* 51–52, A (1996–7): 85–98, pl. 29–36.

Tzanakaki, Katerina. 'Εισαγωγές αττικών ερυθρόμορφων αγγείων στην Κρήτη'. In *Πεπραγμένα του Θ' Κρητολογικού Συνεδρίου. Ελούντα 1–6 Οκτωβρίου 2001. Τόμος Α5. Αρχαία Ελληνική και Ρωμαϊκή περίοδος*, 82–94. Chania: Philologikos Syllogos 'Chrysostomos', 2006.

Villanueva-Puig, M.-C. 'Les vases attiques du VIe et Ve siècle trouvés en contexte funéraire à Athènes'. In *Le vase grec et ses destins*, edited by P. Rouillard and A. Verbanck-Piérard, 63–66. Munich: Biering & Bringmann, 2003.

Villard, François. *Les vases grecs*. Paris: Musée du Louvre, 1948.

———. *La céramique grecque de Marseille. Essai d'histoire économique*. Paris: Ecole française de Rome, 1960.

Vlachopoulos, I., ed. *Αρχαιολογία του Αιγαίου*. Athens: Miletos, 2005.

Wachter, Rudolph. *Non Attic Greek Vase Inscriptions*. Oxford: Oxford University Press, 2001.

Walters, H. B. *A History of Ancient Pottery*. Vol. 1. London: John Murray, 1905.

Wiel-Martin, Federica. 'Vasi reali e vasi raffigurati nelle tombe dipinte di epoca arcaica'. In *Pittura parietale, pittura vascolare. Ricerce in corso tra Etruria e Campania*, edited by F. Gilotta, 9–18. Napoli: Arte Tipografica Editrice, 2005.

———. *La ceramica attica a figure rosse di Adria. La famiglia Bocchi e l'archeologia*. Padova: CLEUP, 2005.

Williams, Dyfri. 'Aegina, Aphaia-Tempel XI. The Pottery from the Second Limestone Temple and the Later History of the Sanctuary'. *Archäologische Anzeiger* (1987): 629–80.

———.'The Invention of the Red-Figured Technique and the Race between Vase Painting and Free Painting'. In *Looking at Greek Vases*, edited by Nigel Spivey and Tobias Rassmussen, 102–18. Cambridge: Cambridge University Press, 1991.

Appendix: The Distribution of Early Attic Red-Figure Pottery in Greece

Since no work on the subject has yet appeared since the pioneer work by Gisela Richter in 1905, I consider it useful to provide a comprehensive picture of the distribution of early attic red-figure in mainland and insular Greece. Only vases dated between

525 BC and 490/480 BC have been taken into account. In the present state of our knowledge, there can be no question of a statistical analysis of the finds.

A good picture may be drawn of finds (either published or currently under study) from Attica (the Agora, the Acropolis and the Kerameikos,[44] as well as Eleusis[45] and Brauron and other sites in Attica),[46] Eretria,[47] Aegina,[48] Corinthia (Perachora, Isthmia, Nemea, Corinth and Acrocorinth),[49] Olympia,[50] Samos,[51] Delos,[52] Rhodes,[53] less so from Megara,[54] Boeotia (Tanagra, Akraiphia, Thebes, Theban Kabirion and Rhitsona),[55] Central Greece (Delphi,[56] Atalanti, and Locris)[57] and Thessaly.[58] Early red figure is absent from Crete,[59] and occurs rarely in the Ionian Islands (especially in Corfu and Lefkas),[60] the Cyclades (except Delos) and the Dodecanese (except Rhodes).[61] In Macedonia and Thrace, despite the fervent archaeological activity of the last thirty years, the number of early red-figured vases known at present is modest and usually confined to coastal sites.[62] Thasos may be singled out for the high quality and relative abundance of early Athenian vases found in two excavation sectors, the Artemision and the 'Tsakonis plot'.[63]

Brotherhoods of Faith and Provident Planning: The Non-public Associations of the Greek World

Vincent Gabrielsen

Introduction

I begin with three short life stories. Strabo tells us the story of his contemporary Hybreas, one of the two notable men in Mylasa. As a young man, Hybreas inherited from his father only a wood-carrying mule and a mule driver. But he gave himself to studying rhetoric with Diotrephes of Antiocheia, and soon earned the post of *agoranomos* at his home city. An incumbent of this post, he then decided to embark on a political career, and by joining 'those frequenting the market', the *agoraioi*, he quickly rose to a position of great power in the city of Mylasa.[1] The Danish storyteller and poet Hans Christian Andersen (whose mother is said to have washed other people's clothes to pay for his education) was equally well 'connected', though hardly to any *agoraioi*. In 1847, during his visit to London, Andersen was elected visiting member of the very prestigious London club, *The Athenaeum*. In his diary, the Dane registered his fascination at the club's cosmopolitan and luxurious interior ('the servants were wearing silken stockings'), but at the same time he also complained that he had to pay the exorbitant price of 3 shillings for one lamb chop and half a glass of mineral water—albeit served on a silver tray.[2] The American founding father, Benjamin Franklin, himself of artisan and farming stock, was in 1727 the founder-member of a club 'of mutual improvement' named the *Junto* (or Leather Apron Club), whose purpose was 'to debate questions of morals, politics and natural philosophy and to exchange knowledge of business affairs', and he later established a number of other clubs for discussion of good causes and new knowledge.[3] Actually, the multitude of new clubs and societies which proliferated in the eighteenth and nineteenth centuries in England, France, America, and elsewhere are now shown to have played a central role in the ongoing modernization process of the social and political life.[4]

Franklin became a Deist, Andersen led the life of an ardent Protestant, and only heavens know to which gods or god our Hybreas from Mylasa offered up his sacrifices—at a guess, Zeus *Agoraios*. At any rate, religious preoccupations most certainly filled some space in these men's lives. All three, moreover, pursued personal goals, or gained advancement, via membership of more or less tightly knit groups of a private nature. Lastly, like many of their contemporaries, all three knew quite well what institutionalized personal relations within 'societies' meant, and what such links could do for them. However, none of these men would have spoken of his connections as 'networks'.

The Greek word *diktyon* means 'net'; *diktyodes* or *diktyoeides* (or *diktyoton*) refer to something with the form of a net; the verb *diktyoomai* means 'to be caught in a net'; *diktyarchountes*, attested in second-century AD Parion (Troas), are officials in an association of 'fishermen', not a 'network' in the modern sense.[5] Still, despite the lack of a special Greek term, most, I believe, will agree that it is both legitimate and highly relevant to direct our attention towards such phenomena in the ancient world that can reasonably be categorized under the general modern description 'networks'. The ones I have chosen for treatment here are the non-public associations of the ancient Greek-speaking world.

The evidence concerning these non-public associations, mainly inscriptions and papyri, has grown noticeably since Franz Poland published his fundamental study on the subject almost a hundred years ago.[6] In recent years, ancient historians have started taking a strong interest in specific types of associations, or in the associations of a particular area, or again in single aspects of associational life—religion being a dominant one.[7] That interest is now shared also by theologians, especially those among them who explore possible connections between the associations of the classical world and the Pauline or early Christian congregations.[8] Characteristically, it is this latter group of scholars, as well as historians of the Roman period,[9] who now stand out for their willingness to venture into thematically far-reaching questions and to give a panoramic view of an eminently complex historical phenomenon.

In this article, using material from various parts of the Greek-speaking world, I shall attempt to offer a rather wide-ranging assessment of the networking capabilities of the ancient non-public associations; though, for reasons that will become apparent shortly, chronologically the stress will be on Hellenistic times. Treatment of the ancient 'brotherhoods' *sui generis* can of course only be undertaken at the cost of leaving out individual features, deviations and local nuances, all of which underline particularity and difference. Still, while all these may prove most useful in a different kind of study, the overall topic to be treated here requires a more generalizing approach: networks and their workings, often straddling several social, juridical, or geographical borders, are less likely to reveal much of themselves in a narrowly focused study that builds on local- or type-specific evidence alone.

I proceed from the contention that there are at least three good reasons for including the non-public associations in an analysis of networks in the Greek-speaking world.

Two of them have mostly to do with current trends. One is the political and theoretical debate which has been going on since the 1980s, and which is mostly fuelled by the efforts of several countries to delegate an assortment of governmental responsibilities to private actors: it is about the ability or efficiency of private organizations to take over specific public functions (e.g. in the field of welfare), and also in fending off the worst depredations of global capitalism;[10] also, one may add the resurgence of 'communitarism' in American political theory—that is, the conception of the community as a mediating force between individual and state. Flowing almost directly from these debates is the next trend: the revived interest among historians of medieval and later times in gaining a better understanding of the history of private associations.[11] Whether ancient historians would want to respond to the first of these trends is a matter of personal choice. But to abstain from engaging in the scholarly discourse generated by the second trend is to insist on an unjustified disciplinary isolationism that denies the experience of the ancient world the opportunity of making its own contribution to the study of a significant historical phenomenon. In 1981, G. E. M. de Ste Croix complained, 'The treatment of corporations in existing works is wholly inadequate, being merely systematic and not historical'.[12] It is time to give the 'historical' its due. My third reason, a more substantive one compared with the other two, provides also the principal justification for my topic. It therefore needs to be laid out in some more detail.

Public and Private Associations

A brief and even superficial comparison between the world of the Classical Greek *polis* and its successor in Hellenistic times is enough to make that reason apparent. In the Classical period, the *dominant* (but not the only) way in which the populations of the Greek *poleis* had come to be organized consisted of membership of units, or associations, that were not only subordinate to the main socio-political unit, the *polis*, but also integral parts of it. Societal organization at that time was almost coterminous with, and certainly under the primacy of, the *political*, and had 'citizenship' as well as the smooth functioning of the public sphere as its pre-eminent concerns. In Aristotle's well-known formulation, 'the *polis* is an association (*koinonia*) of citizens centred around the constitution (*politeia*)'.[13] This larger association rested on, and drew nourishment from, a variety of smaller associations (*koinoniai*), all of which were politically oriented microcosms that cooperated to energize the life of their parent unit, the *polis*.[14] Regardless of whether they had a territorial basis (e.g. demes, *komai*, *ktoinai*, etc.), or were constituted on a predominantly non-territorial criterion (e.g. *phylai*, *phratries*, *patrai*, *syngeneiai*, *hekatostyes*, etc.), they all were *public subdivisions*. One of their main functions was to tie their members to the big *polis*-association (*koinonia*).[15] A significant side effect of all this was that, with the occasional exception of military service,[16] and of course with the partial exception of those few groups of foreign worshippers that were slowly emerging in the later fifth century (see below), non-citizens remained not only political outsiders,[17] but also a rather amorphous and—associationally speaking—parentless group, floating in free space.[18]

In Hellenistic times, citizen-manned, public organizations were still in place.[19] Whether or not their importance overall was in decline (one supposedly starting either early on or from *c*.130 BC) is an issue pending systematic investigation. Yet any preliminary assessment of that issue, including the present one, will have to take into account a number of clear indications to the effect that in several Hellenistic *poleis* public subdivisions continued to hold a central place, both within traditional political structures and, as something new, within those freshly created structures that came to define the relationship between *poleis* and monarchs.[20] Thus, in quite a number of Hellenistic *poleis*, not decline but renewal and reorientation are evidenced as the prevailing features.

In any case, a notable novelty is that the associational sinews that traditionally and predominantly held together the public sphere had now attained a vibrant and assertive counterpart which filled a fairly large part of the picture. These are the *private*, or, as has become customary to call them in English, *voluntary* associations.[21] One of their general characteristics—and the main criterion that justifies our uniting into a single treatment groups of various descriptions—is that they normally had no formal (i.e. constitutional) affiliation to the *polis*, even though they all remained subject to *polis* law.[22] Another general characteristic is that they distanced themselves from prevailing juridical distinctions between status categories, freely admitting as members both citizens and all categories of non-citizens—that is, foreigners, women and in some cases slaves as well.[23] In short, population groups other than the politically privileged one, the citizens, had finally begun organizing themselves—at first sporadically and at a low key (see the case of Athens below), then systematically and on a large scale—into discrete but clearly recognizable *koinoniai*. With this, political 'outsiders' gradually found a home in collectives possessing a distinctly familiar organizational form. They thus acquired for real societal space of their own.

It is almost exclusively from epigraphic evidence that we know of a variety of types and individual names. Beginning for good in the later fourth century BC, the non-public associations of the Greek-speaking world placed themselves into one among several broader categories or types: for instance, into *therapeutai, speirai, systemata, synetheis, syntechniai, synergasiai*, or into what were much more common in Hellenistic times, *eranoi* and *thiasoi*. While the latter two are attested from the start (see p. 191 below), some of the others make their appearance only gradually and as late as in Imperial times, so that by about the beginning of the first century BC a fairly rich typology had come into existence. Within each of these types, then, individual associations are distinguishable by their name, more often than not a composite one. Typically (and disregarding local idiosyncrasies), it consists of at least one, and frequently more, theophoric elements—for example, Apolloniastai, Hermaistai, Athenaistai, Sabaziastai (or Sabazioi) and so on. Such theophoric elements are then often joined by an ethnic or a geographic designation. Occupational descriptions, on the other hand, are both late to enter the names and few and far between initially (cf. p. 195 below). Finally, the theophoric, ethnic, or other elements may be accompanied by a personal name, which usually is the name of the *ktistes*, the founder or president of the association.[24]

Typology and individual names can thus be perceived as two distinct tiers. Each of them began for real to grow richer and more varied in the course of the Hellenistic period.[25] But even before that, their individual manifestations (i.e. the specific types and names) did much to underline the *distinctiveness, diversity* and *exclusiveness* of the groups which they designated; types and especially names, in short, were among the features that ensured that no association was like another. But at the same time these groups were also endowed with a counter-tendency, one strongly underlining *generic affiliation, inclusiveness*, and, above all, *sameness*. A principal factor producing these latter—organizational make-up—will be mentioned shortly. Here, suffice it to note a related one—namely, the existence of a third and higher tier which embraced all types and all names. For each non-public association described itself (and was described by others) also by use of the generic description, *koinon*—that is, 'commonalty' or 'association', pure and simple; beginning at the highest level, the hierarchy of tiers was, for example, *koinon, eranos*, specific name of association. As is known, especially in Hellenistic times, the term *koinon* was often used to describe also the main political community itself—whether a federal state, an *ethnos*, a *polis*, or even a kingdom (e.g. *to koinon ton Makedonon*)—*and* its public subdivisions.[26] Thus, in addition to imposing a certain measure of sameness among the plethora of non-public associations, this higher tier (i.e. the *koinon*) situated these associations en bloc right at the side of the public ones, the two becoming in effect terminologically indistinguishable.[27]

As they emerged, multiplied and spread like fire, primarily within urban centres, they enlarged tremendously the field within which people could act, connect, do business and communicate in a particular and considerably more effective way—that is, as members of one or more organizations. This resurgent 'club-ism' is, I think, a change worth examining closer. What might have been the factors that caused it?

And what can we say of its consequences? Some modern historians refer to this change as the 'associational phenomenon' (*il fenomeno associativo*),[28] and emphasize the way it enriched the religious, social, and cultural life of the ancient world.[29] A large body of inscriptions confirms their view. Almost everywhere, city life (most conspicuously, its civic and religious spheres) gained fresh vigour from such an unprecedented wave of organized conviviality, commensality, mutual aid, and cult worship; in the cities of the Roman East, it is now argued, public ceremonial and local politics were likewise affected to a noticeable degree.[30] To all these, we should finally add the contribution of associations to the architectural embellishment of urban centres and their *necropoleis*, the remains of which are still visible in several places.[31]

Nevertheless, we can and should go further than this. Historically, I propose, this mushrooming of private associations was the carrier of a societal transformation of the first order; not of the sudden and sensational kind—a revolution, as such—but rather one of those quiet yet block-building transformations that seem to fit, albeit in their own way, the newly introduced concept of 'industri*ous* revolutions'.[32] It resulted in institutional innovations which, far from undermining the *polis*, came to enrich it with new networks and fields of activity. Associational proliferation and activity created a huge repository of institutional potential, whose special properties were to connect, communicate, and energize. Whenever they were released (and as far as we can see, they started to be so in the period *c.*330–300 BC), these properties profoundly affected forms of organization within the three principal fields of religion, economy, and politics. Within the space of this short article, I shall attempt to exemplify that potential, though not, as it would have been ideal, with equal weight on all three of these fields; rather, the stress will be on the political. Here, I can only be suggestive, offering some interpretations rather than documenting a case.

Expanding the Civic Space

In his analysis of the relevant documents, Franz Poland observes that, in the way they were organized, the private associations were true imitations of the *polis*; indeed, he even notes that, to some extent, the real *polis* and its private replicas exhibit a parallel institutional development.[33] This very correct observation, however, needs now to be followed by a crucial question: why would anyone about to create an alternative to the public organizations choose to make his private 'club' the spitting image of the quintessential public organization, the *polis*? The minimalist answer—'He knew of nothing else'—is inadequate; for there *were* alternatives to the kind of *polis* (namely, the democratic one), which nearly all non-public associations mimicked,[34] including those that might exist in oligarchic governed *poleis*. The view to be preferred is, instead, that the choice of model was conscious from the start, the result of careful deliberation; that is, the founders of the early private associations in particular, and at least a majority amongst the founding members, not only had a clear conception of, but also agreed on, which organizational structure was to be preferred and which were to be discarded. Their choice seems to have rested on a conviction about the model's manifestly enduring and very positive characteristics, so as to be copied almost wholesale; rather than acting merely at random or instinctively, founders and their memberships largely built their organizations through careful planning and with a long-term perspective in mind. *Provident thinking*, in other words, was there from the beginning; the accuracy with which the *polis* was copied, as well as the rejection of any one of the utopias in circulation or a combination thereof (see note 34), is one clear reflection of this. Certainly, in a few cases, copying reached dimensions that may seem extraordinary. Yet, for the most part, it remained within reasonable limits, making its intended point that the copyist was the mirror image of, but at the same time something different from, the *polis*.

All associations, for instance, had equipped themselves with a 'constitution', *nomoi* (i.e. their by-laws), and a 'civic bureaucracy'—that is, an assortment of financial, priestly and other officials (*tamiai*, *epistatai*, *epimeletai*, etc.), all assisted by secretaries (*grammateis*). This whole group of offices (*archai*) was headed by the president of the *koinon*: the *archeranistes*, or *archithiasites*, or *archisynagogos*, or, more rarely, the *proeranestria* (if a female president).[35] As far as one can tell, *polis* terminology and *polis* ideology (indeed, the ideology of the democratic *polis*) permeates every single aspect of their activities and overall make-up. The members—collectively termed 'the multitude' (*to plethos*) and divided into 'private individuals' (*idiotai*) and 'office-holders' (*archontes*)—act like 'citizens' in nearly all respects. Above all, they hold assemblies (*ekklesiai*, *syllogoi*), at which they pass formal resolutions (*psephismata*) in the name of the whole body—for instance, 'it was resolved by the *koinon* of [name of respective association]'. These resolutions, borrowing heavily from the structure, formulas and rhetoric of public decrees, are (sometimes) published on stone slabs. Stelai recording the honours bestowed either on distinguished members or on benefactors (*euergetai*) outside the membership are equally replete with real-life-*polis* vocabulary, and count for much of the associations' epigraphic output. Membership is chiefly defined by a set of duties and privileges. A standard privilege in many associations is that deceased members are buried at the expense of the association, and within its own cemetery, where commemorative events—punctuated with crowning ceremonies, other awards of honours and feasting—take place on a regular basis. Rather than terminating membership, death-cult made it continue in the afterlife, embracing the brotherhood's 'burial-grounds-sharing' members (*homotaphoi*); death-cult worship, too, elevated especially prominent members (typically the founder) to the status of hero. Many non-public *koina*, therefore, possessed property, as a minimum a burial plot (*taphiai*, *topoi*) and a sanctuary (*hieron*), or a clubhouse (*andron*).[36] Thus endowed with its own constitution, officials, institutions, and a kind of a territorial base, each of these private brotherhoods posed as a 'little republic'. That such a minute copying was occurring more or less simultaneously in various places strongly suggests one thing: the provident planning—indeed, the rational thinking—that lay behind the emergence and development of the model itself, the *polis*,[37] was imported together with the model's institutional trappings.

So, at the same time as new *poleis* (i.e. newly founded cities) were emerging and flourishing across the Hellenistic East, 'little private republics' were cropping up in the old and new parts of the Greek-speaking world. As a result, by about 150 BC, city-states and federations (and kingdoms, too) found in their midst a resourceful and energetic 'other' that consisted of a profusion of private clubs. What was the relationship between the power of the State and the private look-alikes that took root within its purview?

This is a very complex question that cannot be treated satisfactorily here. Suffice it simply to note that, even in those instances in which the prevailing relationship of symbiosis or cooperation between State and non-public associations gave way to tension or conflict (see final section below), the overall result remained that described by John Davies more than twenty years ago—namely, that the *polis*, among other things because of the profusion of private associations, was 'transformed and revitalized'.[38] That relationship had deep roots in the past, since private associations and in particular wide-ranging personal networks of various descriptions had contributed greatly to the earlier transformation and revitalization of the *polis*. Simply put, in expanding its field of activities as a civic society and consolidating its *public* sphere (i.e. our 'common thing', *to demosion*), the early Archaic *polis*, while it did create brand new institutions, also cannibalized a good number of *privately organized* networks, which, after some refurbishment and perhaps even re-naming, it then presented as its own (i.e. as 'public'). The terms *proxenos* and *proxenia* may offer one illustration of how the State converted into public the private connections of a resourceful foreigner by establishing a special, privileged relation with him (cannibalization of a private network); and also how, in order to put that relationship into a public institutional setting, it borrowed heavily from the ways of the older *xenia* custom (cannibalization of a pre-existing social institution that spawned personal networks via 'ritualized friendships').[39] Wherever the evidence permits it, as it does in the case of Athens, we can trace parts of the process through which the Archaic or early Classical city-state sought to consolidate its 'statish-ness'—that is, its gradual imposition of an overarching *demosion*—by means of defining itself vis-à-vis its constituent parts (i.e. the public units), and also by means of distancing itself from its private competitors, thus establishing its superiority over them all. One text is especially relevant and deserves citing in full.

> If a deme or members of a phratry or *orgeones* of heroes or members of a *genos* or messmates or funerary associates (*homotaphoi*) or *thiasotai* or those sailing away for booty or for trade make arrangements in these matters amongst themselves, they shall be valid (*kyrion einai*) unless forbidden by the laws (*demosia grammata*).[40]

This is the well-known clause from Solon's law, known from the Roman jurist Gaius, who is quoted in Justinian's *Digest* 47.22.4. One interesting thing about this law is that it still lumps together bodies, which were to become clearly separated due to the growing distinction between 'public' and 'private'. Even more interesting, however, is the fact that it gives us the approximate point of time at which the concept of 'by-laws', as something separate from '*the* laws', was beginning to emerge in Athens. What is seen happening here in one *polis* was part of a larger current leading to increasingly greater political centralization, a greater measure of 'statish-ness' in the Greek world. Under this more centralized form of State, personal networks—built on kinship, friendship, cult worship, marriage, patronage, professional occupation (including the quest for profit), or even armed violence—did not really disappear. Quite a number of them submerged under (though without necessarily being absorbed by) the 'political' and its *demosion*, where they maintained some form of existence, until they began surfacing again in the late fourth century. Again, of all the places in the Greek-speaking world in which private associations proliferated in Hellenistic times, only Athens offers sufficiently rich pre-300 BC evidence so as to enable us to view the development within a broader historical perspective. What can be seen with adequate clarity is that the breakthrough is a real one, not just the product of our unevenly distributed evidence.[41]

The earliest securely datable piece of evidence for a *koinon* of *eranistai* at Athens, a dedication to Zeus Philios, is from 324/3 BC,[42] the earliest known decrees issued by such a body from *c.*300 BC.[43] Also from the year 324/3 BC is the first securely datable document of a regular association, which is attested already in *c.*342/1 BC and which describes itself neither as an *eranos* nor as a *thiasos*, but simply as 'the *koinon* of the *Eikadeis*'.[44] Except for two previous appearances (see below), *koina* of the *thiasos* type begin to enter the picture for good in the later fourth century—their earliest known decree dating from 302/1 BC,[45] the earliest securely dated document with named foreigners from 301/0 BC.[46] It is in the third century, however, that the number of non-citizen *koina* increases dramatically. So it all seems to begin not long before 340–330 BC, for then to gather momentum during the following three decades. What was there before that can somehow be related to these new, private organizations?

First of all, already in the fifth century, there were a variety of all-citizen (or at least citizen-dominated) clubs: for example, those of 'fellow-diners', 'fellow-drinkers', and 'hero cult worshippers', all existing side by side with the socially suspect performers of particular rites, the hell-fire clubs of rich youngsters (calling themselves *Autolekythoi*, *Ithyphaloi*, and *Triballoi*), the more serious religious groups of *Nouminiastai*, *Tetradistai*, and *Dekadistai*, and a few others of a similar description.[47] Neither these nor the politically oriented and more cohesive *hetaireiai* (set up 'for mutual support in lawsuits and in elections', Thuc. 8.54.4), nor again the so-called philosophical schools, enjoyed anything like a firm organizational structure, or (perhaps apart from the latter) a noteworthy degree of permanency.[48] The same applies to the only groups of *eranistai* to make themselves visible in our sources before 324/3 BC, the ad hoc groups of lenders who provided an interest-free loan to a particular individual.[49] Finally (and still since the fifth century), there were two other kinds of citizen groups, each with a somewhat firmer organization than most of the above *and* a strong preoccupation with hero-cult worship, but also with an attachment to the *polis* organization as subgroups of the Athenian phratries. One of these consisted of groups of *orgeones*, the other of groups of *thiasoi*.[50]

Compared with the public associations, each of these was feeble. But together they constituted a substantial, strongly religion-oriented organizational undergrowth that was both thickening and transforming during the fourth century. It was mainly out of this undergrowth that the private, more tightly organized and differently composed groups that proliferated in the post-300 period would gradually arise. In this transition, religion, it seems, was the main door into and a supreme carrier of associational efflorescence. A full-fledged, independent (i.e. private) *thiasos* association appears on record for the first time in the fifth century (the '*thiasos* of *Etionidai*' worshipping Herakles), and again in the early fourth century, each time marking the emergence of an alternative to the still-dominant phratry-dependent *thiasoi*.[51] By 400 BC, too, the Thracian émigré community at Athens had used the cult of their native goddess Bendis (now recognized as an Athenian public cult) to institute the first non-citizen groups of *orgeones* of Bendis, which thus came to co-exist with the citizen *orgeones* worshipping the goddess; in a document of a later date, the Thracian group (of the Piraeus) referred to its establishment as an official act of the Athenian People, who, in accordance with an oracle from Dodona, had granted the Thracians, 'alone of the *ethne*', the privilege to own real estate and to institute a shrine.[52] Not only did similar groups of non-citizen *orgeones* crop up in due course (e.g. those of the Great Mother), eventually appropriating the term *thiasos* so as to become a '*thiasos* of *orgeones*',[53] but the Thracian *orgeones* of Bendis themselves saw it fit to vote measures that aimed at expanding their membership, 'so that there may be as many *orgeones* of the shrine as possible'.[54]

Thus, in the late fifth and early fourth centuries, new blueprints for 'club-ism' were being created, in time to be taken over and improved by other consumers of organizational life. Regarding formal organization and mode of associational governance, however, the institutional structures of the democratic *polis* continued to be the uncontested model. In 333/2 BC, Cypriot merchants from Kition, by decree of the Athenian People, obtained permission to establish a shrine of Aphrodite and to own the plot of land on which to build it; even prior to their getting a positive official response, the Kitian merchants are seen to be acting as if they already possessed associational identity, fittingly giving their original petition the form of an enactment which they collectively had passed 'in a lawful manner' (*hoi Kitioi emporoi edoxan ennoma*). In all this, they clearly were using blueprints created by others—for instance, such groups as that of Thracian émigrés just mentioned, and certainly a specific group of Egyptians, who in 333/2 had already been granted official permit to institute a shrine for Isis.[55] A few years later (in 324/3), as we saw above, the first genuine private association appears on the extant record. In two of his works on ethics, Aristotle, not having Athens particularly in mind, gave *eranoi* and *thiasoi* as examples of private associations which are organized 'for sacrifice and social intercourse', and associations of seafarers as examples of 'profit-motivated' (*chrematistikai*) organizations.[56] During the second half of the fourth century BC, at Athens and elsewhere, a new kind of *eranoi* had come to join the *thiasoi* as *koinoniai* of the non-public kind.

This significant transformation in the post-Alexander period—for which I coined the concept of 'industrious revolution'—had therefore two aspects. A backward-looking aspect, in as much as the full-fledged personal network was being brought back in again. And a forward-looking aspect, in as much as private clubs were using, within their purview, the very techniques of the *polis* in order to challenge the *polis*' own elitist ideology of exclusivity and, more importantly, its near-monopoly of organizational life. In short, individuals at large—including the privileged category of the citizens—were claiming back their privately organized and run networks. They wanted to be able again to set up, command, and put to good use resourceful *koinoniai* that were neither organically connected to nor governed by the power of the State.

A Powerful 'Other'

As soon as they began achieving this, hitherto under-exploited reserves of intellectual energies, professional skills and collectively endowed loyalties started to become utilized, or stood ready to be utilized. In the long run, the result was a noticeable enlargement of the total associational space and, by virtue of this, a much greater degree of interconnectedness, locally and especially trans-regionally. The State and its constituent parts had now to share a number of time-honoured prerogatives with a potentially powerful 'other' that duplicated their functions. Neither the re-emergence of this 'other', nor the specific organizational form it took, came about by accident or as a by-product of something else; they resulted from careful planning. So did also the momentous spreading of the phenomenon after about 300 BC. What were the most significant consequences of this *koinonio*-mania? Here, I briefly sketch four such consequences.

Firstly, a most immediate one was that direct and close connections became established between individuals who had hitherto been separated not so much by 'physical distance', though such distance did matter, but by 'social distance', traditionally kept in place by strictly upheld distinctions in legal status, gender, wealth, or ethnic origin. The 'associational culture' spread vertically, both up and down the social and gender ladders, to cater for the specific needs of men and women. City elites, quickly recognizing the potential of this sprawling club-ism, linked themselves to it, either by making specific associations the targets of their personal benefactions (*euergesiai*), or by becoming card-carrying members, or, again, by founding their own clubs—as Benjamin Franklin and others were to do centuries later. Governments, on their part, befriended this private 'associational culture' as soon as they realized what kind of resources could be tapped from it (but see also point 4 below). One example of this will suffice: elsewhere, I discuss the evidence showing the role played by private associations in one Hellenistic state's endeavours to secure manpower for its army and especially navy. Further evidence suggests that this trend applied to a larger geographical area. What apparently exercised a special appeal were the advantages of having well-organized and mostly foreign reserves of fighting personnel 'in residence', at a time when manpower demand was extremely high almost everywhere.[57]

The reverse movement (i.e. fighters clubbing for outgoing activity) is attested twice, in Aristotle's description of *koinoniai* of 'fellow soldiers' (*systratiotai*)—who united 'for what is advantageous in warfare, whether money or victory or the seizure of a city' (Arist. *Eth.Nic.* 8.9.5 [1160a]; cf. note 56 above)—and in the earlier mention, by Solon's law, of 'those going away for booty' as a body making its own binding written agreements. With this, we have already moved on to instances showing the downward spread of the 'associational culture'. Less affluent men and women, and even slaves, could find a number of their needs catered for through membership. Promises such as the one made by the 'law' of a third-century *thiasos* could not but exercise a special attraction to the privileged and the unprivileged alike: 'and if someone is wronged, he is to be helped by them all and by all their friends.'[58] Those standing ready to help are of course one's associates.

Second, another consequence issued from the fact that the 'little private republics' proved highly industrious in the production and diffusion of 'sameness'; they made it spread laterally across a large geographical expanse, where it seems to have travelled mostly by merchant ship. But at the same time, private associations also took 'sameness' out of its cradle—the public sphere—and into the one sphere, the non-public one, which had stood out for its lack of it. The result was a considerable expansion of the space, within which a particular set of values and a special mode of communication, the *civic* ones, reigned. The members of brotherhoods, meaning also the members of *different* brotherhoods all over the Greek-speaking world, had acquired now a common language that enrolled them into one (almost universally valid) political culture, a uniform way of thinking, articulating their wishes and acting within well-defined collectives. Their issuing of decrees, their cult worship, their obsession with receiving benefactions and granting honours, their statues, crowns, and other kinds of dedicatory objects, all this and more besides united all those who were members of small, private units under one world view and under a single mode of social expression—a harmonizing, or one should rather say, homogenizing trend that customarily was, and *is*, dear to imperial powers.[59] Yet the basic values they fostered were apparently also dear to everybody else, since all brotherhoods practised an undemanding, ground-level conviviality that embodied notions of equality, freedom, participation, and all else that has the sweet scent of *demokratia*. As receptacles of personal networks, associations contributed to altering the image of their home political communities, making them look more egalitarian, socially more porous and juridically less secluded. In short, a principal part of the transformation I am describing consisted precisely of this intrusion of *civic* and, indeed, democratic ideals, culture and modes of expression into regularly held, private gatherings.

A third consequence underscores the fundamental importance of religion in a major field of economic activity. Two observations bring this out clearly. First, private *koina* with a name openly advertising their special attachment to a professional occupation appear relatively late—around 150 BC, to become numerous only from the first century onwards—and those who do so are initially (from *c.*150 to 60 BC) dominated by professionals in a specific area, seaborne commerce.[60] Second, theophoric names, as well as theophoric elements *in* names, not only dominate almost entirely until *c.*150 BC, but they go on being prominent features also after they have been joined by professional designations.[61] All this has been correctly interpreted as reflecting not the unimportance of profession as a focal point in associational life, but the far greater importance of religion; rather than evidencing the actual preoccupations of brotherhoods (or the vocational status of their members) at different periods of time, naming practices attest to carefully made choices in the mode of self-identification.[62] Religion, in short, remained a solid mainstay at all times. Why? What, besides individually felt religiosity, might have been the reason or reasons why *koina* uniting 'shippers' (*naukleroi*) and 'merchants' (*emporoi*) not only went on using names that identified them as devotees of various deities (*Poseidoniastai*, *Apolloniastai*, *Hermaistai*, *Athenaistai*), but also placed great emphasis on their acting religiously, and being seen as acting so?

Only one aspect of this vast issue will be mentioned here briefly: the propensity of manifest religious devotion, that is, a person's demonstration of 'faith' as a means of creating broad-based 'faith in other people', or, to use its secular name, 'trust'. Religiosity was expressed in the punctilious observance of cult and the unrestrained demonstration of devotion, and these were regarded as the defining characteristics of the 'pious' man (*eusebes*); and pious men were, as a rule, perceived to be upright, conscientious, and dependable people.[63] 'For thirty years I have kept safe the gold of foreigners and citizens, always with exemplary honesty', reads the second-century epigram on a banker's tombstone from Rhodes, and it further inform us that this banker's 'exemplary honesty' owed to his possession of a 'divinely sanctioned sense of justice' (*hosia dikaiosyne*).[64] This could not be but a dependable man.

In another Hellenistic trade centre, Athens, a group of *thiasotai* solemnly pledged in their 'law' to assist each other, 'so that it may be visible to all that we are piously disposed (*euseboumen*) both towards the gods and towards our friends' (*IG* II^2 1275, early third century, lines 8–10; cf. above p. 194). Piety towards the gods was more or less a given; but piety towards men, something ordinarily in short supply, could never be taken for granted. At Athens, like elsewhere, professionals worshipping the Great Gods (the *Dioskouroi*) united to form the 'House of Shippers' (*oikos naukleron*), an association whose members, besides repeatedly stressing their piety (*eusebeia*) and strict observance of their cultic obligations, formally held assemblies (*agora kyria en toi oikoi*) at which they passed decrees (one of 112/111 and another of 111/110 BC survive in a fragmentary state).[65] The very similar groups formed by merchants' diasporas on Delos—the *naukleroi*, *emporoi*, and *ekdocheis*, who placed themselves under the religious patronage of Poseidon, Apollon, and Hermes—have been the subject of detailed study.[66] On Rhodes, similar groups of professionals were in the late first century BC making experiments (which were to prove successful) in lessening 'occupational distance', as they appear to have forged relationships of trust and cooperation between quite untraditional partners: an affluent diasporic group of 'shippers' (*naukleroi*) had merged with another such group of 'farmers' (*georgoi*) to form a single, fairly articulate body based in Lindos; there, they attended to their common professional interests as well as their pious devotion to the local Athena, and all the while they, still as a group, were rising to a privileged civic position within their home city.[67] Such clubs endowed their members with an identity that brought closer three (usually) separate notions—that of the 'worshipper', the 'citizen' and the 'pursuer of profit' (i.e. the members of what Aristotle called *chrematistikai koinoniai*, see note 58 above).

From 112/111 BC, too, dates the decree passed at Athens by a *synodos* of 'shippers and merchants' that worshipped Zeus Xeinios. This decree's existence and general content is known only indirectly, through a *polis* decree issued by the Athenian Council as a response to the petition of the *synodos* for permission to post 'a shield with the painted image of its own *proxenos*' by his office building—their *proxenos* turns out to be a named Athenian, who at the same time was an elected state official, the overseer (*epimeletes*) of the harbour of Piraeus. What kind of distance is lessened here is revealed not so much by the fact that this private, Piraeus-based, Zeus-worshipping group of 'shippers and merchants', much like a foreign state, had a *proxenos* in Athens; or that this *proxenos* occupied a post of key importance for their trade; or, again, that the Council did grant them permission to post the picture within a public area. More striking than all of these is the fact that, by validating the association's decree through the formal act of *epikyrosis* ('ratification'), the Council interacted with this private club constitutionally, as if it was one of the subdivisions of the Athenian state.[68]

The strong and persisting emphasis on religion turned the membership of every association from notional 'citizens' to a group of ardent 'worshippers'. Whatever the origin, special qualities or ritual demands of individual cults, the fact that they were being actively worshipped by private brotherhoods made the religiosity of their membership constantly visible to the outside world. As a result, the equation between an 'associate' and a 'pious man' was becoming all the more self-evident. As repositories of religiosity and piety, associations created plenty of 'faith in other people', trust, particularly the variety which was—and in a way still is—in very high demand—namely, *long-distance trust*. This was another way in which the private associations proved to be industrious. Here, 'sameness' was being generated and spread by the near addiction of every 'little republic' to religion. Trust over long distances was not least needed by diasporas of every description—commercial trust and information-sharing trust being the most vital ones. All in all, the strong emphasis on religion, too, seems to have been the result of provident planning.

Finally, there was the ever-present flirting with power, a factor that lay dormant for most of the time, but which, when awakened, tended to become quite serious indeed. Certain cases of excessive *polis* mimicking, though they do stand out for the sheer pretensions, must be set aside as completely harmless. The whole membership of a club founded by a native of Kyzikos residing on Rhodes was organized into mock public subdivisions—that is, into three *phylai* (just like the Rhodian state), each headed by a *phylarchos* and a *gymnasiarchos*, and all three competing in annually held events (*agones*) that were presided over by elected *agonothetai*; the *phylai* of this second-century BC association were named after the founder, his wife and their granddaughter.[69] Such trends persisted. In second-century AD Philadelphia (Lydia), associations are attested that styled themselves 'the sacred *phyle* of wool-weavers' (*he hiera phyle ton eriourgon*) or 'the sacred *phyle* of shoemakers' (*he hiera phyle ton skyteon*), while in nearby Saittai there was a '*phyle* of linen-weavers' (*he phyle ton linourgon*).[70] From a much earlier date, finally, private clubs had equipped themselves with entire sets of *polis* legal processes—terminology and all—for use within their own, closed-circuit judicial systems; a good example is offered by a decree of the *Haliadai* and *Haliastai koinon* in second-century Rhodes.[71] Cannibalization, it seems, had now changed direction.

Our evidence for the 'serious' cases comes from widely dispersed dates. In the early second century AD, there are the anxieties expressed in Pliny the Younger's correspondence with Emperor Trajan about the *collegia* in the Roman east.[72] A petition for the establishment of a *collegium* of fire-fighters (*fabri*) in Nikomedia was denied for fear of fostering seditious ideas: 'Whatever title we give them, and whatever our object in giving it,' writes Trajan, 'men who assemble for a common purpose will all the same become a political association before long' (Pliny, *Ep.Tra.* 10.33, 34). Regarding the petition from the free city of Amisos (in Pontus) to establish a private *eranos*-association, Pliny is advised not to try to prevent this, 'especially if the contributions [sc. of their members] are not used for riotous and unlawful assemblies' (Pliny, *Ep.Tra.* 10.92, 93; cf. 10.96, 97). Fears for the subversive character of Greek-inspired associations (said to be involved in a conspiracy against the Roman state) are attested as far back in time as 186 BC, when the Senate introduced strict regulations about the activities of Dionysiac cult groups (i.e. banning their gatherings) in Rome and Italy, a measure to be followed in 64 BC by another senatorial ban of associations allegedly involved in anti-constitutional activity;[73] later (towards the end of the second century AD), Ephesos was hit by a bakers' riot.[74] Centuries earlier, Thucydides had put on record the aggressive activism exhibited by a number of fiercely anti-democratic *hetaireiai* in late-fifth-century Athens, upper-class led and manned clubs, which, with assistance from young bullies, played a part in the overthrow of democracy (see note 48 above).

What brings such chronologically distant phenomena and events together under one historical structure is a particular characteristic, perhaps the associational culture's most important one: its ability to unite powerful networks of organizations and interests that can be used—and indeed *were* used—to achieve political mobilization: 'fellow spirits' could meet, share information, make plans, summon courage and act. Governments, which had befriended club-ism because of its resourcefulness in the first place, had also good reasons to feel uneasy with its superb capabilities of mobilizing people and stirring up action.[75]

Chronologically sandwiched between the Classical Athenian *hetaireia* and the *collegia* in the Roman imperial provinces is the extensive, famed and powerful network of the Dionysiac Technitai, the professional actors and other performing artists, who from about 300 BC were responsible for the major Greek festivals and competitions. Since they, and particularly the evidence relating to them, are treated in depth in two recent and almost simultaneously published monographs, I shall limit my account to highlighting only a few, essential features.[76]

First of all, *Dionysiakoi Technitai* is the common name of four principal units, each with a distinct associational status (a *koinon*), and each with its own sub-branches: (1) the *Isthmian-Nemean* unit, with its Peloponnesian, Boiotian, Euboian and Macedonian branches; (2) the *Athenian* unit, with its strong Delphic connection; (3) the *Egyptian* unit, with its Cypriot branch; and finally, the largest of all four, (4) the *Ionian-Hellespontine* unit, with its special attachment to the cities of Teos and Pergamon. The second feature is that, thanks to their specialization in (as well as responsibility for) a politically and religiously vital field—that is, the Pan-Hellenic festivals and competitions—these multi-branched *koina* developed very close bonds of cooperation (e.g. participation in ambassadorial missions) and patronage with their host governments and with ruling powers.

The third feature, finally, relates to the remarkable amount of political clout that was attained by the largest of these associations, the Teos-based *Ionian-Hellespontine koinon*. Briefly, its members (a) enjoyed a special kind of citizenship within Teos that is nearly unparalleled in the Classical Greek or Hellenistic world; (b) they had their legal cases tried not by the ordinary Teian courts, but by a special court, the *koinodikion*, which was composed of Dionysiac Technitai and Teian citizens; (c) a recently discovered tetradrachm in the Attic standard (with a Dionysos head on the obverse and the legend *ton peri ton Dionyson techniton* on the reverse; see also Psoma, this volume) attests to the fact that the Technitai issued their own currency;[77] and finally, as inscriptional evidence shows, (c) they claimed as their own, at least 'their own' in a fiscal sense, part of the territory of Teos—that is, the harbours and the urban centre, *polis*. Not unexpectedly, a conflict broke out between the *koinon* of the Technitai and the city of Teos. When King Eumenes II of Pergamon intervened with a settlement proposal, he urged the two parties to merge by means of a *synoikismos*—that is, the kind of merger normally carried out by political communities only. If our evidence ever provided an example of a private body which made a serious attempt at 'state-building', this is it![78]

Certainly, this particular historical event, the association of the Technitai of Asia Minor itself, and possibly also the entire organization of the *Dionysiakoi Technitai*, are all highly exceptional. Nevertheless, they go some way towards illustrating how much power-potential could be stored in the associational crucible, and also how far those willing to activate that potential could get. Most other *koina*, however, avoided such 'overstretching', going on being industrious in a much less spectacular, albeit no less effective, way.

Notes

[1] Strabo 14.2.24 [659]. It cannot be decided whether *agoraioi* here means 'political leaders' (i.e. 'speakers in the agora') or (as in Diod. 20.82.4–5, 84.5–6, 97.5) 'marketeers' (i.e. 'those buying and selling'). In inscriptions, the word usually refers to those frequenting the market

place, perhaps a loosely organized group: *IGBulg* I^2 15(2) and (3), cf. *IGBulg* III (1) 1415. *Agoraios* as a cult epithet: Theoi Agoraioi: *SEG* 42.561; Herakles Agoraios: *IGBulg* V 5636; Zeus Agoraios: *IG* X (2) 2, 252; *IG* XII (8) 361; *IC* I (9) 1. I would like to thank the two anonymous readers who commented on the paper and made a number of useful suggestions and corrections. Below, I consider the two specific points they raise regarding my main argument.

[2] Hans Christian Andersen, *Dagbøger, 1825–1875*, esp. vol. 3, edited by H. Vang Lauridsen and T. Gad (1974), 234–35 (25/26 July 1847).

[3] Benjamin Franklin, *Autobiography*; L.W. Labaree *et al.*, *Papers of Benjamin Franklin*, 280. Cf. Clark, *British Clubs and Societies*, 390, 401, 413.

[4] England and America: Allen, *Clubs of Augustan London*; Clark, *British Clubs and Societies*. Elsewhere: Françoise, *Sociabilité et société bourgeoise*.

[5] *I. Parion, Die Inschriften von Parion*, 5; cf. Poland, *Geschichte*, 360 n. ++.

[6] Poland, *Geschichte*, criticizing his predecessor, Ziebarth, *Das griechiche Vereinswesen*. See also Rasmussen, *Peri ton Eranon*; Foucart, *Des associations religieuses*; Tod, 'Clubs and Societies in the Greek World'; Fisher, 'Greek Associations'. Associations of Hellenistic and Roman Egypt: San Nicolò, *Ägyptisches Vereinswesen*; de Cenival, *Les associations religieuses en Égypte*.

[7] Athens: Ferguson, 'Attic Orgeones', 'Orgeonika'; Parker, *Athenian Religion*, 333–42; Lambert, *Phratries of Attica*; Jones, *Associations of Classical Athens*, 221–67, 307–10; Leiwo, 'Religion or Other Reasons?' Rhodes: Pugliese Carratelli, 'Per la storia delle associazioni'; Gabrielsen, *The Naval Aristocracy*, 123–29; Gabrielsen, 'The Rhodian Associations'. Delos: Bruneau, *Recherches sur les cultes*; Rauh, *The Sacred Bonds of Commerce*. Individual aspects: e.g., Radin, *Legislation of Greeks and Romans*; Vondeling, *Eranos*; Arnaoutoglou, 'Between *koinon* and *idion*'; Arnaoutoglou, *Thusias heneka*.

[8] Kloppenborg and Wilson, *Voluntary Associations*; Harland, *Associations, Synagogues and Congregations*; Harland, 'Spheres of Contention'; Ascough, *Paul's Macedonian Associations*.

[9] Roman World and Greek East: Liebenam, *Zur Geschichte*; Waltzing, *Étude historique*; Ausbüttel, *Untersuchungen zu den Vereinen*; van Nijf, *The Civic World*; Perry, *The Roman Collegia*. Physical environment of associations: Bollmann, *Römische Vereinhauser*.

[10] Green, *Reinventing Civil Society*, with Ferdinand Mount's review in *Times Literary Supplement* 15 October 1993, 12–14.

[11] Thrupp, 'The Gilds'; MacKenney, *Tradesmen and Traders*.

[12] De Ste.-Croix, *Class Struggle*, 597.

[13] Arist. *Pol.* 1274b32–1276b15. Cf. Hansen, *Polis and City-State*, 133–34; Hansen, *Polis*, 110.

[14] Arist. *Eth.Nic.* 1160a: 'But all associations (*koinoniai*) are parts of the state (*politike koinonia*).'

[15] General studies: Jones, *Public Organization*; Murray, 'Rationality and the Greek City'; Hansen, *Polis*, 102, 114–15. Studies on specific public subdivisions: Osborne, *Demos*; Whitehead, *Demes of Attica*; Siewert, *Die Trittyen Attikas*; Trail, *Demes and Trittys*; Gabrielsen, *Naval Aristocracy of Hellenistic Rhodes*, 29–31, 116–20 (demes and *ktoinai* of Rhodes); Roussel, *Tribu et cité* (*phylai*); Lambert, *Phratries of Attica*; Gabrielsen, *Naval Aristocracy of Hellenistic Rhodes*, 141–49; Cordano, *Le tessere pubbliche* (phratries, *patrai* in Attica, Rhodes, Camarina and elsewhere); Shipley, *A History of Samos*, 184–85 (*hekatostyes* in Samos); Jones, *Associations of Classical Athens* (public associations of classical Athens). On the status of the Athenian *gene*: Bourriot, *Recherches*, 25–26, 199–235, 526–47, with 1180–1346, 1349–61; Roussel, *Tribu et cité*, 65–78; Smith, 'The Clans of Athens', 51–61; Parker, *Athenian Religion*, 56–66, 284–327; Jones, *Associations of Classical Athens*, 242–49 (except fragment 3 of [Arist.] *Ath.Pol.*, whose value is dubious, none of the other Athenian groups attested as *gene* were public associations proper); Gabrielsen, *Naval Aristocracy of Hellenistic Rhodes*, 149–51 (*diagoniai* of Rhodes).

[16] Military service by metics and mercenaries: Parke, *Greek Mercenary Soldiers*; Whitehead, *Ideology*, 82–86; McKechnie, *Outsiders in the Greek Cities*, 79–100; Bettalli, *I mercenari*; Burckhardt, *Bürger und Soldaten*. Navy: Gabrielsen, 'Socio-Economic Classes', 204–12.

[17] Athens: Whitehead, *Ideology*. Greek world: Whitehead, 'Immigrant Communities'; McKechnie, *Outsiders in the Greek Cities*; Lonis, *L'étranger*, esp. Gauthier, 'Météques'. Fugitives: Seibert, *Die politischen Flüchtlinge.*

[18] In his mid-fourth-century BC treatise on how Athens can increase her revenues, Xenophon includes proposals for improving the care (*epimeleia*) to be shown to those foreigners who lived permanently in Athens and also to those who came on short business visits there (Xen. *Vect.* 2–3). Xenophon has many good recommendations, including the building of lodging houses for ship owners and houses and shops for retail traders (3.12–13), but he has not a single recommendation regarding organizations that would ensure a stronger attachment of non-citizens to the Athenian *polis.*

[19] Heuss, *Stadt und Herrscher*; Jones, *The Greek City*, 95–112, 157–69; Orth, *Königlicher Machtanspruch*, 179–80; Gauthier, 'Les cités hellénistiques' (1984); Gauthier, 'Les cités hellénistiques' (1993); Gruen, 'The Polis'; Billows, 'Cities'. Cf. also Savali, 'I neocittadini'.

[20] See e.g. Billows, 'Cities', 209; *contra* Green, 'The Polis', 155–70 (general decline compared with the classical city-states); cf. also Hansen, *Polis*, 133. Jones (*Associations of Classical Athens*, 302–06) argues for a decline of the public associations of Classical Athens after the Macedonian takeover; so also Lambert, *Phratries of Attica*, 275, for the Athenian phratries. Possible differences in *polis*-organization before and after *c.*130 BC: Gauthier, 'Les cités hellénistiques' (1993), 212. Public subdivisions still serving traditional purposes: Blümel, 'Vertrag zwischen Latmos und Pidasa'; Jones, 'The Union of Latmos and Pidasa': *phylai* and *phratoriai* into which citizens are distributed in the treaty (*sympoliteia*) between Latmos and Pidasa (323–313 BC). According to the *politikos nomos*, citizens of Alexandria were registered in demes: *P. Hal*, lines 245ff. In Eumenes II's grant of *polis* status to Tyriaion (Phrygia, 180s BC), the distribution of citizens to *phylai* is mentioned together with 'laws', 'the council' and 'the magistrates': *SEG* 47.1745; Jonnes and Ricl, 'A New Royal Inscription'. Public subdivisions attain a new role in the relationship between *polis* and monarch, particularly in the structure of cultic honours: *SEG* 41.1003 II, lines 19–24 (*c.*203 BC), and Sahin, 'Ein neues Dekret', 13–18 (*symmoriai* of Teos for Antiochos III and his family); *SEG* 41.75, cf. Habicht, *Gottmenschentum*, 152 (deme of Rhamnous for Antigonos Gonatas); cf. also *I. Ilion* 31 and *OGIS* 11.24.26 (*phylai* of Ilion and of Priene, respectively, and their relation to royal cult). *I.Iasos* 5 (*c.*196 BC) is a decree of an Iasian *phyle* which is independently honouring Antiochos III and his family. Finally, *phylai* and other public subdivisions appear to be of vital importance for the functioning of the *polis* of Rhodes: Gabrielsen, *Naval Aristocracy*, 24–31, 112–23, 141–54.

[21] The term 'voluntary' association seems to be a nineteenth-century coinage: Black, *The Association*, 1, 28–29; Langford, *A Polite and Commercial People*, 553.

[22] Poland, *Geschichte*, 330–37.

[23] Poland, *Geschichte*, 289–98, 328–29.

[24] Poland, *Geschichte*, 74–75, 78, 366.

[25] Poland, *Geschichte*, 5–172; Waltzing, *Étude historique*, IV, 1–242, esp. 236–42.

[26] Treaux, '*Koinon*', 39–46; cf. Poland, *Geschichte*, 163–67. Federations: Larsen, *Greek Federal States*, xiv, 4–10, 193, 196; Corsten, *Vom Stamm zum Bund*, 15–16.

[27] Note e.g. the inclusion of *IG* XII, 1 155*d* (a second-century BC decree of the private *koinon* of *Haliadai* and *Haliastai* from Rhodes) among the public decrees in Rhodes and Lewis, *The Decrees*, 266.

[28] De Robertis, *Storia delle corporazione*; cf. van Nijf, *Civic World*, 9–11.

[29] A recent assessment: the papers in Kloppenborg and Wilson, *Voluntary Associations.*

[30] Van Nijf, *Civic World*, 243–47; Harland, *Associations*; Porena, 'Forme di partecipazione'.

[31] The *necropoleis* of Rhodes: Fraser, *Rhodian Funerary Monuments*; Patsiada, 'Rhodiake taphike architektonike'. The house of the *Poseidoniastai* on Delos: Bruneau and Ducat, *Guide de Délos*, 116, with fig. 20.

[32] De Vries, 'The Industrial Revolution'. De Vries introduced the concept 'industrious revolutions' in the context of economic history, and in order to stress the substantial contribution to the economy made by a long series of changes in the pattern of household consumption and household organization under the 'old regime'. Here, 'industrious revolutions' is used in a broader context than the specifically economic one, but still in such a way as to underline the contribution of small-scale and piecemeal institutional innovations that are pursued and accomplished by a plurality of actors or societal units. Even though they are dispersed over a wider area, these units interact with each other, sometimes directly, at other times indirectly, but always (or mostly) in such a way that innovations introduced by one of them to a degree build on those already made by others; the main unit to which the changing patterns of consumption are here related is not the 'household', but the 'privately organized group'; and what is 'consumed' are not foodstuffs and other material goods, but the variety of advantages and benefits ingrained in associational life.

[33] Poland, *Geschichte*, 337–38.

[34] Alternatives to existing *poleis* were posited by several ancient thinkers, among others Aristophanes (*Birds*, *Ecclesiazousai*), and Plato (*Republic*, *Kritias*, *Laws*): on these and further utopias, see now the papers collected in Hansen, *The Imaginary Polis*, esp. the editor's introduction on pp. 9–24; and, of course, alongside *poleis* with a democratic constitution there were also those with an oligarchic constitution as well as those governed by a tyrant (or *strategos*)—for example, Syracuse under Hieron (478–66) and in 404 BC, when one Dionysios was elected *strategos* (Diod. 13.95.1), cf. the proposal to elect Praxagora as *strategos* in Aristophanes' *Eccl.* 247–48. On the 'non-*polis*' regions of Greece, discussed in the Manchester Seminar Series 'Alternatives to the Democratic *Polis*', see Davies, 'The "Origins"', 27, with his n. 21.

[35] See the detailed study of Poland, *Geschichte*, 337–423; more generally, Wilson, 'Voluntary Associations'; Gabrielsen, 'The Rhodian Associations'.

[36] Detailed modern treatment of this aspect: van Nijf, *The Civic World*, 38–55.

[37] Murray, 'Cities of Reason'; Davies, 'The "Origins"', together with the other papers collected in Mitchell and Rhodes, *Development of the Polis*, esp. Osborne, 'Law and Laws' (pp. 74–82).

[38] Davies, 'Cultural, Social and Economic Features', 304–15 (section 'The Polis Transformed and Revitalized').

[39] *Xenia*: Herman, *Ritualised Friendship*. *Proxenia*: Gschnitzer, '*Proxenos*'; Marek, *Die Proxenie*; cf. Wallace, 'Early Greek *Proxenoi*', 189–208. A concrete example consists of the personal, overseas network of *xeniai* shared by leading Spartiates, in the late Archaic and early Classical periods, which, as S. Hodkinson correctly notes, 'gives a real meaning to Irad Malkin's recent phrase, "the Spartan Mediterranean"': Hodkinson in Mitchell and Rhodes, *Development of the Polis*, 93. This is one part of my brief answer to the scepticism expressed by one of the two anonymous readers of this article about my argument, that in expanding its public sphere (*to demosion*) the early *polis* cannibalized a number of private networks. The other (and briefer) part simply consists of a reference to the concise analysis offered by Davies, 'The "Origins"', esp. 29–31.

[40] For the text given here, see Ferguson, 'Attic Orgeones', 62–66, and Whitehead, *Demes of Attica*, 13. Extensive discussion (with further references) in Jones, *Associations of Classical Athens*, 33–45, 311–20, who argues for a somewhat different text (his p. 34). Arnaoutoglou ('Between *koinon* and *idion*', 72–3) doubts that this law clause is from Solon's time and that it can tell us anything about the relationship between state and associations in classical times; the same doubt is also expressed by one of the two anonymous readers. Decisive proof of the text's early (i.e. Archaic) origin is of course unattainable, but I agree with Ferguson, 'Attic Orgeones'; Whitehead, *Demes of Attica*; and most recently Ismard, 'Les associations', that, rather than being a later fabrication, the text transmits an early Athenian law-clause. What matters, however, is that the political centralization, as well as the subordination of public units and private

associations to *polis* law, which is reflected in the text, is a process amply documented by independent, contemporary evidence—see, for example, the papers in Hansen, *Ancient Greek City-State*; and Osborne, 'Law and Laws'.

[41] Jones, *Associations of Classical Athens*, 5–7, 13, 222–23, 302–4, 307–10. Cf. Parker, *Athenian Religion*, 335, 338. The point I make in the text holds true even if one assumes that private associations—despite of our lack of evidence to that effect—did exist in considerable numbers also in the fifth century—the reason why they are not documented in the fifth century supposedly being the non-survival of the relevant documents. However, if they really existed before *c.*400 in the form and numbers in which they are known after *c.*350 BC, then we also have to assume that we are dealing with units that were signally introvert, epigraphically absolutely mute and historiographically utterly unnoticed; I consider that assumption to be erroneous. The suggestion of one anonymous reader, that private associations appeared for the first time at Athens and from there spread to the rest of the Greek-speaking world, seems to me be unlikely for two reasons: (a) Aristotle (*Eth.Nic.* 1160a20–22) speaks of *eranoi* and *thiasoi* in general—that is, without having specifically Athens in mind, and at a time when especially the *eranos*-type of association is still something relatively new at Athens (cf. below); (b) at about that time, the Athenians used *eranos* about two different institutions, the *eranos*-loans (e.g. Dem. 53.8) and the private *eranos*-associations (e.g. *IG* II^2 2935, of 324/3 BC), see note 49 below. Thus, one can equally posit the opposite scenario—that is, that developments in fourth-century Athens were being influenced by developments elsewhere.

[42] *IG* II^2 2935. A fuller list of documents in Parker, *Athenian Religion*, Appendix 4, 333–42; and Jones, *Associations of Classical Athens*, Appendix 1, 307–10.

[43] *IG* II^2 1265 (of a year between 320 and 296 BC) and *SEG* 41.82 (*c.*300–280? BC).

[44] *IG* II^2 1258 (324/3 BC). Earlier mention: *Agora* XIX, no. P26 (of *c.*342/1 BC). The decree of the *Eikadeis* stood within the shrine of Apollon Parnessios. *IG* II^2 2631 and 2632 are two *horoi* of the property of this *koinon*. By 'regular' association I mean that this one was organized like the other private *koina*.

[45] *IG* II^2 2347 ('post med. s. IV') evidences the activities of the association. *IG* II^2 1261 (*SEG* 16.108) carries three decrees of the *thiasos*, dated to 302/1, 301/0 and 300/229 BC.

[46] *IG* II^2 1262 (301/0), 1263 (300/299) 1271 (298/7), 1273 (281/0), cf. 2943, 2352.

[47] For an overview, see Parker, *Athenian Religion*, 333–36; see also O. Murray in Murray, *Sympotica*, 149–61.

[48] *Hetairiai*: Hyp. 4.8 (*c.*330–324 BC) refers to a law forbidding the formation of a *hetairikon* for the purpose of overthrowing the democracy. See generally, Calhoun, *Athenian Clubs*; Sartori, *Le eterie*; Welwei, 'Polisbildung'; Jones, *Associations of Classical Athens*, 223–27. Philosophical schools as *koinoniai*: Arist. *Eth.Nic.* 9.17 (1164b); Theophrastus apud Diog. Laert. 5.53, with Lynch, *Aristotle's School*, who, however (p. 127), rejects the dominant view that the Peripatos and other Athenian schools were organized as *thiasoi* devoted to the cult of Muses; cf. Jones, *Associations*, 227–34; Mason, '*Philosophiae*' (philosophical schools in later periods).

[49] On these *eranoi*: Finley, *Studies in Land and Credit*, 100–6, 289–93; Vondeling, *Eranos*; Millett, *Lending and Borrowing*, 153–59. On the *eranoi* attested in the *horoi*-inscriptions Finley (*Studies*, 102–3) writes: 'there is no genuine evidence of any kind for the alternative view that they are societies'. This statement cannot be correct for the post-330 BC period. [Arist.] *Ath.Pol.* 52.2 mentions lawsuits concerning *eranoi* and *koinoniai* (*eranikai* and *koinonikai dikai*). Since a clear separation is made here between terms which in the case of non-public associations proper were nearly synonymous (i.e. *eranoi* associations *were* also called *koina*), the *eranoi* here are very probably the 'friendly loan groups', see also Jones, *Associations*, 16 n.62, 222–23, 307–8, *pace* Ziebarth, *Das griechiche Vereinswesen*, 179. As to the *koinonikai dikai*, they may concern any kind of 'jointly own property'—see e.g., Dem 14.16: *ta koinonika* (here a category of property exempt from the trierarchy).

[50] *Orgeones*: special connection to phratries, *FGrHist* 328: Philochoros F 35a; see Andrewes, 'Philochoros'; Isaios 2.14. See Roussel, *Tribu et cité*, 134; Lambert, *Phratries of Attica*, 74–77; Jones, *Associations*, 215, 250; and esp. Ferguson, 'Attic Orgeones', 73–95, nos. 1–12 (his 'class A' documents concerning *orgeones of citizens*); and Ferguson, 'Orgeonika', 130–63. *Thiasoi*: *IG* II^2 1237 (as sub-groups of the *Dekeleieis*). See Lambert, *Phratries of Attica*, 81–93; Jones, *Associations*, 216–20. Jones (*Associations*, 218) suggests that 'the segmental relation of subgroup to phratry was a loose one and that in fact admission to or participation in the activities of the subgroup was tolerated in the cases of persons who would not qualify as bona fide *phrateres*'—these persons he identifies as women and non-citizens (219–20). However, the sources adduced in support of this are late, and the argument about a loose segmental relation is not entirely convincing.

[51] *IG* I^3 1016 (mid-fifth century BC); *IG* II^2 2343 (early fourth century BC), on both documents, see Jones, *Associations*, 308.

[52] Pl. *Resp.* 1.327a: 'Thracian' and 'local' (i.e. citizen) worshippers of Bendis, *c.*411 BC; *IG* II^2 1283 (of 261/0 BC): decree of the Thracian *orgeones* in Piraeus, referring to the time at which their group and their shrine were established (lines 4–6). Ferguson ('Orgeonika', 132, cf. Jones, *Associations*, 257) dates the existence of this shrine to 429/8 BC. For documents of the citizen *orgeones* of Bendis: *IG* II^2 1361, 1324, see Ferguson, 'Attic Orgeones', 98–99. See also Simms, 'Cult of the Thracian Goddess Bendis', 59–76; Jones, *Associations*, 43, 256–57. The epigraphic evidence for the worship of Bendis in the Piraeus is catalogued by Garlan, *The Piraeus*, 231–33, nos. 37–46.

[53] *Orgeones* of the Great Mother: *IG* II^2 4609 (late fourth century BC, our earliest possible evidence for the existence of the group), 1314, 1315, 1316 (272/1); see Ferguson, 'Attic Orgeones', 107–15. *Orgeones* attested for the first time as forming a *thiasos* (in the Hellenistic sense of the word): Dow and Gill, 'Greek Cult Table', esp. 112, a palimpsest text underneath the *orgeones* degee *IG* II^2 1246 (early third century BC) and dated to a year after 316 BC.

[54] *IG* II^2 1361.20–23 (late C4 BC); cf. Parker, *Athenian Religion*, 337–38, with 170–1.

[55] *IG* II^2 337, esp. lines 33–5, 42–5. Parker, *Athenian Religion*, 337: 'Both groups [sc. the Kittian *emporoi* and the Egyptian worshippers of Isis] must have constituted a *koinon* or *thiasos* of some kind.'

[56] In *Eth.Eud.* 7.9.3 (1241b), after mentioning the *koinoniai* of body and soul, master and slave, etc., Aristotle writes: 'the other types of *koinoniai* are a part of the *koinoniai* of the *polis*, such as (those of) *phrateres* or *orgeones* [OCT: *orgeon*; Dietsche: *orgeonon*] or business partnerships (*hai chrematistikai* [sc. *koinoniai*]'. Arist. *Eth.Nic.* 8.9.4–6 (1160a): (4) 'But all *koinoniai* resemble parts, as it were, of the *koinonia* of the *polis*. For people travel for some advantage and to secure some of the necessities of life. And the *koinonia* of the *polis*, too, it seems, was originally formed and continues to endure for the sake of advantage. For it is at this that lawmakers take their aim; and they say that justice consists in what is advantageous to the community. (5) Now, the other *koinoniai* aim at some particular advantage (*sympheron*): for example, sailors (*ploteres*) at what is advantageous on a voyage for income from goods (*pros ergasian chrematon*) or some such thing; fellow soldiers (*systratiotai*) for what is advantageous in warfare (*tou kata ton polemon*), whether money or victory or the seizure of a city; and likewise *phyletai* and *demotai*. [And some *koinoniai* seem to come into existence for the sake of pleasure (*hedone*) – *koinoniai* of *thiasotai* and *eranistai*. For these (are organized) for the sake of sacrifice and social intercourse, respectively (*thysias heneka kai synousias*)]'; transl. Jones, *Associations*, 28. The text within brackets is regarded as authentic, but possibly in the wrong place.

[57] Gabrielsen, 'Rhodian Associations, 222, for comparison with the kleruchic *katoikiai*. See more generally Chaniotis, *War in the Hellenistic War*, 78–101.

[58] *IG* II^2 1275 (early third century BC).

[59] This point is well brought out, for imperial Rome, by Cracco-Ruggini, 'La vita associative'.

[60] Poland, *Geschichte*, 106–27, esp. 113; van Nijf, *The Civic World*, esp. 8. and index s.v. 'occupations'. Those involved in seaborne commerce: Vélissaropoulos, *Les naucleres grecs*, 91–124. The *koinon* of the merchants from Kition mentioned above might have been one of the earliest examples with a name revealing professional identity.

[61] Foucart, *Des associations religieuses*, 1–22; Poland, *Geschichte*, 173–74, 246–47; Pugliese Carratelli, 'Per la storia'.

[62] Davies, 'Cultural, Social and Economic Features', 283: 'That such groups took cult form and theophoric names not only reveals what sort of self-identification was most important (or most acceptable in the host town) but also suggests that many of the similarly named associations attested on Rhodes from the later third century onwards had comparable origins and functions.' Cf. van Nijf, *The Civic World*, 8, 40.

[63] Burkert, *Greek Religion*, 272–75.

[64] Maiuri, *Nuova silloge epigrafica*, 29–32, no. 19 (Rhodes).

[65] B. D. Meritt, *Hesperia* 30 (1961) 229–30, nos. 28 and 29. Improved text: Robert, 'Deux décrets', 7–14 [= L. Robert, *OMS* VII, 713–20], esp. p. 7: 'piously' (*eusebos*) in line 7 of the first decree (securely restored) and line 6 of the second. Robert cites further evidence for such *oikoi* of *naukleroi* in Athens (e.g. *IG* II^2 2350) and elsewhere; cf. Vélissaropoulos, *Les naucleres grecs*, 104–6.

[66] Vélissaropoulos, *Les naucleres grecs*, 106–10, and esp. Rauh, *Sacred Bonds*.

[67] The first group to appear on record is that of the 'farmers' residing at Lindos: *I.Lindos* 229 (137 BC), lines 3–4 (line 4 ought to be read as: *kai ge[orgeunton en] tai [Lindiai]*). The merger is attested in *I.Lindos* 384b (ca. 9 BC), lines 15–17: *[toi katoikeu]ntes en Lindia poli kai georgeuntes/ [kai] areuntes en ta Lindia philodoxias* etc., which now can be fairly securely restored to read: *[toi katoikeu]ntes en Lindia poli kai georgeuntes/ [kai naukl]areuntes en ta Lindia philodoxias* etc.; cf. *I.Peraia* (IK 38) 514 (163/4 AD). Privileged position of the *georgeuntes* group: *I.Lindos* 349 (38 BC), i.e. honours awarded to a priest of Athana Lindia exclusively by public bodies and officials, with the sole exception of the *georgeuntes*, who, furthermore, are listed *before* the twelve Lindian demes. That they were a single group of wealthy foreign landowners (or managers of large agricultural estates) residing in Lindos: Gabrielsen, *Naval Aristocracy*, 129.

[68] *IG* II^2 1012 (*Syll.*^{3} 706), esp. lines 22–23: *ten boulen epikyrosai heautoi psephisma*, to be compared to the fourth-century resolution regarding the petition of the Kitian *emporoi* (*IG* II^2 337, cf. above p. 192), in which the Athenian Assembly simply acknowledges that the petition is made 'in a lawful manner'. On the State's ratification (*epikyrosis*) of decrees issued by its public subdivisions: Gabrielsen, 'Subdivisions of the State'. On the decree of this *synodos*, see also Vélissaropoulos, *Les naucleres grecs*, 104.

[69] *IG* XII, 1, 127, cf. Gabrielsen, 'Rhodian Associations', 231–32.

[70] *IGR* 4, 1632, cf. Waltzing, *Étude historique* III, 51, no. 146 (*eriourgoi*); Waltzing, *Étude historique* III, 52, no 147 (*skyteis*). Saittai: Kolb, 'Sitzstufeninschriften', esp. 115, 117, nos. 36–41; *SEG* 40. 1063. See Poland, *Geschichte*, 154; van Nijf, *The Civic World*, 20, 184, 233.

[71] *IG* XII, 1 155 d (I i), lines 90–103. I am indebted to Dr Lene Rubinstein for drawing my attention to the degree to which *polis* legal procedures are imitated in these lines.

[72] On Roman law and *collegia*: Cotter, 'The Collegia'; Arnaoutoglou, 'Roman Law' (citing further works), who rightly points out that Roman restrictions of, or bans on, associations were temporary and localized.

[73] *Senatus Consultum de Bacchanalibus*: *ILS* 18 (=*FIRA* I^2 30), cf. Pailler, *Bacchanales*; Beard, North, and Price, *Religions of Rome*, 92–8. The ban of 64 BC (lifted in 58 BC): Asconius, *Pis.* 7; Asconius, *Corn.* 67; cf. de Robertis, *Storia delle corporazione*, 83–108; Ausbüttel, *Untersuchungen*, 85–86.

[74] Buckler, 'Labour Disputes'; *I.Ephesos* 215. For a silversmiths' riot when Paul was visiting Ephesos: *Acts* 19.23–8; cf. van Nijf, *The Civic World*, 15, 238–39.

[75] Aneziri, *Die Vereine*, 159–64, 194 n. 120: honours voted to kings, prominent Romans, public officials in kingdoms or cities and royal courtiers. Cf. Le Guen, *Les associations de technites* II, 80–81.

[76] Le Guen, *Les associations de technites*; Aneziri, *Die Vereine*.

[77] Lorber and Hoover, 'An Unpublished Tetradrachm', 59–68 (pls. 15–17), who propose the date *c.*155–154 BC. I am indebted to Dr Panos Iossif, who brought the coin and the relevant publication to my attention.

[78] On all of the above points, except (c), see Welles, *RC* 53, now Le Guen, *Les associations* I, 243–50, no. 47; and Aneziri, *Die Vereine*, 387–91, no. D12: Letter of Eumenes II (before 158 BC), esp. IIA, lines 3–5 (*koinodikion*), IIIA, lines 5–8 (proposal that the parties conclude a *syntheke* with a view to a *synoikismos*, cf. e.g., Welles *RC* 3/4, for Antigonos I's arrangement of a *synoikismos* between Lebedos and Teos), IIC, lines 10–15 (fiscal division between the parties of the harbours, *polis* and *chorai* of Teos). The power-political aspects are particularly emphasized by Le Guen, *Les associations* II, 77–82, 100–02. See also Aneziri, *Die Vereine*, 155, 187–88, 292, 302–4, 307–16 (99–100: *koinodikion*).

References

Allen, R. J. *The Clubs of Augustan London*. Cambridge, Mass: Harvard University Press, 1933.

Andersen, H. C. *Dagbøger, 1825–1875*, 12 vols, edited by H. Vang Lauridsen, T. Gad, and K. Weber. Det Danske Sprog-og Litteraturselskab. Copenhagen: Gad, 1971–76.

Andrewes, A. 'Philochoros on Phratries'. *Journal of Hellenic Studies* 81 (1961): 1–15.

Aneziri, S. *Die Vereine der dionysischen Techniten im Kontext der hellenistischen Gesellschaft. Untersuchungen zur Geschichte, Organization und Wirkung der hellenistischen Technitenvereine*. Historia Einzenschriften, no. 162. Stuttgart: Franz Steiner Verlag, 2003.

Arnaoutoglou, I. 'Between *koinon* and *idion*: Legal and Social Dimensions of Religious Associations in Ancient Athens'. In *Kosmos. Essays in Order, Conflict and Community in Classical Greece*, edited by P. Cartledge, P. Millett, and S. von Reden, 68–83. Cambridge: Cambridge University Press, 1998.

———. 'Roman Law and *Collegia* in Asia Minor'. *Revue internationale des droits de l'Antiquité* 49. (2002): 27–44.

———. *Thusias heneka kai synousias. Private Religious Associations in Hellenstic Athens*. Yearbook of the Research Centre for the History of Greek Law, no. 37, Suppl., no. 4. Athens: Academy of Athens, 2003.

Ascough, R. *Paul's Macedonian Associations: The Social Context of* Philipians *and* 1 Thessalonians. Wissenschaftliche Untersuchungen zum Neuen Testament, no. 161. Tübingen: Mohr Siebeck, 2004.

Ausbüttel, E. M. *Untersuchungen zu den Vereinen im Westen des römischen Reiches*. Frankfurter althistorische Studien, no. 11. Kallmünz: Franz Steiner Verlag, 1982.

Beard, M., J. North, and S. Price. *Religions of Rome*. Vol. 1: *A History*. Cambridge: Cambridge University Press, 1998.

Bettalli, M. *I mercenari nel mondo Greco*. Vol. 1. Pisa: ETS, 1995.

Billows, R. 'Cities'. In *A Companion to the Hellenistic World*, edited by A. Erskine, 196–215. Blackwell Companions to the Ancient World. Oxford: Blackwell Publishing, 2005.

Black, E. C. *The Association: British Extraparliamentary Political Organization, 1769–1793*. Cambridge, Mass.: Harvard University Press, 1963.

Blümel, W. 'Vertrag zwischen Latmos und Pidasa'. *Epigraphica Anatolica* 29 (1997): 135–42.

Bollmann, B. *Römische Vereinhauser. Untersuchungen zu den Scholae der römischen Berufs- kult-, und Augustalen-kollegien in Italien.* Mainz: Philipp von Zabern, 1998.

Bourriot, F. *Recherches sur la nature du génos. Études d'histoire sociale athénienne—periode archaique et classique.* Lille and Paris: Atelier Reproduction des Theses, 1976.

Bruneau, P. *Recherches sur les cultes de Délos à l'époque hellénistique et à l'époque impériale.* BEFAR, no. 217. Paris: De Boccard, 1970.

Bruneau, P. and J. Ducat. *Guide de Délos.* 2nd ed. Paris: De Boccard, 1966.

Buckler, W. H. 'Labour Disputes in the Province of Asia'. In *Anatolian Studies Presented to Sir William Mitchell Ramsay*, edited by W. H. Buckler and W. M. Calder, 27–50. Manchester: Manchester University Press, 1923.

Burckhardt, L. *Bürger und Soldaten: Aspekte der politischen und militärischen Rolle athenischer Bürger im Kriegswesen des 4. Jahrhunderts v. Chr.* Historia Einzelschriften, no. 101. Stuttgart: Franz Steiner Verlag, 1996.

Burkert, W. *Greek Religion: Archaic and Classical.* Oxford: Blackwell Publishers, 1987.

Calhoun, G. M. *Athenian Clubs in Politics and Litigation.* Bulletin of the University of Texas. Humanistic Series, no. 14. Austin, 1913. Reprinted. New York: Burt Franklin, 1970.

Chaniotis, A. *War in the Hellenistic War: A Social and Cultural History.* Oxford: Blackwell Publishers, 2005.

Clark, P. *British Clubs and Societies, 1580–1800: The Origins of an Associational World.* Oxford: Oxford University Press, 2000.

Cordano, F. *Le tessere pubbliche dal tempio de Atene a Camarina.* Rome: Istituto Italiano per la storia antica, 1992.

Corsten, T. *Vom Stamm zum Bund: Gründung und territoriale Organisation griechischer Bundesstaaten.* Studien zur Geschichte Nordwest-Griechenlands, no. 4. Munich: Oberhummer Gesellschaft, 1999.

Cotter, W. 'The Collegia and Roman Law: State Restrictions on Voluntary Associations, 64 BCE—200 CE'. In *Voluntary Associations in the Graeco-Roman World*, edited by J. S. Kloppenborg and S. G. Wilson, 74–89. London and New York: Routledge, 1996.

Cracco-Ruggini, L. 'La vita associative nelle città dell'Oriente Greco: tradizioni locali e influenze romane'. In *Assimilation et résistance à la culture gréco-romaine dans le monde ancient. Travaux du VIème Congrès International d'Études classiques. Madrid, Septembre 1974*, edited by D. M. Pippidi, 463–91. Bucarest and Paris: Les Belles Lettres, 1976.

Davies, J. K. 'Cultural, Social and Economic Features of the Hellenistic World'. In *The Cambridge Ancient History.* Vol. 7, part I: *The Hellenistic World*, edited by F. Walbank, A. E. Astin, M. W. Frederiksen, and R. M. Ogilvie, 257–320. Cambridge: Cambridge University Press, 1984.

———. 'The "Origins of the Greek *Polis*"'. In *The Development of the Polis in Archaic Greece*, edited by L. G. Mitchell and P. J. Rhodes, 24–38. London and New York: Routledge, 1997.

de Cenival, F. *Les associations religieuses en Égypte d'après les documents démotiques.* Cairo: Institut français d'archéologie orientale du Caire, 1972.

de Robertis, F. M. *Storia delle corporazione e del regime associativo nel mondo romano.* Bari: Adriatica, 1971.

de Ste. Croix, G. E. M. *The Class Struggle in the Ancient World from the Archaic Age to the Arab Conquests.* London: Duckworth, 1981.

de Vries, J. 'The Industrial Revolution and the Industrious Revolution'. *Journal of Economic History* 54, no. 2 (1994): 249–70.

Dow, S. and D. H. Gill. 'The Greek Cult Table'. *American Journal of Archaelogy* 59 (1965): 104–14.

Ferguson, W. S. 'The Attic Orgeones'. *Harvard Theological Review* 37 (1944): 61–140.

———.'Orgeonika'. *Commemorative Studies in Honor of Theodore Leslie Shear.* Hesperia Suppl., no. 8, 130–63. Princeton, NJ: American School of Classical Studies at Athens, 1949.

Finley, M. I. *Studies in Land and Credit in Ancient Athens, 500–200 BC*. New Brunswick, NJ: Rutgers University Press, 1952.

Fisher, N. R. E. 'Greek Associations, Symposia, and Clubs'. In *Civilizations of the Ancient Mediterranean: Greece and Rome*, edited by M. Gant and R. Kitzinger, 1167–97. New York: Charles Scribner Sons, 1988.

Foucart, P. F. *Des associations religieuses chez les grecs. Thiases, éranes, orgéones*. Paris, 1873. Reprinted. New York: Arno Press, 1975.

Françoise, E., ed. *Sociabilité et société bourgeoise en France, en Allemagne, et en Suisse 1750–1850*. Paris: Recherche Sur Les Civilisations, 1986.

Franklin, B. *Autobiography*, edited by J. A. Leo Leway and P. M. Zall. New York: Norton, 1986.

Fraser, P. M. *Rhodian Funerary Monuments*. Oxford: Oxford University Press, 1977.

Gabrielsen, V. 'Subdivisions of the State and their Decrees in Hellenistic Rhodes'. *Classica et Mediaevalia* 45 (1994): 117–35.

———. *The Naval Aristocracy of Hellenistic Rhodes*. Studies in Hellenistic Civilization, no. 6. Aarhus: Aarhus University Press, 1997.

———. 'The Rhodian Associations and Economic Activity'. In *Hellenistic Economies*, edited by Z. H. Archibald, J. K. Davies, V. Gabrielsen, and G. J. Oliver, 215–44. London: Routledge, 2001.

———. 'Socio-Economic Classes and Ancient Greek Warfare'. In *Ancient History Matters. Studies Presented to Jens Erik Skydsgaard on His Seventieth Birthday*. Analecta Romana Instituti Danici, Suppl., no. 30, edited by K. Ascani, V. Gabrielsen, K. Kvist, and A. Holm Rasmussen, 203–20. Rome: L'Erma di Bretschneider, 2002.

Garlan, R. *The Piraeus from the Fifth to the First Century BC*. London: Duckworth, 1987.

Gauthier, P. 'Les cités hellénistiques: épigraphie et histoire des institutions et des régimes politiques'. In *Acts of the Eighth International Conference on Greek and Latin Epigraphy, Athens, 3–9 October, 1982*, 82–107. Athens: Musée Épigraphique, 1984.

———.'Métèques, périèques et *paroikoi*: bilan et points d'interrogation'. In *L'étranger dans le mond grec*, 2 vols, edited by R. Lonis, 23–46. Nancy: Presses Universitaires de Nancy, 1988.

———.'Les cités hellénistiques'. In *The Ancient Greek City-State*, Acts of the Copenhagen Polis Centre, no. 2. Det Kongelige Danske Videnskabernes Selskab, Historisk-filosofiske Meddelelser, no. 67, edited by M. H. Hansen, 211–31. Copenhagen: The Royal Danish Academy of Sciences and Letters, 1993.

Green, D. G. *Reinventing Civil Society: The Rediscovery of Welfare Without Politics*. London: IEA Health and Welfare Unit, 1993.

Green, P. *Alexander to Actium: The Hellenistic Age*. Berkeley: University of California Press, 1990.

Gruen, E. S. 'The Polis in the Hellenistic World'. In *Nomodeiktes. Greek Studies in Honor of Martin Ostwald*, edited by R. M. Rosen and J. Farrel, 339–54. Ann Arbor: University of Michigan Press, 1993.

Gschnitzer F. s.v.'*Proxenos*'. *Real-Encyclopädie der klassischen Altertumswissenschaft* Suppl. XIII (1973): 629–730.

Habicht, C. *Gottmenschentum und griechische Städte*. 2nd ed. Zetemata, no. 14. Munich: Beck, 1970.

Hansen, M. H., ed. *The Ancient Greek City-State. Symposium on the Occasion of the 250th Anniversary of the Royal Danish Academy of Sciences and Letters, July 1–4 1992*. Historisk-filosofiske Meddelelser, no. 67. Det Kongelige Danske Videnskabernes Selskab. Copenhagen: The Royal Danish Academy of Sciences and Letters, 1993.

———. *Polis and City-State. An Ancient Concept and its Modern Equivalent*. Acts of the Copenhagen Polis Centre, no. 5. Det Kongelige Danske Videnskabernes Selskab, Historisk-filosofiske Meddelelser, no. 78. Copenhagen: The Royal Danish Academy of Sciences and Letters, 1998.

———. *The Imaginary Polis. Symposium, January 7–10, 2004*. Acts of the Copenhagen Polis Centre, no. 7. Historisk-filosofiske Meddelelser, no. 91. Det Kongelige Danske Videnskabernes Selskab. Copenhagen: The Royal Danish Academy of Sciences and Letters, 2005.

———. *Polis: An Introduction to the Ancient Greek City-State*. Oxford: Oxford University Press, 2006.

Harland, P. A. *Associations, Synagogues and Congregations: Claiming a Place in Ancient Mediterranean Society*. Minneapolis: Fortress Press, 2003.

———. 'Spheres of Contention, Claims of Pre-eminence: Rivalries among Associations in Sardis and Smyrna'. In *Religious Rivalries and the Struggle for Success in Sardis and Smyrna*. Studies in Judaism and Christianity, no. 18, edited by R. Ascough, 53–63. Waterloo: Wilfrid Laurier University Press, 2005.

Herman, G. *Ritualised Friendship and the Greek City*. Cambridge: Cambridge University Press, 1987.

Heuss, A. *Stadt und Herrscher des Hellenismus in ihren staats- und völkerrechtlichen Beziehungen*. Klio Beiheft. no. 39. Leipzig: Akademie Verlag, 1937.

Inschriften von Parion, edited by P. Frisch. *Die Inschriften von Parion*. Inschriften griechischer Städte aus Kleinasien, no. 25. Bonn: Dr. Rudolf Habelt, 1983.

Ismard, P. 'Les associations en Attique de Solon à Clisthène'. In *Individus, groupes et politique à Athénes de Solon à Mithridate. Actes du colloque international, Tours, 7–8 mars 2005*, edited by J.-C. Couvenhes and S. Milanezi, 17–34. Tours: Presses Universitaires François Rabelais, 2007.

Jones, A. H. M. *The Greek City from Alexander to Justinian*. Oxford: Oxford University Press, 1940.

Jones, C. P. 'The Union of Latmos and Pidasa'. *Epigraphica Anatolica* 31 (1999): 1–7.

Jones, N. F. *Public Organization in Ancient Greece. A Documentary Study*. Memoires of the American Philosophical Society, no. 176. Philadelphia: American Philosophical Society, 1987.

———. *The Associations of Classical Athens: The Response to Democracy*. Oxford: Oxford University Press, 1999.

Jonnes, L. and M. Ricl. 'A New Royal Inscription from Phrygia Paroreios: Eumenes II Grants Tyriaion the Status of a Polis'. *Epigraphica Anatolica* 29 (1997): 1–29.

Kloppenborg, J. S. and S. G. Wilson, eds. *Voluntary Associations in the Graeco-Roman World*. London and New York: Routledge, 1996.

Kolb, F. 'Sitzstufeninschriften aus dem Stadion von Saittai (Lydien)'. *Epigraphica Anatolica* 15 (1990): 107–19.

Labaree L. W., E. R. Cohn, B. B. Oberg, and W. B. Willcox. *The Papers of Benjamin Franklin*, Vol. 9. New Haven, Conn.: Yale University Press, 1996.

Lambert, S. D. *The Phratries of Attica*. Ann Arbor: University of Michigan Press, 1993.

Langford, P. *A Polite and Commercial People: England 1727–1783*. Oxford: Oxford University Press, 1989.

Larsen, J. A. O. *Greek Federal States: Their Institutions and History*. Oxford: Clarendon Press, 1968.

Le Guen, B. *Les associations de technites dionysiaques à l'époque hellénistique*. Vol. 1: *Corpus documentaire*, II: *Synthèse*. Études d'Archéologie Classique 11–12. Nancy: De Boccard, 2001.

Leiwo, M. 'Religion or Other Reasons? Private Associations in Athens'. In *Early Hellenistic Athens. Symptoms of a Change*, Papers and Monographs of the Finnish Institute at Athens, no. 6, edited by F. Rösén, 103–17. Athens: Finnish Institute at Athens, 1997.

Liebenam, W. *Zur Geschichte und Organisation des römischen Vereinswesen: 3 Untersuchungen*. Leipzig: B. G. Teubner, 1890.

Lonis, R., ed. *L'étranger dans le mond grec*, 2 vols. Nancy: Presses Universitaires de Nancy, 1988–92.

Lorber, C. C., and O. D. Hoover. 'An Unpublished Tetradrachm Issued by the Artists of Dionysos'. *The Numismatic Chronicle* 163 (2003): 59–68.

Lynch, J. P. *Aristotle's School: A Study of a Greek Educational Institution*. Berkeley: University of California Press, 1972.

MacKenney, R. R. *Tradesmen and Traders: The World of the Guilds in Venice and Europe, c. 1250–c. 1640*. Totowa: Barnes and Noble Books, 1987.

G. Maier. 'Eranos als Kreditinstitut.' Diss., Erlangen-Nürenberg, 1969.

Maiuri, A. *Nuova silloge epigrafica di Rodi e Cos.* Florence: Le Monnier, 1925.

Marek, C. *Die Proxenie.* Frankfurt am Main: Peter Lang Verlagsgruppe, 1984.

Mason, S. N. '*Philosophiae*: Greco-Roman, Judean and Christian'. In *Voluntary Associations in the Graeco-Roman World*, edited by J. S. Kloppenborg and S. G. Wilson, 31–58. London and New York: Routledge, 1996.

McKechnie, P. *Outsiders in the Greek Cities in the Fourth Century BC.* London and New York: Routledge, 1989.

Meritt, B. D. 'Greek Inscriptions'. *Hesperia* 30 (1961): 229–32, nos. 28 and 29.

Millett, P. *Lending and Borrowing in Ancient Athens.* Cambridge: Cambridge University Press, 1991.

Mitchell, L. G., and P. J. Rhodes, eds. *The Development of the Polis in Archaic Greece.* London and New York: Routledge, 1997.

Murray, O. 'Cities of Reason'. In *The Greek City from Homer to Alexander*, edited by O. Murray and S. Price, 1–25. Oxford: Clarendon Press, 1990.

Murray, O. *Sympotica: A Symposium on the Symposion.* Oxford: Clarendon Press, 1990.

———.'Rationality and the Greek City: The Evidence from Kamarina'. In *The Polis as an Urban Centre and as a Political Community*, Acts of the Copenhagen Polis Centre, no. 4. Det Kongelige Danske Videnskabernes Selskab, Historisk-filosofiske Meddelelser, no. 75, edited by M. H. Hansen, 493–504. Copenhagen: The Royal Danish Academy of Sciences and Letters, 1997.

Orth, W. *Königlicher Machtanspruch und städtische Freiheit. Untersuchungen zu den politischen Beziehungen zwischen den ersten Seleukidenherrschern (Seleukos I., Antiochos I., Antiochos II) und den Städten des westlichen Kleinasien.* Münchener Beiträge zur Papyrusforschung und antiken Rechtsgeschichte, no. 71. Munich: Beck, 1977.

Osborne, R. *Demos: The Discovery of Classical Attika.* Cambridge: Cambridge University Press, 1985.

———. 'Law and Laws: How do we Join Up the Dots?'. In *The Development of the Polis in Archaic Greece*, edited by L. G. Mitchell and P. J. Rhodes, 74–82. London and New York: Routledge, 1997.

Pailler, J.-M. *Bacchanales. La répression de 186 avant J. -C. à Rome et en Italie.* BEFAR, no. 270. Rome: De Boccard, 1988.

Parke, W. H. *Greek Mercenary Soldiers.* Oxford: Oxford University Press, 1933.

Parker, R. *Athenian Religion.* Oxford: Clarendon Press, 1996.

Patsiada, V. 'Rhodiake taphike architektonike'. In *Rhodos 2.400 Chronia: He polis tes Rhodou apo ten Idrysi tes mechri ten katalepse apo tous Tourkous (1523)*, Vol. 1, 127–36. Athens: Ypourgeio Politismou, 1999.

Perry, J. S. *The Roman Collegia: The Modern Evolution of an Ancient Concept.* Mnemosyne Suppl., no. 277. Leiden: Brill, 2006.

Poland, F. *Geschichte des griechischen Vereinswesens.* Fürstlich Jablonowskichen Gesellschaft zu Leipzig. Historisch-Nationalökonomischen Sektion 23: 38. Leipzig: Teubner, 1909.

Porena, P. 'Forme di partecipazione politica cittadina e contatti con il potere imperiale'. In *Politica e Partecipazione nelle Città dell'Impero Romano*, edited by F. Amarelli, 13–92. Rome: L'Erma di Bretschneider, 2005.

Pugliese Carratelli, G. 'Per la storia delle associazioni in Rodi antica'. *Annuario della Scuola Archeologica di Atene e delle Missioni Italiani in Oriente* 22 n.s. 1–2 (1939–40): 147–200.

Radin, M. *The Legislation of Greeks and Romans on Corporations.* New York: Columbia, 1909.

Randazzo, S. 'Senatus consultum quo illicita collegia arcentur (D. 47, 22, 1, 1)'. *Bollettino dell' Istituto di Diritto Romano* 33–4 (1991–92): 49–88.

Rasmussen, F. L. K. *Peri ton Eranon.* Copenhagen, 1833.

Rauh, N. K. *The Sacred Bonds of Commerce: Religion, Economy, and Trade Society in Hellenistic and Roman Delos.* Amsterdam: J.C. Gieben, 1993.

Rhodes, P. J. and D. M. Lewis. *The Decrees of the Greek States.* Oxford: Oxford University Press, 1997.

Robert, L. 'Deux décrets d'une association à Athènes'. *Archailogike Ephemeris* (1969): 7–14.

Roussel, D. *Tribu et cité. Études sur les groupes sociaux dans les cités grecques aux époques archaique et classique*. Paris: Les Belles Lettres, 1976.

Sahin, S. 'Ein neues Dekret der Symmoria zu Ehren ihre Prostatai in Teos'. *Epigraphica Anatolica* 5 (1985): 13–18.

san Nicolò, M. *Ägyptisches Vereinswesen zur Zeit der Ptolemäer und Römer*, 2 vols. Münchener Beiträge zur Papyrus Forschung und antiken Rechtsgeschichte, no. 2. Munich: Beck, 1913–15.

Sartori, F. *Le eterie nella vita politica ateniese del VI e V secolo a.C.* Rome: L'Erma di Bretschneider, 1957.

Savali, I. 'I neocittadini nelle città ellenistiche'. *Historia* 34 (1985): 387–431.

Seibert, J. *Die politischen Flüchtlinge und Verbannten in der griechischen Geschichte.*, 2 vols. Darmstadt: Wissenschaftliches Buchgesellschaft, 1979.

Shipley, G. *A History of Samos 800–188 BC*. Oxford: Oxford University Press, 1987.

Siewert, P. *Die Trittyen Attikas und die Herresreformen des Kleisthenes*. Vestigia, no. 33. Muchen: Beck, 1982.

Simms, R. R. 'The Cult of the Thracian Goddess Bendis in Athens and Attica'. *Ancient World* 18 (1988): 59–76.

Smith, R. C. 'The Clans of Athens and the Historiography of the Archaic Period'. *Échos du Monde Classique. Classical Views* 4 (1985): 51–61.

Thrupp, S. 'The Gilds'. In *The Cambridge Economic History of Europe, III: Economic Organization and Policies in the Middle Ages*, edited by M. Postan, E. E. Rich, and E. Miller, 230–80, 624–34. Cambridge: Cambridge University Press, 1963.

Tod, M. N. 'Clubs and Societies in the Greek World, Lecture Three'. In *Sidelights on Greek History: Three Lectures*, edited by M. N. Tod, 71–93. Oxford: Basil Blackwell, 1932.

Trail, J. S. *Demes and Trittys. Epigraphical and Topographical Studies in the Organization of Attica*. Toronto: Athenians, 1986.

Treaux, J. '*Koinon*'. *Revue des Études Anciennes* 89 (1987): 39–46.

van Nijf, O. M. *The Civic World of Professional Associations in the Roman East*. Dutch Monographs on Ancient History and Archaeology, no. 17. Amsterdam: Gieben, 1997.

Vélissaropoulos, J. *Les nauclères grecs: Recherches sur les institutions maritimes en Grèce et dans l'Orient hellénisé*. Hautes Études du Monde Gréco-Romain, no. 9 Geneva and Paris: Droz and Champion, 1980.

Vondeling, J. *Eranos*. Groningen: J. B. Wolters, 1961.

Wallace, M. B. 'Early Greek *Proxenoi*'. *Phoenix* 24 (1970): 189–208.

Waltzing, J.-P. *Étude historique sur les corporations professionelles chez les Romains depuis les origins jusqu'à la chute de l'Empire d'Occident*, 4 vols. Louvain: Peeters, 1895–1900.

Welwei, K.-W. 'Polisbildung, Hetairos-Gruppen und Hetairien'. *Gymnasium* 99 (1992): 481–500.

Whitehead, D. *The Ideology of the Athenian Metic*. Proceedings of the Cambridge Philological Society, Suppl., no. 14. Cambridge: Cambridge Philological Association, 1977.

———. 'Immigrant Communities in the Classical Polis: Some Principles for a Synoptic Treatment'. *L'Antiquité Classique* 53 (1984): 47–59.

———. *The Demes of Attica 508/7—ca. 250 BC*. Princeton, NJ: Princeton University Press, 1986.

Wilson, S. G. 'Voluntary Associations: An Overview'. In *Voluntary Associations in the Graeco-Roman World*, edited by J. S. Kloppenborg and S. G. Wilson. London and New York: Routledge, 1996.

Ziebarth, E. *Das griechiche Vereinswesen*. Fürtstlich Jablonowskichen Gesellschaft zu Leipzig, Historisch-Nationalökonomischen Sektion 21: 34. Leipzig: Hirzel, 1896.

Thessalians Abroad, the Case of Pharsalos

Maria Stamatopoulou

It has been some years since the publication of Tataki's excellent monograph *Macedonians Abroad*, which compiled the epigraphic, archaeological, and literary evidence for the presence of Macedonians outside their territory. The book in question has been valuable not only for scholars examining Macedonian history, but also for anyone interested in Hellenistic culture and history. Its proven usefulness inspired me to embark upon a similar project regarding the Thessalians, with the aim of understanding what the evidence tells us about links of the Thessalians with the outside world, and the types and frequency of these contacts, together with the regional and chronological patterns. Understanding the conduct, modes of self-representation, and self-promotion, as well as the interaction of Thessalian leading families at a regional and Pan-Hellenic level, might enable us to make better sense of Thessalian history and material culture in the first millennium BC.

I have chosen Pharsalos as a case study because it was one of the leading Thessalian cities in antiquity, the capital of the tetrad Phthiotis, and because there is ample and varied evidence about the city.[1]

Pharsalos is situated in southern Thessaly, on the southern border of the valley of the river Enipeus, in a very favourable and strategic location (Figure 1). The city, with its acropolis on a steep, naturally defensible hill, controlled the crossroads of major routes that connected the Ionian sea with the Aegean and northern Greece with the south. Moreover the city commanded a very fertile *chora* and rich pasturelands. Literary sources stress the quality of the Pharsalian cavalry, which was considered the best

Figure 1 Map of Thessaly, after Reinders, *New Halos*, 22, fig. 4.

in Thessaly,[2] and which was a major factor determining its foreign policy. Pharsalos, similarly to other Thessalian cities, was a politically organized community (*polis* in the political sense) already in the fifth century.[3] References in Thucydides (iv.78.2) and Aristotle (*Pol.* 1306a11–12) suggest that it was governed by a strict oligarchy.

As with the rest of Thessaly, Pharsalos was famous in antiquity for its wealth, and notorious for the extravagant lifestyle of its leading families.[4] While Theopompus's attack on the Pharsalians' laziness, extravagance and licentious behaviour may have been motivated by his hatred towards Philip II and the Pharsalians' collaboration with the Macedonian king,[5] the continuous references in literary sources to the Thessalians' love of luxury and extravagance suggest that there may be some truth in the matter.

This article will examine what the existing evidence shows about Pharsalians in the international arena, the kind of networks they formed with the outside world, and any meaningful trends or differences from other Thessalian cities.

The sources for Thessalian history prior to the end of the sixth century are problematic and the archaeology of the region during these periods is not yet well known.[6] However, the two earliest references to Pharsalians abroad are characteristic of what will follow (Table 1, nos. 1–2): a) Kleomachos, leader of a cavalry contingent in aid of the Chalkidians during the Lelantine war (Plut. *Mor.* 760e–761b). The historicity of the Lelantine war has been a matter of debate, as has the character of the Thessalian help to the Chalkidians (state or private).[7] Given the supposed early date of the Lelantine war, it is more likely that Kleomachos's military aid to Chalkis should be seen in the context of ritualized guest-friendship, *xenia,* rather than a state initiative.[8] b) Phaidros, victor in the men's *stadion* in the 56th Olympiad (*c.*556). It is well known that participation in Pan-Hellenic games was one of the predominant means of self-representation and promotion of elites during the Archaic period.[9] These two features—military aid by members of the Pharsalian elite as a result of a personal friendship with their peers abroad and participation in athletic competitions—are the two key activities in which we will see Pharsalians abroad.

Evidence increases in the fifth century BC, a period during which Thessaly must have already been organized into tetrads (*moirai*). Epigraphic, literary, and archaeological evidence reveals that there were *poleis* with civic organizations and institutions, as well as considerable investment in the elaboration and monumentalization of public space.[10] The fifth century BC was marked by the increase in Athenian power and the struggle between Athens and Sparta to control the north. Thessalian foreign policy appears inconsistent, with shifts in alliances and absence of organized common action, since politics was shaped by the competing interests of influential individuals within the cities.[11]

During this period, two major families predominate in terms of the Pharsalian presence abroad: the Daochids and the Menonids.[12] As concerns the latter, we know from Demosthenes (23.199; 13.23) that Menon I contributed to Kimon's military campaign at Eion the enormous sum of 12 talents of silver and a cavalry contingent of 200 or 300 *penestai*, for which he was awarded citizenship by the Athenians (Table 1, no. 6). He is often identified with Menon, son of Menekleides, of the *deme* Gargettos, whose name is preserved on a large number of *ostraka* found in the Athenian

Table 1 List of pharsalians attested abroad, sixth–first centuries BC.

	Date	*Name*	*Present in*	*Role*	*Evidence*
	7th-6th centuries				
1	???	**Kleomachos**	Lelantine War	Commander of cavalry in aid of Chalkis	Plut. *Mor.* 760e–761b; *LGPN* III.B s.v. Kleomachos (19)
2	556	**Phaidros**	Olympia	Olympic victor in men's stadion, 56th Ol.	Moretti, *Olympionikai*, no. 107; *LGPN* III.B. s.v. Phaidros (3)
	5th century				
3	?484	**Agias, s. of Aknonios, f. of Daochos I**	Panhellenic games	*Periodonikes*, Olympic victor in pankration, ? 74th Ol.; 5 times victor at Nemea, 3 at Pythia, 5 at Isthmia in the pankration	Moretti, *Olympionikai*, no. 192; *FD* III.4 460, 2; *IG* IX,2 249; *LGPN* III.B. s.v. Agias (30)
4	?484	**Telemachos, s. of Aknonios**	Panhellenic games	*Periodonikes;* Olympic victor in wrestling, ?74th Ol.; same number of victories as Agias	Moretti, *Olympionikai*, no. 190; *FD* III.4 460, 3; *LGPN* III.B s.v. Telemachos (9)
5	?484	**Agelaos I, s. of Aknonios**	Delphi, Pythia	victor at the Pythia, youths' stadion	*FD* III.4 460, 4; *LGPN* III.B s.v. Agelaos (24)
6	476/5	**Menon I**	Eion/Athens	Military aid to Kimon at Eion; granted citizenship by the Athenians	Dem. 23.199; 13.23; *LGPN* III.B s.v. Menon (52)
7	454/3	**Orestes, s. of Echekratidas**	exile in Athens	Athenians organized expedition against Pharsalos to restore him	Thuc. i.111.1
8	mid-5th c.	**Epitimos**	Athens	Panathenaia? Died in penthathlon competition	Plut. *Vit.Per.* 36; *LGPN* III.B s.v. Epitimos (8)
9	431	**Menon II**	Athens	Head of Pharsalian horse; aid to Athens according to old alliance	Thuc. ii.22; *LGPN* III.B s.v. Menon (53)
10	424	**Strophakos**	Chalkidike	*Proxenos* of the Chalkidians; friend of Brasidas	Thuc. iv. 78.1; *LGPN* III.B s.v. Strophakos (1)
11	411	**Thucydides, s. of Menon**	Athens	*Proxenos* of the Athenians in Pharsalos; helped deter a civil war in Athens	Thuc. viii.92.9; Marcellin. *Vit. Thuc.* 28 = Polemon Fr. 5; *LGPN* III.B s.v. Thoukydidas (1)
12	Late 5th-early 4th c.	**Menon III, s. of Alexidamos**	Athens, Asia Minor	Protagonist of Platonic dialogue, *Meno* (402 BC); head of Thessalian mercenaries *(xenagos)* in *Anabasis*, *patrikos xenos* of the Persian king; pupil of Gorgias	Pl. *Meno;* Xen. *An.* 1.4, 13–16, 2.5.31, 2.6.21–29, 2.6.68; Ktesias, *FGH* 688 F27; *LGPN* III.B s.v. Menon (54)

	4th century				
13	394-387	**Iphitos**	Athens	Honorary decree	*IG* II^2 26,7; *SEG* XLVI 118; *LGPN* III.B s.v. Iphitos
14	374-369	**Polydamas**	Sparta	Hereditary *proxenos* of Sparta; mediator in Pharsalos during *stasis*	Xen. *Hell.* 6.1.2–4, 6.4.34; *LGPN* III.B. s.v. Polydamas (3)
15	361	**?Agelaos II**	Athens	Treaty of Thessalian Koinon with Athens against Alexander of Pherai; *archon* of Thessalians	*IG* II^2 116, 23, 34; Theopomp. Fr. 18 Jacoby; *LGPN* III.B s.v Agelaos (26)
16	361/60	**Sosibios**	Delphi	Donated a drachm at Delphi for the reconstruction of Temple of Apollo	*CID* II 4 I, 31; *LGPN* III.B s.v. Sosibios (47)
17	before 353/2	**?Kallippo[s]**	Athens	Honorary decree?	*IG* II^2 186,7; *LGPN* III.B s.v. Kallippos (64)
18	before 353/2	**?Aoīos**	Athens	Honorary decree?	*IG* II^2 186,8–9; *LGPN* III.B s.v. Aoīos
19	before 353/2	**?Kaphis**	Athens	Honorary decree?	*IG* II^2 186,9; *LGPN* III.B. s.v. Kaphis (11)
20	before 353/2	**?[X]enokrate[s]**	Athens	Honorary decree?	*IG* II^2 186,10; *LGPN* III.B s.v. Xenokrates (65)
21	?351-347	**Anaxandros**	Delphi	*Proxenos* of Delphi	*Hesp.* 16.61; *SEG* XVI 312 l. 3; *LGPN* III.B s.v. Anaxandros (9)
22	346-344	**A.o . . .**	Delphi	*Polemarch* mentioned in Pharsalian dedication	*SEG* I 210; Marcadé, *Signatures* I, 35,2; Vatin, 'Pharsaliens', 452–453; Helly, *État*, 258–260; Habicht, 'Polemarchen', 254–256
23	346-344	**Aiatos**	Delphi	*Polemarch* mentioned in Pharsalian dedication	*SEG* I 210; Marcadé, *Signatures* I, 35,2; Vatin, 'Pharsaliens', 452–453; Helly, *État*, 258–260; Habicht, 'Polemarchen', 254–256; *LGPN* III.B s.v. Aiatos (2)
24	346-344	**Eupolemos**	Delphi	*Polemarch* mentioned in Pharsalian dedication	*SEG* I 210; Marcadé, *Signatures* I, 35,2; Vatin, 'Pharsaliens', 452–453; Helly, *État*, 258–260; Habicht, 'Polemarchen', 254–256; *LGPN* III.B s.v. Eupolemos (19)
25	?339	**Nikomachos, s. of Eua . . .**	Delphi	Proxeny decree	Vatin, 'Pharsaliens', 446 l. 1; *SEG* XXII 462 l. 1; *LGPN* III.B s.v. Nikomachos (99)
26	344-338	**Polemoklēs s. of Astolaos**	Delphi	Proxeny decree	Vatin, 'Pharsaliens', 446 l. 1; *SEG* XXII 462 line 1; *LGPN* III.B s.v. Polemoklēs

(*continued*)

Table 1 – *Continued*

	Date	Name	Present in	Role	Evidence
27	345-327	**Kottyphos**	Delphi	*Hieromnemon; strategos* of the federal army in the war against Amphissa	Aesch. III 124,128; *Schol. Aeschinem* Orat.3, schol. 124; Dem. 18.151; *CID* II 34 I, 27; 36 I 22; 36 II [12], 34; 43, 16, 41; 44, [4]; 71,3,49; 72,[13]; 74 I, 31, [63]; 74 II, 18,76; *LGPN* III.B s.v. Kottyphos (7)
28	345-337	**Euaion, s. of Euthymos?**	Delphi	*Naopoios* at Delphi between 345-337 BC; *argyrologos* in 342/1 BC	*CID* II 11A,4; II B, [5] *(argyrologos* in 342/1 BC); 31, 72 *(naopoios)*, 82 *(naopoios - prostateuon)* 33, 8; 74 I 65; *LGPN* III.B s.v. Euaion (4)
29	345	**Gennaios, s. of Agatharchos**	Delphi	*Naopoios* during archonship of Damoxenos	*CID* II 31, 72; 119, 3; *LGPN* III.B s.v. Gennaios (21)
30	after 336/5 BC	**unknown Pharsalian**	Athens	Decree honouring a Pharsalian	*IG* II^2.430; *SEG* XLV 71
31	337-334	**Panairos, s. of Panairos**	Delphi	*Tamias*	*CID* II 74 I, 42; 77 I, [3]; *LGPN* III.B s.v. Panairos (5–6)
32	337/6-332	**Daochos II**	Delphi, Anaphe	*Tetrarchos* of the Thessalians; *hieromnemon* at Delphi; *proxenos* of Anaphe; 'friend of Philip II'; dedicator of 'Daochos monument	Dem. 18.295; Polyb. 18.14.4; *FD* III.4, 460 no. 7,3; *CID* II 64,17; 74 I, 2, 31; 74 II, 22, 75; 75 II, 10, 26; 76 I, 17; 79A I, 5; 79A II, [7], 36; 80, [6]; 82, 18; *CID* IV 9, 2; *LGPN* III.B s.v. Daochos (2)
33	337/6-332	**Thrasydaos**	Delphi	*Hieromnemon;* friend of Philip II	Dem. 18.295. *CID* II 8 II,11; 64,17; 74 I, 2, 31; 74 II, 22, 75; 75 II, 10, 26; 76 I, 17; 79A I, 5; 79A II, [7], [36]; 80, [5]; 82, 18. *CID* IV 9,2; *LGPN* III.B s.v. Thrasydaos (3)
34	337-328	**Timoleon, s. of Hellanikos**	Delphi	*Naopoios* at Delphi	*CID* II 74 I, 65; 121 III, [4]; *LGPN* III.B s.v. Timoleoun (2)
35	334-324	**Molyson s. of Paidinas**	Delphi	*Naopoios,* contemporary to Daochos and Thrasydaos; contemporary to Politas (spring 325)	*CID* II 32,30; 79 A I, 9; 95 I,6; 97,16; 98 A, 5; 102 II B, 15; 119,8; 121 III,17; *LGPN* III.B s.v. Molyson

36	329-326	**Politas**	Delphi	*Hieromnemon*	*CID* II 32,42; 69,19; 71 [42]; 72, [4]; 89, [2], 6; 94, 2; 96, [3]; 97, 56; 99 B, [12]; 100 I, [1]; 102 I fr.A, [4]; *LGPN* III.B s.v. Politas (27)
37	327-324	**Daimonios, s. of Menekrates**	Delphi	*Naopoios*	*CID* II 97, 33; 99 A 5; 102 II B, [20]; 119,7; *LGPN* III.B s.v. Daimonios (3)
38	328-324	**Hippodamas, s. of Agathon**	Delphi	*Naopoios*	*CID* II 95, 9; 97, 19; 102 IIB, 19; 119, 9; 121 III, [3]; *LGPN* III.B s.v. Hippodamas (4)
39	c. 330-300	**Panaitios**	Dodona	Dedicator of a silver kylix to Zeus Naios at Dodona	*SEG* LII 571; *LGPN* III.B s.v. Panaitios (3)
40	366-338	**Melanippos**	Oropos	Victor at men's *pentathlon* at Amphiaraeia	*IOropos* 520, 28; *LGPN* III.B s.v. Melanippos (7)
41	last quarter of 4th c.	**Kyrsilos**	Alexander's army	Member of Alexander's army; campaign in Armenia	*FGrH* 130; Berve, *Alexanderreich,* no. 457; *LGPN* III.B s.v. Kyrsilos (3)
42	323 BC - late 4th c.	**Ariston**	Alexander's army	*Hetairos?* in Alexander's army	Berve, *Alexanderreich,* no. 139; *LGPN* III.B s.v. Ariston (287)
43	d. 321 BC	**Menon IV**	Athens?; Asia Minor	*Hipparch* of the Thessalian horse in the Lamian War; his daughter married to king Aiakides of Epeiros, mother of Pyrrhus. The name possibly attested in Athenian decree honouring Thessalian exiles (321/0 or 320/19 BC)	Diod.Sic. 18.12.3, 18.15, 17.6, 38.5–6; Plut. *Vit.Pyrrh.* 1.4; Dem 27. Berve, *Alexanderreich,* no. 517; *SEG* XLV 83; *LGPN* III.B s.v. Menon (55)
	3rd century				
44	3rd c	**Diopeithes, s. of Dionysios**	Demetrias	Tombstone	Arvanitopoulos, *Thess.Mnem.*, no. 259; *LGPN* III.B s.v. Diopeithes (4)
45	3rd c	**Philedō**	Demetrias	Tombstone	Arvanitopoulos, *Thess.Mnem.*, no. 97; *LGPN* III.B s.v. Philedō
46	3rd c-2nd c BC	**Philinna, d. of Sosimenes**	Demetrias	Tombstone	Arvanitopoulos, *Prakt.* 1912, 186, no. 46; *LGPN* III.B s.v. Philinna (10)
47	284	**Philomelos**	Olympia	Olympic victor in *stadion*, 124th Ol.	Moretti, *Olympionikai,* no. 534; *LGPN* III.B s.v. Philomelos (19)
48	281	**Thōrax**	Thrace/Asia Minor	Soldier in Diadochs' army	App. *Syr.* 64; *LGPN* III.B s.v. Thōrax (7)

(*continued*)

Table 1 – *Continued*

	Date	Name	Present in	Role	Evidence
49	250-215	**Gorgos, s. of Simos**	Krannon	Honorary decree by the city of Krannon (*isopoliteia* etc)	*SEG* XXIII 437; Mastrokostas, *REA* 56 (1964) 312–315; *BE* (1972) 233; *LGPN* III.B s.v. Gorgos (19)
50	230-200	**Eurypylos, of Archippides**	Delphi	*Theadorokos* of Pharsalos for Pythia	Plassart, 'Théodoroques', 1. III, 21; *LGPN* III.B s.v. Eurypylos (3)
51	?228-216 BC	**Krytinas**	Amphilochia	*Epimeletes,* inscribed on a building block of the walls at Amphilochia	*ADeltion* 22 (1967) *B2,* 321; *BE* 1970, 343; *LGPN* III.B s.v. Krytinas
52	207/6 or 206/5	**]iolos**	Delphi	*Hieromnemon* during the archonship of Polykleitos-under Aitolians	*CID* IV 87, 3
53	209-203	**Polleas**	Delphi	*Hieromnemon,* during the archonship of Damokrates-under Aitolians	*CID* IV 95, 8; *LGPN* III.B s.v. Polleas (2)
54	c. 200	**Eutychos**	Delphi	*Priest* during the archonship of Hybreas	*SGDI* 2072, 16; *LGPN* III.B s.v. Eutychos (51)
	2nd century				
55	c. 150	**Metou[pos]**	Krannon	Honorary decree	*SEG* XXXV 583 l. 3; *LGPN* III.B s.v. Metoupos (5)
56	2nd half of the 2nd c	**Mnasippos, son of Oiolykos**	unknown city	Relief bearing a list of names with their ethnics divided according to region	Habicht, 'Prosopographie', 121–122; *BMI* 1154 a, 19; *LGPN* III.B s.v. Mnasippos (27)

Kerameikos, dated to the second quarter of the fifth century.[13] The identification has been recently challenged,[14] but it is clear from independent contemporary evidence that Menon belonged to a major Pharsalian family, which followed a consistently pro-Athenian policy in the fifth century (see below).[15] All in all, the emphasis on the use of Menon's own funds and *penestai* should perhaps be seen as indicative of a private initiative, not a state action.

Soon after the Battle of Tanagra, in *c.*454/3, Thucydides mentions that the Athenians organized an unsuccessful expedition against Pharsalos to restore Orestes, son of the *basileus* Echekratidas (Thuc. i.111.1; Diod.Sic. xi.83.3–4—the association of the Echekratids with Pharsalos is disputed).[16] What is interesting in Thucydides' reference is the ability of the city to deter such an attack,[17] and the willingness of the Athenians to instigate such operations in the north, in order to restore their exiled ally.

In 431, at the beginning of the Peloponnesian war, the Thessalians sent help to Athens *kata to palaion xymmachikon* (Thuc. ii.22.3). The cities involved were Larisa, Pharsalos, Krannon, Gyrton, Pyrasos, and Pherai, all leading Thessalian cities. Interestingly, the Pharsalian cavalry was headed by Menon II.[18]

Quite telling for the nature of help offered by Pharsalians to foreign powers, such as Athens and Sparta, during this period is the well-known episode of Brasidas' march through Thessaly in 424. Thucydides clearly states that it was the help of Brasidas' *epitedeioi* in Thessalian cities that secured his safe passage despite the pro-Athenian sympathies of the Thessalian *plēthos* (Thuc. iv.78.1).[19]

Another member of Menon's family, Thucydides (Table 1, no. 11), was *proxenos* of the Athenians; according to the homonymous historian, the Pharsalian played a paramount role in preventing civil war in Athens in 411, when he was present in the city (Thuc. viii.92.8; cf. Marcellin. *Vit.Thuc.* 28).[20] Finally, Menon III, son of Alexidamos, probably a relative of Thucydides, was also present in Athens in the last decade of the fifth century, as is shown by the homonymous Platonic dialogue (Table 1, no. 12).[21] Tutored by Gorgias, Menon III socialized with Socrates and his circle while in Athens. His presence there has been interpreted by some scholars as an attempt to seek help against Lycophron of Pherai.[22] He was a *patrikos xenos* of the Persian king (Pl. *Meno* 78d3) and of Ariaeus, an *erastes* of the Aleuad Aristippos (Pl. *Meno* 70b3, b5, 76b5), famous for his beauty and wealth and notorious for his arrogance, extravagance, and promiscuity (Pl. *Meno* 76b5, 82b1; Xen. *An.* 2.6.21–28, 68).[23] Menon III figures prominently in Xenophon's *Anabasis* as the *xenagos* (leader) of the Thessalian mercenaries sent by his friend Aristippos of Larissa to Cyrus (although Xenophon's account is highly polemic).

Thus, throughout the fifth century, Pharsalians appear to be actively involved with both Athens and Sparta. The reasons for foreign interest in Pharsalos are self-evident: Pharsalian nobles had considerable wealth and an excellent cavalry. In addition, Pharsalos and its elites, as potential leaders of Phthiotis and perhaps also at times of the perioikic region of Achaia Phthiotis, could be useful in securing export of grain from the Thessalian plains,[24] or in securing the safe passage of goods through Pharsalian territory towards the sea. This was the reason why in the mid-fourth century, Philip II offered the harbour city of Halos to his faithful associates in Pharsalos (Dem. 11.1).[25] Any city controlling southern Thessalian affairs could be a very useful ally for a state with maritime or other interests in the northern Aegean.

What about Pharsalians' interests? The continued presence of the Menonids in Athenian contexts would imply that the link with the city was a family one; Kimon and Menon I probably shared a guest friendship between aristocrats (the former had a son named Thessalos).[26] Menon's significant service to the Athenian victory was probably recognized at state level with citizenship. Thucydides' proxeny in the late fifth century led Walbank to suppose that it was perhaps a hereditary one.[27] On the other hand, one of Brasidas' *epitedioi*, Strophakos (Table 1, no. 10), was a *proxenos* of the Chalkidians,[28] who at the time were opposing Athens. From the Pharsalian point of view, it is hardly surprising that elite families had bonds with their Athenian and/or Spartan peers. In the 420s the presence of the Lacedaimonians at Herakleia on Trachis was good reason for making Thessalians, especially in the south, wary of Spartan expansion. The presence in Athens of Thucydides in 411 and Menon III in the final years of the century has been explained in the light of the unrest in Thessaly during this period that peaked with the rise in power of Lycophron at Pherai. Allying themselves with the likely winners appears to be a constant feature of Thessalian military and foreign policy, and Menon III seems to exemplify this attitude.[29]

Like *xenia* bonds, the participation in Pan-Hellenic athletic contests was a prerogative of the elites until at least the late fifth century. Similarly to other Thessalians, the Pharsalians participated in the games with considerable success, although they do not seem to have been interested in commemorating their victories through dedications. The best known Pharsalian athletes are members of Daochos' family (Table 1, nos. 3–5): Agias, Telemachos and Agelaos, sons of Aknonios, of which the first two were *periodonikai*. Perhaps we should also include Epitimos among Pharsalian athletes (Table 1, no. 8). He is known from a single reference in Plutarch (Plut. *Vit. Per.* 36.5.2) as a contemporary of Perikles, who died in Athens during a *pentathlon* competition; it has been suggested that he might have participated in the Panathenaic games.[30]

So far we have relied on epigraphic and literary evidence, without referring to the archaeology of the city. It must be pointed out that relatively little is known about Pharsalos, mainly because the modern town lies above the ancient city, and thus far excavations either have been conducted before the mid-twentieth century or were rescue in character.[31] It appears, however, that, like other Thessalian cities, Pharsalos invested in public architecture and demarcation of public space in the course of the late sixth and fifth centuries BC.[32] There is good evidence that the acropolis of Pharsalos was fortified in the early fifth century BC;[33] scattered architectural members suggest that there were monumental stone Doric buildings.[34] Very little material from the city itself dates to this period, mostly votive deposits discovered at various spots on the northwest slopes of the acropolis hill, containing figurines, associated with the cult of Demeter.[35] Inscriptions to Zeus Thaulios have been found in the western edge of the city, on the Aghia Paraskevi hill above the springs of Apidanos.[36]

The city issued silver coins after the Persian wars, probably in the mid-fifth century, bearing Athena's head on the obverse and part of a horse or a horseman on the reverse. The iconography of these coins show that Pharsalos' coinage was distinct from those of Methylion, Skotoussa, and the rest of Pelasgiotis.[37]

Although little archaeological material from the city has been fully published, it is clear that the Pharsalians imported extensively from Athens and Corinth, especially pottery.[38] Funerary sculpture of relatively high quality shows stylistic affinities with the northern Aegean.[39] The most impressive finds are the two *tholos* tombs in the western cemetery of the city, both dated to the late sixth/early fifth centuries. The best-known one, the so-called Verdelis tomb, built over a LH IIB tomb, consisted of a built *tholos* (4.5 m diam) covered by a mound that was lined by a stone enclosure, accessible through a built dromos (7.5 m long). The location of the tomb, its type and some of the finds suggest that forging a link with the past was a factor in the choice of grave type.[40]

Fourth Century

If we look at the fourth-century evidence (Table 1, nos. 13–43), it becomes apparent that not much changed in the frequency and type of Pharsalian representation abroad before the middle of the century. Pharsalians appear as *proxenoi* of Athens, Sparta or Thebes, and sparingly in the amphictyonic lists.

It is widely accepted that towards the end of the fifth century and in the early fourth century Thessaly was in turmoil. Athens' defeat in the Peloponnesian war, Macedonia's increasing presence in the south and, especially, Lykophron's rise in power at Pherai acted as catalysts in Thessalian affairs.[41] Pharsalos' geographical position in the southern part of Phthiotis made it highly desirable for hegemonic cities like Sparta and Thebes, and its close proximity to Pherai affected its political life. During this period shifting alliances characterized the stance of the Pharsalians and major military events occurred near its territory.

In 395 there was a Spartan garrison in Pharsalos,[42] but by the time of Agesilaos' return from Asia the Pharsalians were allied with the Thebans (Xen. *Hell.* 4.3.3; Plut. *Vit. Ages.* 16.5) and their cavalry suffered a humiliating defeat by the Spartan king, in which the Pharsalian *hipparchos* Polycharmos died (Xen. *Hell.* 4.3.8). In the 370s Polydamas of Pharsalos, a *proxenos* of Sparta (Xen. *Hell.* 6.1.2–5), addressed the Spartans asking for aid against Jason. During Pelopidas' campaigns to the north, against Alexander of Pherai, which culminated in the Battle at Kynos Kephalai, the Pharsalians were allied with the Thebans.[43]

Among the few Pharsalians appearing outside Thessaly during this period, Iphitos, who is otherwise unknown, is mentioned in an Athenian honorary decree of the early fourth century BC (*IG* II^2 26, 7), whereas Polydamas was hereditary *proxenos* of Sparta (Xen. *Hell.* 6.1.4). Their presence abroad is linked to the uneasiness caused in the Thessalian cities by the rapid increase in Pheraean power. Polydamas appears to have been widely admired and respected not just in Pharsalos, where he was entrusted with mediating during a *stasis*, but also in Thessaly as a whole (Xen. *Hell.* 6.1.2–4); he was killed by Polyphron.[44] In the well-preserved treaty of 361 BC between the Thessalians and Athens against Alexander of Pherai, an Agelaos appears as *archon* of the Thessalians (Table 1, no. 15); it is likely that he was a member of the Daochid family.[45] Finally in a very fragmentary inscription from Athens dated to the first half of the fourth century a number of Pharsalians received honours by the city of Athens (*IG* II^2 186); due to the very fragmentary state of the inscription it is not possible to determine who among them were definitely Pharsalians (Table 1, nos. 17–20).

By contrast the second half of the fourth century sees a dramatic increase in Pharsalian presence abroad, especially at Delphi where Pharsalian nobles, similarly to other Thessalians, feature prominently in honorary decrees (Table 1, nos. 21, 25–26),[46] amphictyonic catalogues and dedications (Table 1, nos. 22–24, 27–29, 31–38).[47]

Eminent Pharsalians regularly served as *hieromnemones*—for example: Kottyphos, who is listed in the oldest surviving list of *hieromnemones*,[48] appears to have served (together with Kolosimmos) between 346 and 337—that is, at the time when the Macedonian king was first admitted in the amphictyony (Table 1, no. 27). He was also leader of the war against Amphisa (Aesch. III 128; *Schol. in Aesch.* III.124.2; Dem. 18.151). Thrasydaos and Daochos II were the Thessalian *hieromnemones* between 336 and 332 (Table 1, nos. 32–33). They had previously participated in colleges of amphictyonic treasurers and were sent in an embassy to Thebes on behalf of Philip (339/8: Plut. *Vit. Dem.*18.1–2; *FGH* 135–6 F20). Both were famously accused by Demosthenes (18.295; cf. Polyb. 18.14.4) as agents of Philip.[49] Thrasydaos was further accused by Theopompus as a man of little intelligence, a flatterer and a tyrant of his own people (*FGH* 115, F 209).[50] Daochos II, as his famous sculptural monument at Delphi testifies, belonged to a most noble and influential Pharsalian family; he is called *tetrarchos* of the Thessalians on the base of his statue at Delphi. His name is also restored in a list of *proxenoi* from Anaphe (*IG* XII, 3 251).

Continuity characterized Pharsalian presence in the amphictyony during the period of Alexander III's reign, as Politas was one of the Thessalian *hieromnemones*, serving from *c.*330–324 (Table 1, no. 36); he seems to have belonged to a family whose members played a prominent role in Delphic and Pharsalian administration.[51]

During the same period various Pharsalians served as *naopoioi* (Table 1, nos. 28–9, 34–5, 37–8):[52] Euaion, son of Euthymos (345–337), who is also mentioned as *argyrologeon* and *prostateuon*; Gennaios, son of Agatharchos (*c.*345); Timoleon, son of Hellanikos (337–328); Molyson, son of Paidinas (334–324); Hippodamas, son of Agathon (328–324); Daimonios, son of Menekrates (327–323). Others, like Panairos, son of Panairos (Table 1, no. 31), served as *tamiai*.[53]

Most of these persons rarely figure in other sources, but the limited epigraphic evidence suggests that they were very prominent citizens, since members of their families appear to have held offices in Pharsalos over a long period of time, or to have made dedications in the city.[54] Some, like the *hieromnemon* Politas and the *naopoios* Hippodamas, son of Agathon (Table 1, nos. 36, 38), might have even belonged to the same family.[55]

Evidence for Pharsalian dedications also shows a marked increase in the second half of the fourth century, both at Delphi and abroad. At Delphi both state and individual dedications are known.[56]

(a) A stone base, preserved in two parts, bearing a horse that was dedicated collectively by the Pharsalians, signed by two sculptors from Atrax, dated to the mid-fourth century BC, mentions three polemarchs from Pharsalos (Table 1, nos. 22–24). The same base had a proxeny decree of Delphi for two citizens of Pharsalos, Nikomachos, son of Eua—, and Polemoklēs, son of Astolaos (Table 1, nos. 25–27), issued under the archonship of Etymondas (date: 339).[57]

(b) A statue of Achilles and Patroklos, dedicated by the Pharsalians, is mentioned by Pausanias (X. 13.5). Lavva has proposed that the same sculptural group was depicted on the reverse of fourth-century coins of the city.[58]

(c) The best-known Pharsalian dedication of any period is the Daochos monument at Delphi.[59] According to recent studies by Jacquemin and Laroche, the group was probably set in a *lesche*-type building at the sanctuary, erected during the second half of the fourth century.[60] The Daochos monument is an extravagant family monument, demonstrating the aristocratic pedigree and the credentials of the *hieromnemon* at Delphi through the display of his distinguished ancestors who had excelled in athletics (Agias, Telemachos, Agelaos), politics (Aknonios, Daochos I), and war (Sisyphos I).

The second half of the fourth century was the peak of Pharsalian presence abroad: there are references to Pharsalians in proxeny decrees from Locri Epiknemidioi (*IG* IX,1 267, line 3), Athens (*SEG* XLV 71 = *IG* II^2 430). They also contributed to the costs of rebuilding the Temple of Apollo at Delphi (*CID* II 8.I.8; Table 1, no.16) and donated money at Argos (*IG* IV 617, line 5).[61] Another Pharsalian, Panaitios, dedicated a silver *kylix* at Dodona in the 330s (Table 1, no. 39), and Melanippos won at the men's *pentathlon* at the Amphiaraeia (Table 1, no. 40).[62]

It is thus evident that the second half of the fourth century was a period of prosperity for Pharsalos. Numismatic and archaeological evidence, although still limited, seems to confirm this inference.[63] The fourth century saw intense building activity in Pharsalos (Figure 2). During this period the city expanded and the new extensive fortification walls (and cross walls) were built. Their perimeter is estimated to be approximately 5 km.[64] A number of bases for dedicatory statues as well as architectural remains and members that could have belonged to a public building were found near the site of the presumed *agora* of the city.[65] Other evidence shows continued investment in funerary display.[66] This was a period when Pharsalians made major dedications in their home town, such as Daochos' dedication of a bronze statue of Agias, made by Lysippos himself,[67] and the city's dedication of a statue of Homer, set near the *agora*.[68]

The posthumous dedication of statues of successful fifth-century athletes, like Agias, the appropriation of Achilles and Patroklos by the Pharsalians, and the wish to stress their earlier successes as in the Daochos monument can be explained in the light

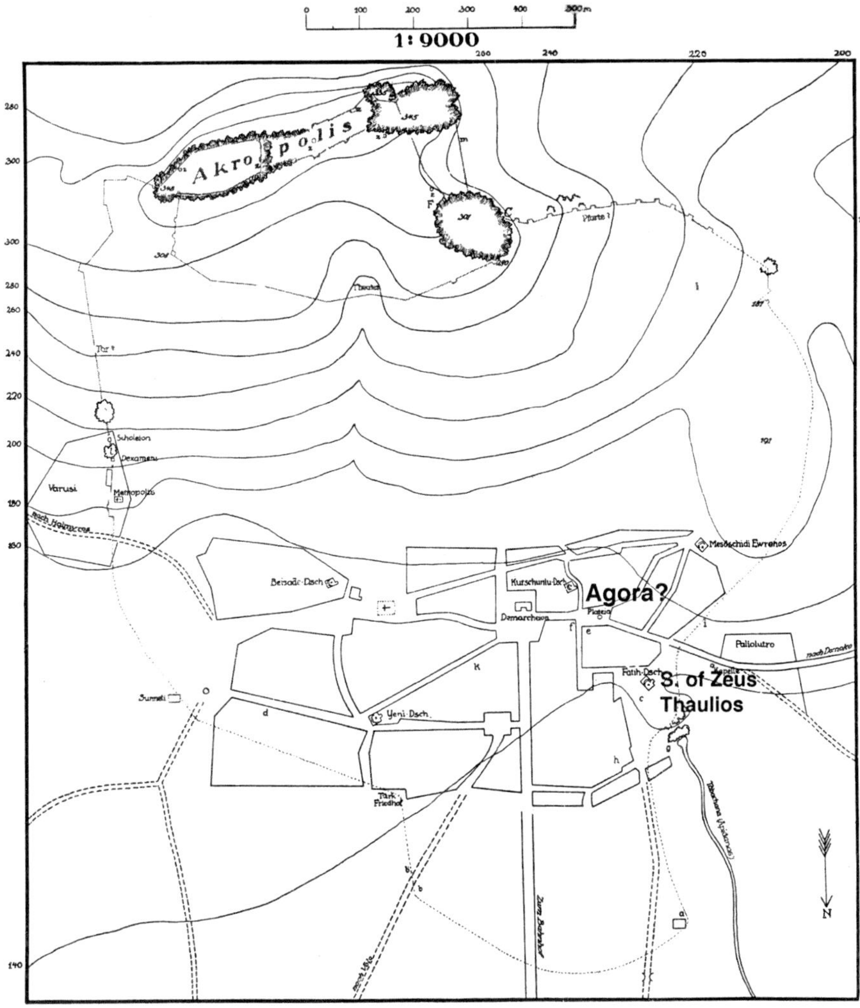

Figure 2 Plan of Pharsalos, adapted from Stählin, *Pharsalos*, plan.

of contemporary political developments and the close relationship of Daochos with the Macedonian kings.[69]

It was this close alliance and collaboration with Philip II that made Pharsalians act as major 'players' at Delphi and in the South during this period. This bond would continue during Alexander III's reign, with Pharsalians still operating at Delphi and actively participating in the Macedonian army. Kyrsilos (Strabo 11.14.12), Ariston (*Histor. Alexandri Magni* 3.31.8), and perhaps Menon IV are named in the sources.[70] The Pharsalian cavalry was held in very high esteem by the Macedonians: it was considered the best in Thessaly and fought on the left side under the command of Parmenion (Arrian. *Alex.Anab.* 3.11).

Third to Second Centuries

In contrast to earlier periods, Pharsalians nearly disappear from the epigraphic record in the third and second centuries (see below). Even taking into account that in some cases they may not have been designated with their city ethnics, as for example in lists of mercenaries,[71] the near absence of Pharsalians at Delphi and Thessaly itself is quite telling, especially when compared with the evidence for citizens of other Thessalian cities, such as Larisa, Pherai, Krannon, Gyrtone, or Kierion.

Some have sought to attribute the decline in Pharsalian representation to its stance during the Lamian War when Pharsalos, like the rest of Thessaly except Pelinna, sided with the Greeks. During the Lamian War the importance of the Pharsalian cavalry, the nearly exclusive presence abroad of Pharsalians that originated from a few noble houses and the change of sides halfway in the war are again evident.[72] After the initial stages of the war, Pharsalos sided with Athens. Menon IV was commander of the Thessalian cavalry, which played an important role in Antipatros' defeat (Diod.Sic. 18.15.1–7, 18.17.4–5, 18.38.3–5). It is uncertain whether Menon IV belonged to the same family as Menon I, who aided Kimon, but he was undoubtedly very distinguished. His daughter married King Aiakides of Epirus, father of Pyrrhos, and his reputation was, according to Plutarch, second only to that of Leosthenes (Plut. *Vit. Pyrrh.* 1.4).[73] Soon after the Battle of Krannon in 322, about fifty exiled Thessalians found refuge in Athens. Some, like Menon, son of Strophakos (*IG* II^2 545 and 2406), bore good Pharsalian names, and might have originated from the city.[74]

The third century is characterized by the expansion of Aitolian power in the north, first in the amphictyony. Pharsalos probably remained under Macedonian control until the 230s. After 278/7, we find no Pharsalians in Delphic lists until the last decade of the third century, when]iolos and Polleas feature as *hieromnemones* (Table 1, nos. 52–53), and an Eutychos is mentioned as a priest during the archonship of Hybreas in *c.*200 BC (Table 1, no. 54);[75] finally, Eurypylos, son of Archippides, is mentioned as the Pharsalian *theadorokos* in a list dated *c.*230–200 (Table 1, no. 50). The absence of Pharsalians and/or other Thessalians from the amphictyonic council is hardly surprising, as it is widely accepted that Macedonians and those under their influence did not participate at Delphi during this period, and until the reorganization of the Council in the early second century.[76]

As already mentioned, it is possible that during the third century some Pharsalians had served as mercenaries in the Successors' armies; however, usually there are no city ethnics in such lists. However, a Pharsalian, Thōrax (Table 1, no. 48), is said to have buried the body of Lysimachos in 281 BC (App. *Syr.* 64).

In Thessaly itself Pharsalians are attested on: a) two honorary decrees by the neighbouring city of Krannon, for Gorgos, son of Simos, dated *c.*250–215 BC, and Meto[upos], who was offered proxeny at about 150 BC (Table 1, nos. 49, 55); b) three painted tombstones in Demetrias (Table 1, nos. 44–46).[77] The small number of Pharsalians on the Demetrias tombstones is not surprising, given that overall city ethnics of Thessalians are extremely rare in Demetrias.[78]

How can we explain this near absence of Pharsalians after the fourth century?

It has been suggested that Greek cities suffered from food shortage during and immediately following the Lamian War.[79] The participation of the Pharsalian cavalry in the war and the military operations near neighbouring Krannon must have affected agricultural production in the region. Moreover, following the foundation of Demetrias in 293, Macedonian interest in Thessaly increasingly shifted towards the new city, which soon became a large, vibrant and cosmopolitan urban centre, with a resident Macedonian army and a royal residence (the 'Anaktoron').

Pharsalos seems to have been under Macedonian control for most of this period, except for a brief interval between the death of Demetrios II in 229 and the early years of the Social War.[80] Even so, Aitolian expansion in Achaia Phthiotis must have affected Pharsalos. The Gauls' raids in the region in the 280s, the control of Herakleia on Trachis by the Aitolians at about the same time, and of large part of Achaia Phthiotis by 239/8 BC; the Macedonian Wars and the Aitolian raids in the region must have been detrimental to its economy.[81] This is clearly indicated by Polybius' comment about the capture of Phthiotic Thebes by Philip V in 217 BC (5.99.4), that as long as Thebes was under Aitolian control, Demetrias, Larisa, and Pharsalos suffered since Aitolians used the city as a basis for raids against them.[82]

The economic crisis of the late third and early second centuries affected not only Pharsalos but also the entire region, as is apparent by the bestowals of citizenship, and often land, in Larisa, Phalanna, and Pharsalos in the late third century. The late-third-century politography decree of Pharsalos (*IThess* 50 = *IG* IX, 2 234) granted citizenship and landholdings to 176 individuals.[83] It seems that there was a debt crisis in the region, especially in Krannon, in the early second century, for which the city asked the Senate's intervention.[84]

Archaeological evidence from the city confirms the epigraphic and literary record. Although excavations are rescue in character, there is no archaeological evidence suggestive of a decline in the third century. During the early Hellenistic period there were at least two *stoas* near the area of the presumed *agora*,[85] and a number of large Hellenistic houses have been found in the city.[86] However, in various parts of the city a clear destruction layer by fire can be dated to the end of the second century,[87] and areas that were formerly residential were used for burials during the early Roman period, indicating that there was a change in the urban plan of the city and perhaps a reduction of the size of the habitation area.[88] It is too early to establish whether this destruction was widespread or whether it can be linked to specific battles, but it does suggest that there was serious disruption in Pharsalos at the time.

Is the absence of Pharsalians in the epigraphic record to be attributed to the wars and to the economic and social crisis in Thessaly at the time? This is partly true, but if we look at other Thessalian cities we see that citizens of Larisa, Pherai, Krannon, Gyrtone or Kierion, appear—even if sporadically—in documents from Delos, Kos, Ambrakia, Asia Minor, and as *strategoi* of the new Thessalian Koinon.[89] It has been shown that in the early second century the Thessalian koinon displayed considerable political activity.[90] Although there is no clear reference in the sources, it appears that

Pharsalos did not join the Koinon following its reorganization after 196 and Perseus' defeat in 169, since there are no Pharsalians among the lists of *strategoi* of the Koinon.[91]

In conclusion, the evidence for Pharsalians' presence outside their territory has shown that they were not very active in forming far-reaching relationships with communities far from Thessaly. Their presence is limited to states south of Thessaly, mainly Athens, Sparta and Delphi. Even in the fourth century, the peak of Pharsalian power, most of the evidence comes from Delphi. During the Hellenistic period, and especially in the second century, Pharsalians virtually disappear from the epigraphic record, an absence that is striking when compared with the evidence from other major Thessalian cities.

As has been repeatedly stressed, during the fifth and early fourth centuries most references to Pharsalian networks with the outside world can be explained as *xenia*-ritualized friendships of certain individuals with prominent Athenians, Spartans, and perhaps Thebans. Moreover, the epigraphic and literary evidence shows that public life and representation abroad was dominated over many generations by members of a few families. This is in agreement with the evidence for the political system in Pharsalos, which was a narrow oligarchy. This phenomenon also characterized public life in other Thessalian cities, such as Larisa and Pherai.[92]

The peak of Pharsalian connections abroad was during the fifth and fourth centuries, when their cavalry was in high demand, and its geographical position made it highly desirable for states with hegemonic aspirations. Pharsalians were proactive in forming lasting networks only in the second half of the fourth century, when they allied themselves with the Macedonian kings. Thereafter, the introvert stance of the Pharsalians was even more pronounced, prompted by military events, the economic crisis, the end of penetism in the late third century,[93] and the exclusion of the city from the Thessalian Koinon. Despite its wealth and love for the high life, Pharsalos never became a leading centre in Thessaly. Larisa, Pherai, and/or Demetrias played that role more effectively and their epigraphic record seems to prove it.

The attempts of the Pharsalian ruling class at networking had worked well throughout the Archaic and Classical periods. The new challenges of the Hellenistic period, however, necessitated a reconsideration of Pharsalos' priorities. The Pharsalian elites often failed to make the right decisions and their demise, manifest in their disappearance from the international arena, resulted in the final decline of their own city.

Acknowledgements

I would like to thank Katerina Panagopoulou and Christy Constantakopoulou for their invitation to participate in their excellent conference and their unique hospitality in Crete. Elaine Matthews very kindly allowed me access to the databases of the *Lexicon of Greek Personal Names* and R. Catling offered significant practical help.

S. Psoma commented on numismatic issues and about Strophakos; and N. Papazarkadas read a first draft of the article and made very useful comments. Prof. Christian Habicht generously read the final draft and offered invaluable corrections and additions. I warmly thank them all. All errors of course are the author's.

Notes

[1] *IACP* cat. no. 413; Stählin, *Pharsalos*; Béquignon, 'Pharsalos'.

[2] *IACP* cat. no. 413, p. 703; Arrian. *Alex.Anab.* 3.11; Spence, *Cavalry*, 23–25; Kroll, 'Archive', 87–8 (for boukephalas).

[3] *IACP* cat. no. 413; Mili, 'Studies in Thessalian Religion', 20–23.

[4] Martin, *Sovereignty*, 60; Sprawski, *Jason of Pherae*, 47, 52–55; Hornblower, *Greek World*, 96–97; Stamatopoulou, 'Thessalian Aristocracy', 327, nn. 122–24.

[5] Flower, *Theopompus of Chios*, 67–70, 121–23; Buckler, 'Actions of Philip II', 386.

[6] Stamatopoulou, 'Thessalian Aristocracy', 315–21.

[7] Helly, *L'État*,16; Parker *Untersuchungen zum Lelantischen Krieg*, 145–47; Stamatopoulou, 'Thessalian Aristocracy', 318 and n.61; Mili, 'Studies in Thessalian Religion', 12, 18, n. 49.

[8] Herman, *Ritualised Friendship*; Herman, 'Treaties'; Stamatopoulou, 'Thessalian Aristocracy', 318, 332, 338 (with regards to Thessaly).

[9] Stamatopoulou, 'Thessalian Aristocracy', 332–37.

[10] Corsten, *Vom Stamm zur Bund*, 178–82; Stamatopoulou, 'Thessalian Aristocracy', 316–17; Mili, 'Studies in Thessalian Religion', 16–18, 20–25.

[11] Hornblower, *Greek World*, 97–99; Sprawski, *Jason*, 35–48; Stamatopoulou, 'Thessalian Aristocracy', 337–39.

[12] I will not discuss the Echekratids as their relation with Pharsalos is disputed. See below, note 16. For the Daochids: Béquignon, 'Pharsalos', 1053–56; Helly, *L'État*, 51–52, 63–66.

[13] Sordi, *Lega*, 100–01; Ducat, *Pénestes*, 24–28; Osborne and Byrne, *Foreign Residents*, nos. 7211–12; Osborne, *Naturalization*, 20–23; Siewert, *Ostrakismos-Testimonien*, 62, 120–28.

[14] Brenne, *Ostrakismos*, 183–90, 240–41; Ruggeri, 'Menone'.

[15] For the Menonids: Béquignon, 'Pharsalos', 1058; Helly, *L'État*, 233–35, 303–06; Ruggeri, 'Menone', 80–82.

[16] Morrison, 'Meno', 61; Sordi, *Lega*, 107; Ruggeri, 'Menone', 82; Helly, *L'État*, 102–06; Mili, 'Studies in Thessalian Religion', 24. Also: Gomme, *Historical Commentary* I, 323–24; Hornblower, *Commentary* I, 178. Hornblower, *Greek World*, 97 accepts the connection with Pharsalos.

[17] The city must have been fortified in the mid-fifth century: note 33.

[18] Hornblower, *Commentary* I, 277.

[19] Hornblower, *Commentary* II, 256–62; Herman, 'Treaties', 97; Sprawski, *Jason*, 26–27; Sprawski, 'Lycophron and Jason', 440–41.

[20] Phillips, 'Men Named Thoukydides', 390–91; Osborne and Byrne, *Foreign Residents*, no. 7210; Gomme *et al.*, *Commentary*, 312–13; Walbank, *Athenian Proxenies*, no. 74.

[21] Sharples, *Plato: Meno.*

[22] Sordi, *Lega*, 139; Brown, 'Menon', 403; Lane Fox, 'Introduction', 16–17; Lane Fox, 'Sex, Gender and the Other', 198.

[23] Sharples, *Plato: Meno*, 18; Brown, 'Menon'; Herman, *Ritualised Friendship*, 47 (for his *xenia* with Arieus); Morrison, 'Meno' 66–68; Sprawski, *Jason*, 39–41.

[24] Xen. *Hell.* 5.4.56: in 377 the Thebans received two ships from Pagasai; 6.1.11: Thessalians exporting surplus production.

[25] Buckler, 'Actions of Philip II', 383; Sánchez, *Amphictionie*, 200; Sordi, *Lega*, 362–63.

[26] Davies, *Athenian Propertied Families*, 307; Herman, *Ritualised Friendship*, 19–21; Herman, 'Patterns of Name Diffusion', 353.

[27] Walbank, *Athenian Proxenies*, 384–85, no. 74; Herman, 'Patterns of Name Diffusion', 361, 363, sees the name as indicative of a *xenia* relationship.

[28] Thuc. iv.78.1; Hornblower, *Commentary* I, 256–58; Hornblower, *Commentary* II, 103; Psoma, *Olynthe*, 216. A Menon son of Strophakos is attested in the list of Thessalian exiles in Athens following the Lamian War: *IG* II^2 545, l.7: Tracy, *Athenian Democracy in Transition*, 87–91; *LGPN* II (ι) and III.B. s.v. Strōphakos. C. Habicht has drawn my attention to another possible Strōphakos; he restores his name on the dedicatory inscription *IThess* I, 77, l.2 (Stro . . .), on a base for a bronze statue found in 1906 in Pharsalos near the presumed area of the ancient *agora* and dated to the early fourth century. I would like to thank him for the suggestion. Finally a Pharsalian Strophakos is attested in the politography decree of the late third century: *IThess* I, 50, l.89; *LGPN* III.B s.v. Strophakos (2).

[29] Dem. 1.22. Sprawski, *Jason*, 29–30. For Spartan interests in Thessaly: Andrewes, 'Two Notes on Lysander', 219–26. Stamatopoulou, 'Thessalian Aristocracy', 338, nn.185–86 for the role of *xenia* relationships in these shifts of alliances.

[30] Stadter, *Commentary*, 328. For Thessalian participation in athletic competitions: Stamatopoulou, 'Thessalian Aristocracy', 330–32, 335–36 (Panathenaia). See also: *IThess* I, 74: for a possible Pythian victor, Trochilos (second half of the fourth century).

[31] Stählin, *Pharsalos*; Stählin, *Das hellenische Thessalien*, 136–42; Béquignon, 'Pharsalos', 1039–46.

[32] Stamatopoulou, 'Thessalian Aristocracy', 323–24.

[33] The style of masonry strongly recalls that of the monumental *tholos* tomb (Verdelis tomb) from the western cemetery of the city: Katakouta and Toufexis, 'Ta teiche', 189–90, 197. Lang, *Archaische Siedlungen*, 279, cat. no.104.

[34] Stählin, *Pharsalos*, 4; *ADeltion* 28 (1973) B2, 333 (G. Raios' plot).

[35] *ADeltion* 19 (1964), B2, 260–61; 21 (1966) 254; 28 (1973) B2, 333 (G. Raios' plot at Agios Nikolaos); 29 (1973–74) B2, 578–79, Phrourio hill, locality Platoma.

[36] *IThess* I 62 (fourth century); Doulgeri-Intzesiloglou, 'Thessalikes epigraphes', 137–54.

[37] Lavva, *Munzprägung*. She (17) sees the last issues of silver coinage of the city in the late fifth century. See review by S. Psoma in *A Survey of Numismatic Research 1996–2001* (2003): 49–50. Also: Kraay, *Archaic and Classical Greek Coins*, 116; Moustaka, *Kulte*, 26–27; Liampi, 'Das Corpus der Obolen', 119.

[38] Stamatopoulou, 'Thessalian Aristocracy', 327 and n. 136; *ADeltion* 27 (1972) B2, 416, pl. 345a (7 Athinas st.).

[39] Biesantz, *Thessalischen Grabreliefs*, K36, K37; *IThess* I 83–88.

[40] Stamatopoulou, 'Thessalian Aristocracy', 328–29; Moustaka, *Kulte*, 60–63. This will be paralleled in the fourth century with the attempt to promote Achilles as a 'national hero'.

[41] Gehrke, *Stasis*, 189–94; Buckler, *Theban Hegemony*, 110–35, 175–82; Morrison, 'Meno'; Sprawski, *Jason*, 38–48; Sprawski, 'Lycophron and Jason'; Rhodes, *History*, 209–15, 248–52; Tuplin, *Failings*, 149–50.

[42] For Spartan interest in Thessaly during this period and Medeios' intervention in Pharsalos: Andrewes, 'Two Notes on Lysander', 221–26; Morrison, 'Meno', 67; Sprawski, *Jason*, 41–47; Rhodes, *History*, 209–15.

[43] Buckler, *Theban Hegemony*, 114–15, 119–35, 175–82; Sánchez, *Amphictionie*, 505.

[44] Gehrke, *Stasis*, 126; Helly, *L'État*, 309; Buckler, *Theban Hegemony*, 110; Martin, *Sovereignty*, 85; Sprawski, *Jason*, 55.

[45] Rhodes and Osborne, *GHI*, no. 44; Tod, *GHI* II no. 147; Corsten, *Vom Stamm zur Bund*, 237; Mili, 'Studies in Thessalian Religion', 28-30. For a *stemma* of the Daochidai: Béquignon, 'Pharsalos', 1055.

[46] Also: Vatin, 'Pharsaliens'.
[47] Lefèvre, *Amphictionie*, 24–29.
[48] Sánchez, *Amphictionie*, 56; Rhodes and Osborne, *GHI*, no. 66: 336.
[49] Sánchez, *Amphictionie*, 56; Walbank, *Historical Commentary on Polybius* II, 567.
[50] Flower, *Theopompus*, 114, 129.
[51] See below, note 54.
[52] On the office and social milieu of the *naopoioi*: Lefèvre, *Amphictionie*, 263; Sánchez, *Amphictionie*, 128–33.
[53] A Panairos was among the *epitedioi* of Brasidas (Thuc. iv.78.1; *LGPN* IIII.B. s.v. Panairos [1]). An early-third-century grave stele from Pharsalos bears the epitaph Menippos son of Panairos: *ADeltion* 43 (1988) B1, 282 (Φ9) = *IThess* I 94; *LGPN* III.B s.v. Panairos (3).
[54] A small votive *naiskos* to Hestia and Heros Symmachos (*IThess* I 68) was dedicated by the son of a Thrasydaos in Pharsalos in the first half of the fourth century BC: Biesantz, *Thessalischen Grabreliefs*, L50/9; *LGPN* III.B s.v. Thrasydaos (2).
[55] The relevant evidence is: a) A proxeny decree of the city of Pharsalos, dated to the early third century BC where a Politas, son of Hippodamas, is listed among the *tagoi* of the city (*IThess* I 53, line 12; *SEG* XXXVI 549; *GLPN* III.B s.v. Politas [28]). b) An early-fourth-century altar from the area of the presumed *agora* of Pharsalos dedicated to Zeus Olympios by a Polycharmos, son of Agathon: *IThess* 58, *ADeltion* 28 (1973) B2, 333, pl. 290 γ-δ (9 N. Koukouflis st; L. Georgiou plot); *SEG* XXVII 22; *GLPN* III.B s.v. Polycharmos (16–17). It is tempting to see a link to the Polycharmos who died in the battle against Agesilaos. (Xen. *Hell.* 4.3.8). If we take into account the well-known *naopoioi* at Delphi, we could perhaps have a family consisting of: i) Agathon I (living in the fifth century), father of Polycharmos, who died in the early fourth century (and dedicated the altar to Zeus in Pharsalos); ii) Agathon II: father of Hippodamas, the *naopoios* in 328–324. This Hippodamas might have been related to the Politas mentioned in the proxeny decree of Pharsalos in the early third century BC. It is also likely that the hieromnemon Politas of the third quarter of the fourth century BC was a member of the same family.
[56] Jacquemin, *Offrandes*, 51.
[57] *SEG* XXII 462; Jacquemin, *Offrandes*, no. 389; Marcadé, *Recueil des signatures*, 35, pl. viii.2; Vatin, 'Pharsaliens'; Helly, *L'État*, 258–60; Habicht, 'Städtische Polemarchen'.
[58] Lavva, *Munzprägung*, 19; Moustaka, *Kulte*, 26–27.
[59] *FD* III.4 134–38, no. 460; Jacquemin, *Offrandes*, no. 391; Lörn, *Griechische Familienweihungen*, 118–23, no. 139; Miller, 'Macedonians', 268–69.
[60] Jacquemin and Laroche, 'Monument de Daochos'.
[61] Lefèvre, *Amphictionie*, 26.
[62] Sideris, 'Bronze Drinking Vases', 168–69, figs. 7–8, 177.
[63] Numismatic evidence: Kraay, *Archaic and Classical Greek Coins*, 117–18; Papaevangelou-Genakos, 'Zetemata metrologias', 44 (for bronze issues); Béquignon, 'Pharsalos', 1043: finds from the modern square, presumably the site of the ancient *agora*.
[64] Katakouta and Toufexis, 'Ta teiche', 197–98; *ADeltion* 29 (1973–74) B2, 567 (M. Kousia's plot), 576–78 (Ag. Paraskevi hill) dated on pottery evidence. Also: *ADeltion* 43 (1988) B1, 268–71.
[65] *ADeltion* 34 (1984) B, 151 (26 Larisis st: A. Doumas' plot).
[66] Monumental *periboloi* with burial in stone sarcophagi or bronze *hydriai*: Verdelis *AE* 1948–49; 1950–51; *Prakt* 1951, 155–57; 1952, 200–03; 1954, 155–57; 1955, 140–45; relief grave *stelai*: Biesantz, *Thessalischen Grabreliefs*, L38.
[67] *IThess* I 57.
[68] *IThess* I 56. For other fourth-century sculpture from Pharsalos: Biesantz, *Thessalischen Grabreliefs*, L34, L50, L52.
[69] Sánchez, *Amphictionie*, 220–68; Sordi, *Lega*, 287–91.

[70] Launey, *Recherches* I, 214; Heckel, *Marshals*, 359–61: no. 4.4. Polydamas (Heckel proposes to identify him as Pharsalian, contrary to the prevailing view: *LGPN* vol. IV).

[71] Launey, *Recherches* II, 1139–45; *SEG* L 1377.

[72] Grainger, *League*, 56–69; Hammond and Griffith, *History of Macedonia* III, 107–17; Tracy, *Athenian Democracy in Transition*, 23–30.

[73] Berve, *Alexanderreich*, no. 517; Lane Fox, 'Introduction', 26.

[74] As C. Habicht rightly pointed out to me in a recent communication (letter of 15 November 2007), not all these persons need be Pharsalians. As he reminded me, Menoitios, son of Archelaos, on line 9 was a Pheraean (the father of the gymnasiarch of Pherai in 309/299: *LGPN* III.B s.v. Menoitios [1]) and Strophakos is also attested as a name for a citizen of Krannon: *SEG* LI 711, B22–23. Habicht, 'Ehrung', 197; Launey, *Recherches* I, 214–15; Tracy, *Athenian Democracy in Transition*, 87–90.

[75] Polleas: Lefèvre, *Amphictionie*, 95; Kramolisch, *Strategen*, 41 n. 96. I would like to thank Prof. Habicht for the reference to Eutychos.

[76] Grainger, *League*, 108, 113; Scholten, *Politics*, 241–51. For the explanation for the presence of Pharsalians among the Aitolian *hieromnemones* in the last decade of the third century BC, a time when Pharsalos was held by Philip V: Walsh, 'Bones of Contention', 41–42; Grainger, *League*, 335–36; Scholten, *Politics*, 44–47; Will, *Histoire* I, 360–63.

[77] Volos Archaeological Museum, Λ97, Λ259 and E445.

[78] Only another six city ethnics are found.

[79] Tracy, *Athenian Democracy in Transition*, 32–34. See also the grain shortage implied by the well-known decree that discusses donation of corn from Cyrene (330–326 BC): Rhodes and Osborne, *GHI*, no. 96.

[80] For the controversy about whether Pharsalos was part of Aitolia between 221 and 205 BC: Fine, 'Problem of the Macedonian Holdings', 148–50; Béquignon, 'Pharsalos', 1066–70; Kramolisch, *Strategen*, 40–41; Walbank, *Historical Commentary on Polybius* I, 627; Walbank, *Historical Commentary on Polybius* II, 355–56; Walsh, 'Bones of Contention'; Scholten, *Politics*, 165–70; Will, *Histoire* II, 169–71, 250–52; *contra* Grainger, *League*, 389–92. The only certainty is that Pharsalos was Macedonian in 217 BC: Polyb. 5.99.4.

[81] Grainger, *League*, 98–104, 238–39, 284–86, 292, 365–74, 381–82; Scholten, *Politics*, 153–55, 168, 215–17; Walbank, *Historical Commentary on Polybius* II, 554, 576–79.

[82] Grainger, *League*, 292; Scholten, *Politics*, 215–17.

[83] Habicht, 'Epigraphic Evidence', 67–71; Chaniotis, *War*, 24, 86, 125, 132. Also: Helly, *L'État*, 302–11; Launey, *Recherches* I, 656–59; Mili, 'Studies in Thessalian Religion', 49, 67–68.

[84] Migeotte, *Souscriptions*, no. 34. Chaniotis, *War*, 121, 125, 132; Grainger, *League*, 517–19; Walsh, 'Disorders'.

[85] Stählin, *Pharsalos*, 8; Béquignon, 'Pharsalos', 1043–44.

[86] *ADeltion* 29 (1973–74) B2, 567 (26 Canada st); 35 (1980) B1, 288–91 (53 Athinas st: I. Argyris' plot); 43 (1988) B1, 271–74: Corner of Athinas and Hondropoulou st (Tsoumani brothers' plot); 49 (1994) B1, 335–37; 51 (1996) B1, 373–74 (21 Patroklou st); 54 (1999) B1, 423 (no. 7, 26 Martiou st.: Ch. Polyzos' plot).

[87] *ADeltion* 51 (1996) B1, 376 (38 N. Koukoufli st.: N. Katsaounis' plot), 376–77: 134 Athinas st (D. Alexopoulos' plot).

[88] *ADeltion* 51 (1996) B1, 377–78: 22 Athinas st (A. and G. Charbis' plot), 16 Kolokotroni st (I. Boubas' plot), on the eastern slopes of the Agia Paraskevi Hill.

[89] For example: Habicht, 'Ehrung', 193–96; Katarachias, 'Thessaloi'; Tziafalias, 'To Ergo', 86 (Atrax: copy of a decree of Lesbos honouring Atrax for giving grain: third century BC), 87 (Larisa: *stele* with decrees of the end of the second century BC); *SEG* XLVIII 1109 (Athenagoras of Larisa honoured by Kos); Kramolisch, *Strategen*.

[90] Pounder and Dimitrova, 'Dedication', 37; Habicht, 'Ehrung', 198–99.

[91] Béquignon, 'Pharsalos', 1070; Kramolisch, *Strategen*, 40–1; Grainger, *League*, 414–18, 445.
[92] Habicht, 'Ehrung', 198–99; Habicht, 'Epigraphic Evidence', 59–61.
[93] Ducat, *Pénestes*, 107–13.

Abbreviations

Arvanitopoulos, *Thess.Mnem.* = Arvanitopoulos, Apostolos S. *Thessalika Mnemeia. Perigraphe ton en toi Mouseioi Volou Grapton Stelon Demetriados-Pagason.* Athens: Hestia, 1909.

CID II = Bousquet, Jean. *Corpus des inscriptions de Delphes.* Vol. II, *Les Comptes du quatrième et du troisième siècle.* Paris: Boccard, 1989.

CID IV = Lefèvre, François. *Corpus des inscriptions de Delphes, tome* IV. *Documents amphictioniques.* Paris and Athens: Boccard, 2002.

FD III.4 = Pouilloux, Jean. *Les inscriptions de la terrasse du temple at de la region nord du sanctuaire. Fouilles de Delphes* III. *Épigraphie.* Vol. 4. Paris: De Boccard, 1976.

IACP = Hansen, M. H., and T. H. Nielsen, eds., *An Inventory of Archaic and Classical Poleis: An Investigation Conducted by the Copenhagen Polis Centre for the Danish National Research Foundation.* Oxford: Oxford University Press, 2004.

IOrop = Petrakos, Vasileios C., *Epigraphes Oropou.* Vivliotheke tes en Athenais Archaiologikēs Etaireias 170. Athens: Archaiologikē Hetaireia, 1997.

IThess I = Décourt, Jean-Claude. *Inscriptions de Thessalie* I. *Les cités de la vallée de l'Énipeus.* Etudes Epigraphiques 3. Athens: CNRS/EFA, 1995.

Rhodes and Osborne, *GHI* = Rhodes, Peter. J., and Robin Osborne, *Greek Historical Inscriptions 404–303 BC.* Oxford: Oxford University Press, 2003.

SGDI = Collitz, Hermann. *Sammlung der griechischen Dialekt-Inscriften* II. Göttingen: van Vanderhoeck and Ruprecht, 1899.

References

Andrewes, Antony. 'Two Notes on Lysander'. *Phoenix* 25 (1971): 206–26.

Arvanitopoulos, Apostolos, S. 'Anaskaphai kai ereunai en Thessalia kai Makedonia kata to etos 1912'. *Praktika tēs en Athēnais Archaiologikēs Hetaireias* (1912): 280–356.

Beck, Hans Georg. *Polis und Koinon. Untersuchungen zur Geschichte und Struktur der griechischen Bundesstaaten im 4. Jahrhunder v. Chr.* Historia Suppl. Vol. 114. Stuttgart: Franz Steiner, 1997.

Béquignon, Yves. 'Pharsalos'. *Real-Encyclopädie der klassischen Alterthumswissenschaft*, Supplementary volume XII (1970): 1038–84.

Berve, Helmut. *Das Alexanderreich aus prosopographische Grundlage.* Munich: C. H. Beck, 1926.

Biesantz, Hagen. *Die thessalischen Grabreliefs.* Mainz: Philip von Zabern, 1965.

Brenne, Stephan. *Ostrakismos und Prominenz in Athen. Attische Bürger des 5.Jhs. v. Chr. auf den Ostraka.* Tyche, Suppl. 3. Vienna: Holzhausen, 2001.

Brown, S. Truesdell. 'Menon of Thessaly'. *Historia* 35 (1986): 387–404.

Buckler, John. *The Theban Hegemony 371–362 BC.* Cambridge, Mass. and London: Harvard University Press, 1980.

———. 'The Actions of Philip II in 347 and 346 BC: A Reply to N. G. L. Hammond'. *Classical Quarterly* 46, no. 2 (1996): 380–86.

Chaniotis, Angelos. *War in the Hellenistic World: A Social and Cultural History.* Malden, Mass: Blackwell Publishing, 2005.

Corsten, Thomas. *Vom Stamm zur Bund. Gründung und territoriale Organisation griechischer Bundesstaaten.* Studien zur Geschichte Nordwest-Griechenlands 4. Munich: Oberhummer Gesellschaft, 1999.

Davies, John. K. *Athenian Propertied Families 600–300 BC.* Oxford: Clarendon Press, 1971.

Doulgeri-Intzesiloglou, Argyroula. 'Thessalikes epigraphes se topiko alphabeto.' Ph.D. diss., University of Thessaloniki, 2000.

Ducat, Jean. *Les Pénestes de Thessalie.* Besançon: Annales littéraires de l'Université de Besançon. Paris: Belles Lettres, 1994.

Fine, Johnvan Antwerp. 'The Problem of the Macedonian Holdings in Epirus and Thessaly in 221 BC'. *Transactions and Proceedings of the American Philological Association* 63 (1932): 126–55.

Flower, Michael A. *Theopompus of Chios. History and Rhetoric in the Fourth Century BC.* Oxford: Clarendon Press, 1994.

Gehrke, Hans-Joachim. *Stasis. Untersuchungen zu den inneren Kriegen in den griechischen Staaten des 5. und 4. Jahrhunderts v. Chr.* Vestigia 35. Munich: C. H. Beck, 1985.

Gomme, Arnold Wycombe. *A Historical Commentary on Thucydides.* Vol. 1. Oxford: Clarendon Press, 1972.

Gomme, Arnold Wycombe, A. Andrewes, and K. J. Dover. *A Historical Commentary on Thucydides.* Vol. V. Oxford: Clarendon Press, 1981.

Grainger, John D. *The League of the Aitolians.* Mnemosyne 200. Leiden, Boston and Cologne: Brill, 1999.

Habicht, Christian. 'Beiträge zur Prosopographie der altgriechischen Welt'. *Chiron* 2 (1972): 103–34.

———. 'Ehrung eines thessalischen Politikers in Athen (IG II^2 933 + Hesperia 29, 1960, 76 no. 154)'. *ZPE* 20 (1976): 193–99.

———. 'Städtische Polemarchen in Thessalien'. *Hermes* 127 (1999): 254–56.

———. 'Epigraphic Evidence for the History of Thessaly under Macedonian Rule'. In *The Hellenistic Monarchies, Selected Papers*, 59–73. Ann Arbor: University of Michigan Press, 2006.

Hammond, N. G. L. and G. T. Griffith. *A History of Macedonia.* III. *336–167 BC.* Oxford: Clarendon Press, 1988.

Heckel, Waldemar. *The Marshals of Alexander's Empire.* London and New York: Routledge, 1992.

Helly, Bruno. *L'État thessalien. Aleuas le Roux, les tétrades et les tagoi.* Lyon: Collection de la Maison de l'Orient Méditerranéen, 1995.

Herman, Gabriel. *Ritualised Friendship and the Greek City.* Cambridge: Cambridge University Press, 1987.

———. 'Treaties and Alliances in the World of Thucydides'. *Proceedings of the Cambridge Philological Society* 216 (1990): 83–102.

———. 'Patterns of Name Diffusion within the Greek World and Beyond'. *Classical Quarterly*, New Series 40 (1990): 349–63.

Hornblower, Simon. *A Commentary on Thucydides.* Vol. I, Books I–III. Oxford: Clarendon Press, 1991.

———. 'The Religious Dimension to the Peloponnesian War, or, What Thucydides does not Tell us'. *Harvard Studies in Classical Philology* 94 (1992): 169–97.

———. *A Commentary on Thucydides.* Vol. II, Books IV–V.24. Oxford: Clarendon Press, 1996.

———. *The Greek World 479–323.* 3rd ed. London: Routledge, 2002.

Jacquemin, Anne. *Offrandes monumentales à Delphes.* Bibliothèque des écoles françaises d'Athènes et de Rome 304. Paris/Athens: Boccard/EFA, 1999.

Jacquemin, A. and D. Laroche. 'Le monument de Daochos ou le trésor des Thessaliens'. *BCH* 125 (2001): Études: 305–32.

Katakouta, S. and G. Toufexis. 'Ta teiche tes Pharsalou'. In *Thessalia. Dekapente chronia archaiologikes ereunas 1975–1990. Apotelesmata kai prooptikes*. Vol. B 189–200. Athens: Kapon, 1994.

Katarachias, K. 'Thessaloi ste Delo kata ton 3o kai 2o ai. P.Ch (Epigraphikes Martyries)'. *ThessHem* 43 (2003): 27–37.

Kraay, Martin. *Archaic and Classical Greek Coins*. London: Methuen, 1976.

Kramolisch, Herwig. *Die Strategen des thessalischen Bundes vom Jahr 196 v. Chr. bis zum Ausgang der Römischen Republic. Demetrias* II. Bonn: Habelt, 1978.

Kroll, John H. 'An Archive of the Athenian Cavalry'. *Hesperia* 46 (1977): 83–140.

Lane Fox, Robin. 'Introduction'. In *The Long March: Xenophon and the Ten Thousand*, edited by R. Lane Fox. 1–45, New Haven, Conn. and London: Yale University Press, 2004.

———. 'Sex, Gender and the Other'. In *The Long March: Xenophon and the Ten Thousand*, edited by R. Lane Fox, 184–214, New Haven, Conn. and London: Yale University Press, 2004.

Lang, Franziska. *Archaische Siedlungen in Griechenland. Struktur und Entwicklung*. Berlin: Akademie Verlag, 1996.

Launey, Marcel. *Recherches sur les armées hellénistiques*. 2nd ed., 2 vols. Bibliothèque des écoles françaises d'Athènes et de Rome 169. Paris: De Boccard, 1987.

Lavva, Stella. *Die Munzprägung von Pharsalos*. Saarbrücken Studien zur Archäologie und alten Geschichte Band 14. Saarbrücken: Saarbrücker Druckerei und Verlag, 2001.

Lefèvre, François. *L'Amphictionie pyléo-delphique: Histoire et institutions*. Bibliothèque des écoles françaises d'Athènes et de Rome 298. Paris/Athens: Boccard/EFA, 1998.

Liampi, Katerina. 'Das Corpus der Obolen und Hemiobolen des thessalischen Bundes und die politische Geschichte Thessaliens im 2. Viertel des 5. Jahrhunderts v. Chr'. In *Hellas und der griechischen Osten. Studien zur Geschichte und Numismatik der griechischen Welt. Festschrift für Peter Robert Franke zum 70. Geburtstag*, edited by W. Leschhorn, A. V. B. Miron, and A. Miron. 99–126. Saarbrücken: SDV Saarbrücken Druckerei, 1996.

Lörn, Christoph. *Griechische Familienweihungen. Untersuchungen einer Repräsentationsform von ihren Anfängen bis zum Ende des 4. Jhs. v. Chr.* Internationale Archäologie 54. Rahden/Westf: Verlag Marie Leidorf, 2000.

Marcadé, Jean. *Recueil des signatures de sculpteurs grecs*. Vol. I. Paris: De Boccard, 1953.

Martin, Thomas R. *Sovereignty and Coinage in Classical Greece*. Princeton, NJ: Princeton University Press, 1985.

Mastrokostas, Euthymios. 'Inscriptions de Locride et de Thessalie'. *REA* 56 (1964): 312–5.

Migeotte, Léopold. *Les souscriptions publiques dans le cités grecques*. Haute-Études du monde Gréco-Romain, 17. Geneva: Librairie Droz and Quebec: Editions du Sphinx, 1992.

Mili, Maria. 'Studies in Thessalian Religion.' Ph.D. diss., University of Oxford, 2005.

Miller (-Colette), Stella, G. 'Macedonians at Delphi'. In *Delphes. Cent ans après la grande fouille. Essai de bilan. Actes du colloque international organise par l'École française d' Athènes, Athènes-Delphes, 17–20 septembre 1992*, edited by A. Jacquemin. BCH Suppl. 36, 263–81. EFA Athènes: Du Boccard, 2000.

Moretti, Luigi. *Olympionikai, i vincitori negli antichi agoni olimpici*. Rome: Academia Nazionale, 1957.

Morrison, J. S. 'Meno of Pharsalus, Polycrates and Ismenias'. *Classical Quarterly* 36 (1942): 57–75.

Moustaka, Aliki. *Kulte und Mythen auf thessalischen Münzen*. Beiträge zur Archäologie 15. Würzburg: K. Triltsch, 1983.

Osborne, Michael J. *Naturalization in Athens*, 2 Vols. Verhandelingen van de Koninklijke Academie voor Wetenschappen, Letteren en Schone Kunsten van België, Klasse der Letteren 101. Brussels: AWLSK, 1981–83.

Osborne, M. J., and S. G. Byrne. *The Foreign Residents of Athens. An Annex to the Lexicon of Greek Personal Names: Attica*. Studia Hellenistica 33. Louvain: Peeters, 1996.

Papaevangelou-Genakos, Cleopatra E. 'Zetemata metrologias thessalikon chalkon nomismaton: He periptose tes Phalannas'. In *Obolos 7. Coins in the Thessalian Region. Mints, Circulation, Iconography, History Ancient, Byzantine, Modern. Proceedings of the Third Scientific Meeting*, 33–50. Athens: Ekdose ton Philon tou Nomismatikou Mouseiou, 2004.

Parker, Victor. *Untersuchungen zum Lelantischen Krieg und verwandten Problemen den frühgriechischen Geschichte.* Historia 109. Stuttgart: F. Steiner, 1997.

Phillips, D. L. 'Men Named Thoukydides and the General of 440/39 BC (Thuc. 1.117.2)'. *Historia* 40 (1991): 385–95.

Plassart, André. 'Inscriptions de Delphes. La Liste des théodoroques'. *BCH* 45 (1921): 1–85.

Pounder, R. L. and N. Dimitrova. 'Dedication by the Thessalian League to the Great Gods in Samothrace'. *Hesperia* 72 (2003): 31–39.

Psoma, Sélène. *Olynthe et les Chalcidiens de Thrace: études de numismatique et d'histoire.* Stuttgart: F. Steiner, 2001.

Reinders, H. R. *New Halos; a Hellenistic Town in Thessalia, Greece.* Utrecht: HES, 1988.

Rhodes, Peter J. *A History of the Classical Greek World 478–323 BC.* Oxford: Blackwell, 2006.

Ruggeri, Claudia. 'Menone, figlio di Menecleide, ateniense, del demo di Gargetto'. *ZPE* 138 (2002): 73–86.

Sánchez, P. *L'Amphictionie des Pyles et de Delphes: recherches sur son rôle historique, des origines au IIe siècle de notre ère.* Historia 148. Stuttgart: F. Steiner, 2001.

Scholten, Joseph B. *The Politics of Plunder. Aitolians and their Koinon in the Early Hellenistic Era 279–217 BC.* Berkeley, Los Angeles and London: University of California Press, 2000.

Sharples, R. W., ed. *Plato: Meno.* London: Aris and Phillips, 1985.

Sideris, Athanasios. 'Bronze Drinking Vases Bearing Dedicatory Inscriptions'. *Eirene* 38 (2002): 167–201.

Siewert, P., ed. *Ostrakismos-Testimonien I. Die Zeugnisse antiker Autoren, der Inschriften und Ostraka über das athenische Scherbengericht aus vorhellenistischer Zeit (487–322 v.Chr.).* Historia 155. Stuttgart: F. Steiner, 2002.

Sordi, Marta. *La lega tessala fino ad Alessandro Magno.* Studi pubblicati dall'Istituto italiano per la storia antica 15. Rome: Istituto Italiano per la Storia Antica, 1958.

Spence, I. G. *The Cavalry of Classical Greece: A Social and Military History.* Oxford: Clarendon Press, 1993.

Sprawski, Slawomir. *Jason of Pherae. A Study on History of Thessaly in Years 431–370 BC.* Electrum 3. Krakow: Jagiellonian University Press, 1999.

———. 'Were Lycophron and Jason Tyrants of Pherae? Xenophon on the History of Thessaly'. In *Xenophon and his World. Papers from a Conference held in Liverpool in July 1999*, edited by C. Tuplin. Historia 172: 437–52. Stuttgart: F. Steiner, 2004.

Stadter, Philip A. *A Commentary on Plutarch's Pericles.* Chapel Hill, N.C. and London: University of North Carolina Press, 1989.

Stählin, Friedrich. *Pharsalos. Topographische und geschichtliche Untersuchungen uber die Hauptstadt der Phthiotis.* Nurnberg: Altes Gymnasium Nurnberg, 1914.

———. *Das hellenische Thessalien; Landeskundige und geschichtliche Beschreibung Thessaliens in der hellenischem und römischem Zeit.* Stuttgart: J. Engelhorn, 1924.

Stamatopoulou, Maria. 'Thessalian Aristocracy and Society in the Age of Epinikian'. In *Pindar's Poetry, Patrons and Festivals from Archaic Greece to the Roman Empire*, edited by S. Hornblower and C. Morgan, 309–41. Oxford: Clarendon Press, 2007.

Tataki, Argyro. B. *Macedonians Abroad. A Contribution to the Prosopography of Ancient Macedonia.* Meletemata 26. Athens: Kentro Hellenikes kai Romaikes Archaiotetos, Ethnikon Idryma Ereunon, 1998.

Tracy, Stephen V. *Athenian Democracy in Transition. Attic Letter-Cutters of 340–290 BC.* Hellenistic Culture and Society 20. Berkeley, London: University of California Press, 1995.

Tuplin, Christopher. *The Failings of Empire. A Reading of Xenophon Hellenica 2.3.11–7.5.27*. Historia 76. Stuttgart: F. Steiner, 1993.

Tziafalias, Athanasios. 'To Ergo tes IE' Eforeias Proistorikon kai Klasikon Archaioteton'. In *To Ergo ton Eforeion Archaioteton kai Neoteron Mnemeion tou YPPO ste Thessalia kai sten eurytere perioche tes (1990–1998). 1st Epistemonike Synantese. Volos, Maios 1998*, 85–96. Volos: 13th Ephorate of Prehistoric and Classical Antiquities, Ministry of Culture, 2000.

Vatin, Claude. 'Pharsaliens à Dèlphes'. *BCH* 88 (1964): 446–54.

Verdelis, Nikolaos M. 'Procheiros anaskaphike ereuna sta Pharsala'. *Archaiologikē Ephēmeris* (1948–49) *Chronika*: 40–2.

———. 'Chalke tephrodochos kalpis ek Pharsalon'. *Archaiologikē Ephēmeris* (1950/51): 80–105.

———. 'Anaskafikai ereunai en Thessalia'. *Praktika tēs en Athēnais Archaiologikēs Hetaireias* (1951): 129–63.

———. 'Anaskafikai ereunai en Thessalia'. *Praktika tēs en Athēnais Archaiologikēs Hetaireias* (1952): 164–204.

———. 'Anaskafikai ereunai en Thessalia'. *Praktika tēs en Athēnais Archaiologikēs Hetaireias* (1953): 120–32.

———. 'Anaskaphe Pharsalou'. *Praktika tēs en Athēnais Archaiologikēs Hetaireias* (1954): 153–9.

———. 'Anaskaphe Pharsalou'. *Praktika tēs en Athēnais Archaiologikēs Hetaireias* (1955): 140–6.

Walbank, F. B. *A Historical Commentary on Polybius*. 3 vols. Oxford: Clarendon Press, 1957–1979.

Walbank, Michael B. *Athenian Proxenies of the Fifth Century BC*. Toronto and Sarasota: Samuel Stevens, 1978.

Walsh, Joseph J. 'Bones of Contention: Pharsalus, Phthiotic Thebes, Larisa Cremaste, Echinus'. *Classical Philology* 88 (1993): 35–46.

———. 'The Disorders of the 170s BC and Roman Intervention in the Class Struggle in Greece'. *Classical Quarterly*, New Series 50 (2000): 300–3.

Westlake, H. D. *Thessaly in the Fourth Century BC*. London: Methuen, 1935.

Will, Édouard. *Histoire politique du monde hellénistique (323–30 av. J.-C.)*. 2nd ed., 2 vols. Annales de l'Est 30. Nancy: University of Nancy, 1979–82.

Profitable Networks: Coinages, *Panegyris* and Dionysiac Artists

Selene Psoma

In the Greek world, coinages were struck mostly in the name of states, either *poleis* or federations, and in that of kings. In a few cases coins carried the name of, and were certainly issued by, military personalities and local dynasts to pay soldiers' wages (*siteresion* and *misthos*). Evidence for a coinage issued by a religious association was recently provided by Lorber and Hoover publishing an Attic-weight tetradrachm with a head of Dionysos on the obverse, a thyrsus in an ivy wreath and the legend τῶν περὶ τὸν Διόνυσον τεχνιτῶν on the reverse (Figure 1).[1] The legend of the wreathed tetradrachm issued by the Dionysiac artists refers, as the authors pointed out, to its issuing authority: the *Koinon* of the Ionian and Hellespontine artists of Dionysos based at Teos. The tetradrachm of the Dionysiac artists bears no marks of control, and that could probably refer to one-time emission intended from the outset.[2]

This coin is added, as Lorber and Hoover noted, to the list of *polis*-like attributes and behaviour of the associations of Dionysiac artists. They further underlined that there 'is no surviving evidence that would suggest a need or a desire for a proprietary

Figure 1 Dionysiac artists of Hellespont and Ionia. 2nd century BCE. Tetradrachm (Attic): 16.87 g. Lorber and Hoover, 'An Unpublished Tetradrachm', 59–68, pl. 15.2.

currency with which to conduct the business of the association' and gave a list of the main expenditures of the Dionysiac artists.[3] Financial occupations of the artists were also noted by the authors, who considered the Dionysiac artists' Attic-weight tetradrachm as a presentation piece for recipients from outside the Attalid Kingdom. Coinages were sometimes issued to celebrate special events, especially in the Roman period;[4] subsequently they had a fiscal function. Numismatic research has shown that coins previously interpreted as medals issued for commemorative reasons were full coinages with purely fiscal character.[5] In this paper, a different approach and fiscal perspective of the coinage of the Dionysiac artists of Teos will be proposed.

A date around 155–145 BCE was proposed for the Dionysiac artists' tetradrachm, as it entered the market in a hoard of wreathed tetradrachms, mostly of Myrina, which, on the basis of their occurrence in Syrian hoards, are dated to that same period (Figure 2).[6] It belongs to the second issuing period of Attic weight minted by cities of the Attalid Kingdom.[7] If we accept that date, it is not difficult to put the coinage in the relevant historical background. From 146/5 or 145/4 dates a letter of Lucius Mummius to the

Figure 2 Myrrina, *c.*160–150 BCE. Tetradrachm (Attic): 16.05 g. Stockholm (*SNG* 2216). From Mørkholm, *Early Hellenistic Coinage*, 620.

artists, by which he seems to have conceded them special rights.[8] Between 145 and 133 BCE, the Ionian and Hellespontine artists of Dionysos moved from Teos to Myonnesos and then to Lebedos (Str. XIV 1, 29).[9] These events could be associated with the coinage.

However, the striking stylistic similarity of the wreath of the Dionysiac artists' tetradrachm with the wreath on Attic-weight Alexanders of the city of Teos that date *c.*188–180 BCE suggests an earlier date, around 170–160 BCE, for the type in question (Figure 3).[10] From that same period dates also an important part of the so-called wreathed coinages on Attic standard, minted in several cities of the Attalid Kingdom.[11] From these years dates also a well-known inscription from Teos that reports 'the escalating dispute between the Hellespontine and Ionian association and its host-city Teos, over competing claims to festival revenues'.[12] The dispute led to the intervention of Eumenes II, whose letter to the artists provides further evidence for them organizing the fair in the urban centre of the city of Teos, while the Teian civic authorities were responsible for the fair organized at the *chora* and the harbours.[13] That same royal letter planned the *synoikismos* of the Teians and the Dionysiac artists that finally failed.[14] It was most probably on these occasions of the *synoikismos* and the celebration of the fairs in the city and of course with royal approval that the artists issued their Attic-weight currency, thus underlying their identity and pride, and with profit in mind. In fact, the *synoikismos*, the organization of the fairs at Teos and the related coinage are all three '*polis*-like attributes and behaviour' of the association of Ionian and Hellespontine Dionysiac artists.[15]

Given this historical background of the artists' coinage, it will be interesting to explain its fiscal function, after examining a number of parallel cases.

Other Coinages Issued by Religious Associations

The coinage of the Dionysiac artists was issued by a religious association per se, and from that point of view it reminds the fourth-century BCE silver coinage on the

Figure 3 Teos. 188-180 BCE. Tetradrachm (Attic) of Alexander types. From Lorber and Hoover, 'An Unpublished Tetradrachm', 59–68, pl. 17.11.

Aeginetan standard issued by the most prestigious religious association of the Greek world—namely, the council of the Delphic Amphictyons ('Αμφικτιόνων), between the autumn of 336 and spring 334 BCE (Figures 4 and 5).[16] Different interpretations have been proposed for that coinage also.[17]

The tetradrachm of the Dionysiac artists was contemporary with Asia Minor coinages on the Attic standard that were carrying the name of a deity in the genitive and no ethnic or other indication of an issuing authority:[18] the coinage of Artemis Pergaia ('Αρτέμιδος Περγαίας; Figure 6) that began during the third century,[19] the coinage of Athena Ilias ('Αθηνᾶς 'Ιλιάδος; Figure 7)[20] and from the second-century BCE Attalid Kingdom, the Athena Nikephoros of Pergamon coinage with the legend 'Αθηνᾶς Νικηφόρου (Figure 8).[21] The coinages with the names of Athena Ilias and Artemis Pergaia were issued, as L. Robert explained on the basis of epigraphic and numismatic evidence, by the confederations of cities that had Athena Ilias and Artemis Pergaia as their protecting deities, respectively.[22] The coinage with Athena's Nikephoros types and name dates before 160 BCE, and has been connected either with the *Nikephoria* of Pergamon reorganized by Eumenes II or with the struggle of Eumenes II with the Galatians.[23] That coinage in silver and bronze was also issued by an association of cities participating in the important cult of Athena Nikephoros, as denoted by the various monograms referring to city ethnics that occur on the reverse of these coins.[24]

Function of Coinages Issued by Religious Associations of Cities

For Louis Robert the coinage in the name of Athena Ilias was connected with the 'economic reality' of the fairs of the *Panathenaia*,[25] and he suggested that the coinage in the name of Artemis Pergaia was also issued by cities participating in the cult and celebrating together the well-known *panegyris* mentioned by Strabo (14.4.2) in the

Figure 4 Amphictyonic Council, 336–334 BCE. Silver Stater (Aeginetan): 12.27 g. Bank Leu 20 (25.iv.1978) 86. From Mørkholm, *Early Hellenistic Coinage*, 204.

Figure 5 Amphictyonic Council, 336–334 BCE. Silver Hemidrachm (Aeginetan): 2.95 g. New York: *SNG Berry* 585. From Mørkholm, *Early Hellenistic Coinage*, 205.

extra muros sanctuary of the goddess.[26] Furthermore, several *panegyreis* were also celebrated at Pergamon, and the coinage of Athena Nikephoros was also connected with the *panegyris* of that goddess.[27]

As far as the Amphictyonic coinage is concerned, epigraphic evidence relates it to payments for building activity, but that cannot exclude[28] its use in the *amphictyonikai (pylatides) agorai*,[29] organized on the occasion of the *pylaiai*, the two annual meetings of the Amphictyonic council to which refer its types:[30] Demeter on the obverse and Apollo seated on the *omphalos* on the reverse.[31] The first one was celebrated in autumn (*oporine pylaia*), during the months of Boathoos or Heraios, and took place at the sanctuary of Demeter at Pylai (Hdt. 7, 201: Thermopylai), in the area of Malis. The second one (*earine pylaia*) was celebrated at Delphi in early spring (in the month of Endyspoitropios).[32] Later, during the second century BCE, the Amphictyonic council decided on the exchange rate of the silver Attic-weight tetradrachm used in the city markets and during the *panegyreis*.[33]

Figure 6 Artemis Pergaia, 255–240 BCE. Tetradrachm (Attic): 16.69 g. London: *BMC Lycia* 119. I. pl. 24. From Mørkholm, *Early Hellenistic Coinage*, 482.

Figure 7 Athena Ilias, 165–150 BCE. Tetradrachm (Attic): 15.93 g. Copenhagen (*SNG*362). From Mørkholm, *Early Hellenistic Coinage*, 652.

Figure 8 Athena Nikephoros, *c.*165 BCE. Tetradrachm (Attic): 16.75 g. Copenhagen. From Mørkholm, *Early Hellenistic Coinage*, 615.

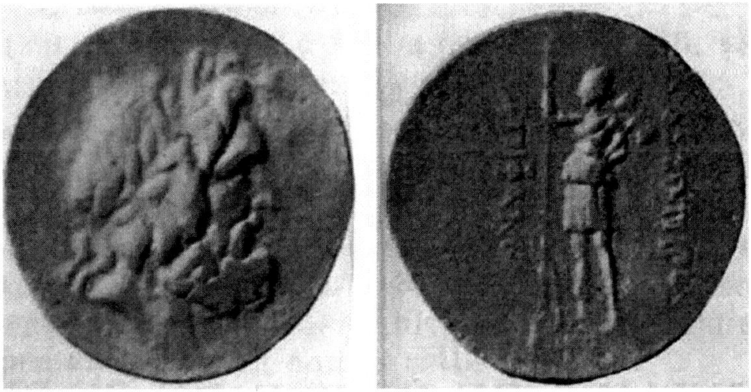

Figure 9 Klazomenai. 2nd century BCE. Tetradrachm (Attic). From Seyrig, 'Un tétradrachme énigmatique'.

An important part of the *panegyreis* was commercial, and fairs were quite important for the financial life of the city and the sanctuary;[34] a number of profit-yielding and fiscal activities took place in civic or rural fairs, which offered the opportunity for selling and buying exceptional merchandise, such as slaves, animals, and luxury objects. Different taxes were paid by the participants to the organizing cities, like the *pentekoste*, the *eponia*, the slaves' tax and the *skenai*' tax.[35] By means of these fiscal activities, the organizers extracted a share of the fair-related profit.

In some cases the *agoranomoi*, or magistrates in charge, also forced the participants to use the city's standards for all transactions.[36] During the second century BCE an official set of standard weights and measures was housed at Eleusis (*IG* II^2 1013, l. 48–49 = Clinton, *Eleusis*, no. 237).[37] Inscriptions and literary sources of a later date provide evidence on the means of exchange during the commercial part of the fairs. According to Dio Chrysostomos (*Res.Chrys.* 3 [*PG* 50.438]; *Pasch.* 4 [*PG* 52.770]), all fair transactions had to be conducted in cash.[38] The Hadrianic letter to the city of Pergamon (*OGIS* 484) noted the exclusive use of bronze currency during the *panegyreis.*[39] The second-century BCE Amphictyonic decree obliged the cities of all the Hellenes to respect the exchange rate of the Attic tetradrachm (*CID* IV 127) at the value of four silver (Attic or reduced Aeginetan) drachms.[40] The decree in question, almost contemporary with the coinages under consideration, mentions both the *panegyris* and the Attic-weight tetradrachm, and in this way establishes a connection between the two.

With this evidence in mind, the *raison d'être* of all the previously discussed silver coinages can be explained in the following way: the coinage issued in the name and types of the god in whose honour the fairs were organized by the corresponding association of cities was the only legal currency (*dokimon nomisma*) for large payments during this event, and all large-scale transactions had to be made with that currency.[41] The organizers of the fairs obliged the participants by law to exchange their currencies on Attic or other weight coinages for the currency in name and types of the honoured god, and probably to pay an *agio* for this procedure.[42] It was in this way that R. Bogaert

Figure 10 Pergamon, *c.*166–161 BCE or earlier. *Cistophorus*: 12.42 g. Copenhagen: *SNG* 413. From Mørkholm, *Early Hellenistic Coinage*, 616.

brilliantly explained the fiscal function of these coinages, which he also related to the *panegyris*.[43] To these silver coinages could be added an important number of bronze coinages that were also issued in the names of gods.[44]

In second-century BCE Asia Minor, where all Attic-weight fair coinages and the coinage of the Dionysiac artists were issued, the Attalids adopted a new currency after Apameia—namely, the *cistophoroi*, introducing new types and a weight of 12.60 g (Figure 10),[45] notably 25% lower than the Attic tetradrachm (17.2 g). Consequently Attalid Asia Minor automatically became a closed monetary zone reserved to the circulation of the *cistophoroi*: in the frontiers of the Attalids one silver Attic-weight tetradrachm was exchanged with one silver *cistophorus*, resulting in a 25% gain.[46] However, the Attalid king (Figure 11) and a number of cities related to the Pergamene kingdom issued their coinages in the Attic-weight standard, establishing a common point between these and the *panegyris* coinages,[47] while epigraphic evidence testifies the use of Attic-weight currency in fairs during the years following Apamea. In one of the eight decrees in honour of Archippe from Kyme (A2: *SEG* XXXIII 1036 l. 22, 24), it is mentioned that seventy Attic drachms were offered to the *boule pandemos* by Archippe, for the purchase of the ox for the sacrifice. Large animals, slaves, and luxury items were to be bought in annual *panegyris*, as literary and epigraphic evidence underlines:[48] it should be noted that the sacred law from Magnesia on the Maeander (*IMagnesia* 98) specifies that the impressive victim (ταῦρος ὡς κάλλιστος) had to be bought by the *oikonomoi* οἱ ἐνεστηκότες καὶ οἱ ἀεὶ καθιστάμενοι ... ἐν τῆι πανηγύρει ἑκάστου ἔτους. The frequent discovery of all these Attic-weight coins in Syrian hoards needs to be underscored.[49] Because of the Attic weight of these coinages, the participants thus were able to set them aside for deals in the corresponding fairs of the following year, or to use them in international transactions outside Attalid Asia Minor, in the kingdom of the Seleucids or in mainland Greece.[50] It is in this way that we propose to explain the coinage of the Dionysiac artists, in close relation with the *panegyris* organized by the artists at Teos, and with the *synoikismos*, both mentioned in the aforementioned letter of Eumenes II.

Figure 11 Eumenes II (197–158 BCE), 175–165 BCE. Tetradrachm (Attic): 16.62 g. Paris, ex Bank Leu 33 (5.v.1983) 364. From Mørkholm, *Early Hellenistic Coinage*, 614.

Dionysiac Artists and International Currencies

The coinage issued on Attic weight by the Dionysiac artists in association with the *synoikismos* and the fairs organized by the artists brings us to another topic—namely, the currency in which the artists were paid. Four inscriptions of Hellenistic date from different parts of the Greek world—from Euboean Eretria to Corcyra in the Ionian Sea, and from Carian Iasos to the Ionian Teos—denote payments made to the Dionysiac artists.

According to the Eretrian decree, which dates from the early third century BCE,[51] the *misthosis* of the Dionysiac artists was to be paid in drachms of Demetrios I (Figure 12). As O. Picard has pointed out, the Dionysiac artists accepted only the Attic-weight drachms of Demetrios Poliorcetes weighing 4.3 g, because the Euboean drachms of that period, with their weight of 3.6 g, were overvalued, and had ostensibly been struck for local circulation (Figure 13).[52]

According to *IG* IX 4, 798 (l. 7, 19, 41–42, 69), from around second-century BCE Corcyra, payments to the artists were calculated in Corinthian currency.[53] The case of Corcyra and the Corinthian currency needs to be seen in the same perspective. The city of Corcyra ceased to issue staters of Corinthian types and standard (8.9 g) around 300 BCE. Smaller denominations on that standard continued for some time (until 229 BCE?).[54] During the third century, the city issued silver coins of 5 g and their halfs (didrachms and drachms) and hemidrachms with its own types (forepart of cow/double stellate pattern), which changed once more at the end of the third century BCE (Dionysos' head/Pegasus). The new standard was close to the reduced Aeginetan standard, which was developed in central Greece and the Peloponnese during the third century BCE.[55] A further change occurred some years later. The new drachm was only 3.3 g (bust of Dione/Pegasus) and the hemidrachms were struck with Pegasus, Aphrodite's head, or Pegasus (Figure 14), Apollo's head.[56]

Figure 12 Demetrios I, *c.*298–294 BCE, mint of Tarsus. Drachm (Attic): 4.16 g. Copenhagen: *SNG* 1189. From Mørkholm, *Early Hellenistic Coinage*, 165.

Figure 13 Euboean League, Chalkis, *c.*304–290/85 BCE. Drachm (local weight): 3.43 g. Stockholm (*SNG* 1432). From Mørkholm, *Early Hellenistic Coinage*, 245.

Figure 14 Corcyra, *c.*229–200 BCE. Drachm (local weight): 2.89 g. Copenhagen (*SNG* Epirus-Acarnania 203). From Mørkholm, *Early Hellenistic Coinage*, 526.

Figure 15 Antiochus III (223–187 BCE), *c.*222–220 BCE, mint of Sardis. Tetradrachm (Attic): 16.64 g. Stockholm. From Mørkholm, *Early Hellenistic Coinage*, 402.

As the Dionysiac artists moved from one place to another,[57] they had no interest in receiving the overvalued silver currency of Corcyra, and preferred the full-weight Corinthian currency, which they could use all around western Greece, and easily exchange for Attic-weight currency.[58]

As far as Asia Minor is concerned, the three relevant epigraphic monuments that are also mentioned by Hoover and Lorber, are *IIasos* 4 and 152 from Iasos,[59] and *CIG* 3069 (*OGIS* 326; MacCabe, *Teos* 26) from Teos. The first inscription from Iasos dating from the mid-second century BCE mentions the drachms of Antiochos a 'long time after Western Asia Minor ceased to be Seleucid'.[60] According to Le Guen, whom Hoover and Lorber seem to follow, 'the reference is anachronistic and the result of quoting a by-law of the *koinon* that had been formulated when Teos was a possession of Antiochus III and not subsequently updated'.[61] Although Le Guen is probably thinking in the right direction, we must take note of the Attic standard of the drachms of Antiochos that had to be paid to the artists by members of their *koinon* who would fail to fulfil their duties (Figure 15).[62] We need also to note that second-century BCE Iasos never issued Attic-weight currency bearing Alexander or civic types, as did other cities of the Attalid kingdom during the second century BCE,[63] and that the Attic-weight drachms of Antiochos the Great would be the only currency of international character available.[64] In the letter of Queen Laodike to the city of Iasos (*IIasos* 4.24: end of third century BC), the Queen's donation was calculated in drachms issued by her husband, Antiochos III: εἰς προῖκας ταῖς τῶν ἀσθενούντων πολιτῶν θυγατράσιν, διδόντες μὴ πλέον Ἀντιοχέων δραχμῶν τριακοσίων ἑκάστηι τῶν συνοικιζομένων.

The third Asia Minor inscription is a Teian honorary decree of the Attalists for Kraton, son of Zotichos, and originally from Chalkedon (*OGIS* 326). It dates from the years following 146/5 and mentions the bequest of Kraton, to be paid after his death to the Attalists in Alexander drachms.[65] This inscription, that is one of the latest from Asia Minor mentioning the Alexander drachms, reveals once more the attachment of the artists to the Attic-weight standard.[66] We also need to underline that second-century

Teos issued Attic-weight Alexanders between 188 and 180 but no Attic-weight coins of civic types (Figure 3).[67] Consequently the *alexandreiai drachmai* of the inscription could be the Attic weight currency of the city.

In second-century Asia Minor, the artists preferred the international, full Attic-weight currency to the local overvalued and much lighter *cistophoroi*. When they decided to strike their own coinage, they chose the Attic standard adopted also by the Attalids and by a number of cities in second-century Attalid Asia Minor for their relations with the Seleucids and other peoples,[68] and by a number of religious associations, due to the profit that its use during the relevant *panegyreis* could generate and because of its international character. In third-century mainland Greece and in second-century Asia Minor, the network that the Dionysiac artists developed participated in two distinct networks—the extended Attic-weight currency network, and the *panegyris* coinages network—thus proving its own adaptability, promoting its identity and providing revenues for itself.[69]

Acknowledgements

The author wishes to thank the organizers of the Congress 'Networks in the Greek World', Katerina Panagopoulou and Christy Constantakopoulou for inviting me to publish my article in the volume. For useful discussions I am indebted to Sophia Aneziri, Vincent Gabrielsen, Christos Gatzolis, Christina Kokkinia, Maria-Gabriella Parissaki, P. Thonemann and my sister Rania, who as a private banker confirmed the network of gain. Yannis Lolos improved the English and Thodoris Kourembanas took the photographs of the coins. Thanks are also due to unnamed referees for their very useful comments. Responsibility for mistakes remains strictly my own.

Notes

[1] Lorber and Hoover, 'An Unpublished Tetradrachm', 59–68, pls. 15–17. Its publication was almost contemporary with the two important monographs on the Dionysiac artists: Le Guen, *Associations, passim*; Aneziri, *Vereine, passim.*

[2] For similar cases, see Psoma, 'Coinages in the Name of Gods'.

[3] Lorber and Hoover, 'An Unpublished Tetradrachm', 65 n.42. For full discussion of revenues and expenses of the artists see, Aneziri, *Vereine*, 169–202.

[4] See Elkins, 'Flavian Colosseum Sestertii', 218–20. Almost the same theory is developed by K. Liampi, 'Die Einführung', 219: 'Gedenkprägung'.

[5] See Kraay, *Archaic and Classical Greek Coins*, 66–67 on Athenian 5th-century decadrachms, and 211 on the Demarateion. On the Demarateion, see now Knoepfler, 'La chronologie du monnayage', 5–40, pls. 1–3.

[6] Lorber and Hoover, 'An Unpublished Tetradrachm', 62.

[7] Le Rider, 'Sur un aspect du comportement monétaire', 42–43.

[8] Aneziri, *Vereine*, 362: D15.

[9] Aneziri, *Vereine*, 81–84; 124.

[10] On these dates see Le Rider, 'Sur un aspect', 40. Lorber and Hoover, 'An Unpublished Tetradrachm', 64: 204–190 BCE based on Price, *Coinage*, no 3211. Mattingly, 'Ma'aret-en-Nu'man hoard',

85 on parallels from Myrina, Cyme and Erythrai: 'the meeting-point of Alexanders and autonomous issues at both mints must lie near 160 BC.'

[11] Le Rider, 'Sur an aspect', 40–42.

[12] Lorber and Hoover, 'An Unpublished Tetradrachm', 64.

[13] *IPerg* 163; Le Guen, *Associations*, no. 47; Aneziri, *Vereine*, D12.

[14] Aneziri, *Vereine*, 100–04.

[15] Lorber and Hoover, 'An Unpublished Tetradrachm', 61.

[16] Bogaert, *Banques et banquiers*, 109–10; Raven, 'Amphictyonic Coinage', 1–22; Kinns, 'Amphictyonic Coinage Reconsidered', 1–22, pls. I–V; Marchetti, 'La fin des émissions', 87–100.

[17] According to Marchetti ('La fin des émissions', 87–100), who does not follow the dates proposed for the beginning of Alexander's Attic weight currency (Troxell, *Studies in the Macedonian Coinage, passim*), it was issued to become the *monnayage par excellence* of the Greek world but the attempt failed, Philip II was killed and Alexander III chose the Attic standard for his 'imperial' coinage.

[18] Not to be confused with a number of coinages that bear legends of the type 'Ηρακλέους Σωτῆρος Θασίων etc. See also Robert, *Monnaies antiques en Troade*, 44. They also differ from southern Italian, Sicilian, and other Greek coinages with the name of the god in the nominative or genitive, accompanying the same god's head or statue: see Liampi, 'Die Einführung', 211 n. 23.

[19] Seyrig, 'Monnaies hellénistiques', *RN* 1963, 38–51, pls. IV–V; Le Rider, 'Un tétradrachme', 75–6; Mørkholm, *Early Hellenistic Coinage*, 30; Colin, *Die Münzen von Perge*, 39, 85–93 (chronology).

[20] Robert, *Monnaies antiques en Troade*, 22–46; Le Rider, 'Un tétradrachme', 75. For the unique, until very recently, silver tetradrachm (Figure 9) with the legend Διὸς Σωτῆρος 'Επιφανοῦς, see Seyrig, 'Un tétradrachme énigmatique', 24–25. Its types point to an attribution to Smyrna: Le Rider, 'Un tétradrachme', 76 and 78: 'Nous ignorons pour le moment d'où est originaire le tétradrachme de Zeus Sôter Epiphane.' Le Rider, 'Sur un aspect', 42: Smyrna. The legend under the exergual line of a second specimen in good condition from the 1987 hoard of Tartous, Syria (*CH* VIII 471, pl. LXIV 1) ΚΛΑΖΟΜΕ) leaves no doubt about its attribution to Klazomenai. For the third known specimen, see Tsangari, *Νόμισμα ἑλληνικόν*, 248, no. 166.

[21] G. Le Rider, 'Un tétradrachme', 66–79.

[22] Robert, *Monnaies antiques en Troade*, 22–46; Mørkholm, *Early Hellenistic Coinage*, 29. See also Özhan and Tombul, 'A New Hellenistic Decree', 109–13. On Perge, see Seyrig, 'Monnaies hellénistiques', 38–51, pls. IV–V; Boehringer, *Zur Chronologie mittelhellenistischer Münzserien*, 57–58, followed by Le Rider, 'Un tétradrachme', 75–76, argued against the common Pamphylian *era* of Seyrig. A return to the common *era* and Seyrig's dating of it was argued by Mørkholm, 'The Era of the Pamphylian Alexanders', 69–75. See also Boehringer, 'Beobachtungen und Überlegungen', 65–75, and pls. 5–7.

[23] For the *Nikephoria*, see Le Rider, 'Un tétradrachme', 70 and n.3, and Jones, 'Diodoros Pasparos', 183–205; Jones, 'Diodoros Pasparos Revisited', 1–14 on the epigraphic evidence concerning the *Nikephoria*. For the connection of the coinage with the Nikephora, see Mørkholm, 'Some Pergamene Coins', 187–92. For its connection with the struggle of Eumenes II with the Galatians, see Price, 'The Larissa 1968 Hoard', 233ff. The *terminus ante quem* of 160 is provided by the presence of two known specimens at the Sitichoro, Thessaly hoard (*IGCH* 237) of *c.*165 BCE (see Price, 'The Larissa 1968 Hoard', 233ff.), and of one specimen in Maaret-en-Nouman hoard (*CH* IX [2002] 511 (Mattingly, 'The Ma'aret-en-Nu'man Hoard', 83 with n.8).

[24] Robert, *Monnaies antiques en Troade*, 46 and n.2. See 46: 'On a relevé sur certaines des marques d'une origine non pergaménienne, suivant quelque système qui parait original.' On these coins,

see von Fritze, *Die Münzen von Pergamon*, 28–30, 32; followed by Le Rider, 'Un tétradrachme', 77 nn.1–2.

[25] Robert, *Monnaies antiques en Troade*, 38.

[26] Robert, *Monnaies antiques en Troade*, 45–46. On that important sanctuary, see now Mackay, 'The Major Sanctuaries', 2050–51 and 2059ff. On the meaning of the word *panegyris* = commercial part of a religious festival, see Robert, Review of *Samothrace*, 67–69; = religious festival, see de Ligt and de Neeve, 'Ancient Periodic Markets', 392–94. See also Chandezon, 'Foires et panégyries', 74–76.

[27] Robert, *Monnaies antiques en Troade*, 46.

[28] Kinns, 'Amphictyonic Coinage', 1–22 with previous bibliography and discussion.

[29] Plu., *Paroim.* 2, 5 (= Zen., *Paroim.* 5, 36) on Tyttigias the *andrapodiskos*; Ath., VI 7 referring to the poet Alexis (*PCG* II fr. 201); [Scymn.], 1–980, 600, who mentions the *Amphictyonike agora*; Thphr., *HP* 9, 10, 2, 6; Str. IX 3, 7 (= 420C); 4, 17 (= 429C); D. Chr., LXXVII 4; Harp., s.v. pylai, *EM*, s.v. pylagoroi; Hsch., s.v. pylatides agorai; Schol. *ad* Aeschin. III 124; Schol. *in* Ar., *Nub.* 623; *in* Pac. 734b; *in* Av. 121, 766 with references to Cratinus' *Pylaia* (*PCG* IV frs. 180–92).

[30] It will be useful to note that the third-century Amphictyonic decree that voted in favour of *asylia* for the Aitolians (*CID* IV 33, 122–31) mentioned the *asphaleia* of those who were going to the *panegyreis.*

[31] B. V. Head was the first to relate the Amphictyonic coinage to these markets on the evidence of its types (*HN*^{2}, 341–42). That coinage was rather short-lived: Mørkholm, *Early Hellenistic Coinage*, 86.

[32] F. Lefèvre, 'Remarques sur le calendrier', 579–94. At Delphi, the year begins in the summer: see Trümpy, *Untersuchungen*, 212–13.

[33] Psoma, 'A propos de drachmes d'argent', 79–88. On that decree, see also Sosin, 'Alexanders and stephanephoroi', 191–208.

[34] On *panegyreis* see now Chandezon, 'Foires et panégyries', 70–100. See also Czech-Schneider, 'Das Apollonheiligtum von Aktion', 76–100.

[35] On that tax see Knoepfler, 'L'intitulé oublié', 287–88 (= *SEG* 38 [1988] 380) of 100 BCE.

[36] Chandezon, 'Foires et panégyries', 81, and Migeotte, 'Les pouvoirs des agoranomes', 293 on *Syll.*^{3} 736, ll. 99–101.

[37] Kroll, *The Athenian Agora*, 29–30.

[38] See de Ligt, *Fairs and Markets*, 58.

[39] Wörrle, *Stadt und Fest*, 212, n.165.

[40] See above, note 33 and Migeotte, 'Les pouvoirs des agoranomes', 292–93.

[41] On the term *argyrion dokimon*, see Picard, 'Sur deux termes des inscriptions', 679–90, followed by D. Knoepfler, 'Tétradrachmes attiques', 235–41 and 213–17. On *episemon nomisma*, see also Martin, 'Silver Coins and Public Slaves', 21–48.

[42] This *agio* was named *epikatallage* in fourth-century inscriptions from Delphi (*FD* III 5, 25 II A11: 344 BCE; *FD* III 5, 58 l. 7 and 15: 331 BCE) and Epidauros (*IG* IV [2] 1, 103 B 41). The term is also noted by Thphr., *Char.* 30. 15. On the exact meaning of that term, see Sosin, 'Agio at Delphi', 67–80, and independently Le Rider, *Naissance de la monnaie*, 261–63.

[43] Bogaert, *Banques et banquiers*, 297.

[44] See Psoma, 'Coinages in the Name of Gods'.

[45] Le Rider, 'Politique monétaire', 164–69.

[46] On the Attalids, see Le Rider, 'Politique monétaire', 181.

[47] On Eumenes II see Mattingly, 'The Ma'aret-en-Nu'man Hoard', 83–84: *c.*165 BC for his rare portrait coins. On the second-century BCE wreathed tetradrachms of the free cities of Asia Minor, see Le Rider, 'Sur un aspect', 37–59, pls. I–III; Kinns, 'Asia Minor', 105–19; Picard, 'Les Romains et les émissions', 245–50. All three reached the same conclusion: these coinages

were issued by the cities of the Attalid kingdom for transactions with the outside world. See also Le Rider, 'Politique monétaire', 164–69 and 179–82; Houghton, 'Seleucid Coinage and Monetary Policy', 57–58.

[48] Chandezon, 'Foires et panégyries', 70–100.

[49] See above, note 47.

[50] See above, note 23.

[51] *IG* XII 9, 207 add S. 176+ *IG* XII Suppl. S. 178 (*SEG* 13, 1956, 462; 30, 1980, 1095). On this inscription, see now Le Guen, *Associations*, no. 1; Aneziri, *Vereine*, 53, n.184.

[52] Picard, *Chalkis et la Confédération eubéenne*, 343–48, on the coinage of Euboea during that period. On that same law, see also Picard, 'Monnaie ὁλοσχερής', 244–45.

[53] *IG* IX 4, 798 (= *IG* IX 1, 694). See also *IMagnesia* 44, l. 32: payments of Corcyra and *IMagnesia* 46 l. 41 of Epidamnos: sums calculated in Corinthian currency.

[54] Fried, *Autonomous Silver Coinage*, Gr. V.

[55] On that standard, see Psoma, 'A propos de drachmes d'argent', 83.

[56] On the coinage of Korkyra during the Hellenistic period, see Mørkholm, *Early Hellenistic Coinage*, 84, 152; Ashton and Warren, 'Hoard of Western Greek and Pseudo-Rhodian Silver', 5–16.

[57] Immediately after the sack of Corinth, the artists of Isthmos, their wives and children were granted by L. Mommius *aleitourgesian*, *anepistathmeian*, *ateleian* kai *aneisforian* (*IG* VII 2413-4; cf. Sherk, *RDGE* 44 l. 1–9). See also Aneziri, 'World Travellers'.

[58] On the Attic and Corinthian standards, see Kroll, 'Observations on Monetary Instruments', 77–91. On exchange rates between different currencies in Metropolitan Greece, see Bousquet, *Etudes sur les comptes de Delphes*, 126ff. On Hellenistic hoards from western Greece containing Corinthian currency: *IGCH* 140, Cephallenia, 1935: c. 300–275 BCE; 171 Mycenae, Argolis, 1895: 250-240 BCE; 182 Therianos, W of Patras, Achaea, 1930: after 265 BCE (Varoucha), *c.*220 BCE (Picard); 200 Kyra Vrysi, near Corinth, 1934: third century BCE; 201 Epirus, before 1825: third century BCE; 207 Butrinto (anc. Buthrotum), 14 km. S of Sarandë, Albania, 1927: late third century BC.; 246 Peloponnesus, 1936: *c.*175–150 BCE; 261 Zougra (anc. Pellene), Achaea, 1859: *c.*146 BCE; *CH* II 74, western Greece, 1973? Third century BCE; IV 31 (VI 45, near Stratos, Acarnania, Greece 1974: 330 BCE); IX 136, Anactorium environs, 1993: 300–240 BCE; 248, Amphilochia, 1995: 160 BCE.

[59] Le Guen, *Associations*, no. 53 l. 21–22; and Aneziri, *Vereine*, D13 l. 21–22.

[60] On the retreat of the Seleucids from Asia Minor after Apameia, see Will, *Histoire politique*, 223–28; Ma, *Antiochos III*, 282–83.

[61] See Lorber and Hoover, 'An Unpublished Tetradrachm', 65, n.43.

[62] On that inscription, see Aneziri, *Vereine*, D13 l. 21–22, and 194–95, and n.125.

[63] Le Rider, 'Sur un aspect', 40.

[64] On the relations of Antiochos III and his wife Laodike with the Carian city of Iasos, see Ma, *Antiochos III*, 89, 112, 161, 180–82, 216–17, 230, 232, 334–35.

[65] Aneziri, *Vereine*, 82, n.332 on the date of Kraton's death.

[66] On Alexander drachms in second- and first-century BCE inscriptions, see Knoepfler, 'Alexandreion nomisma', 46. On Attic and other monetary units in second- and first-century BCE Asia Minor inscriptions, see Psoma, 'Monetary Units in Asia Minor Inscriptions'.

[67] Le Rider, 'Sur un aspect', 40 on Alexanders of Teos, and 43 on the Ionian cities issuing coins with civic types.

[68] The Seleucid kingdom, Pamphylia, and the Propontis received coins of these series too, and we cannot exclude, for reasons already explained by G. Le Rider, their arrival in Delos or Rhodes. See Le Rider, 'Sur un aspect', 51.

[69] See Gabrielsen, 'Brotherhoods of Faith'.

Abbreviations

BullEpigr = *Bulletin Epigraphique.*

Coin Hoards I–IX. London: Royal Numismatic Society, 1975–2001.

CID IV = Documents Amphictyoniques. In *Corpus des Inscriptions de Delphes IV* edited by F. Lefèvre. Athens: École Française d'Athènes, 2002.

FD III 5 = E. Bourguet, *Les comptes du IVeme siècle,* Fouilles de Delphes III 5. Paris: De Boccard, 1932.

IGSK = *Inschriften griechischer Städte aus Kleinasien.*

IMagnesia = O. Kern, *Die Inschriften von Magnesia am Maeander.* Berlin: W. Spemann, 1900.

IMagn.Sip = Th. Inken. *Die Inschriften von Magnesia am Sipylos.* In *IGSK* 8. Bonn: Dr. R. Hambelt Verlag, 1978.

IPerg(amon) = M. Fränkel, *Die Inschriften von Pergamon,* I–II. Berlin, W. Spemann 1890–95.

ISmyrna = G. Petzl, *Die Inschriften von Smyrna* I–II 1/ 2. In *IGSK* 21–22 1/2. Bonn: Dr. R. Hambelt Verlag, 1982–90.

IG = *Inscriptiones Graecae.* Berlin: Walter de Gruyter, 1873.

IGCH = *Inventory of Greek Coin Hoards* edited by Margaret Thompson, Otto Mørkholm, Colin M. Kraay. New York: American Numismatic Society, 1973.

PCG = *Poetae Comici Graeci* edited by R. Kassel and C. Austin. Berolini et Novi Eboraci: W. de Gruyter, 1983–91.

SEG = *Supplamentum Epigraphicum Graecum.* Leiden: Gieben, 1923–2005.

Sherk, RDGE = R. T. Sherk, *Roman Documents from the Greek East. Senatus Consulta and Epistulae to the Age of Augustus.* Baltimore: Johns Hopkins Press, 1969.

Syll. [3] = *Sylloge Inscriptionum Graecarum* edited by G. Dittenberger. Leipzig, S. Hirzelium: 1915–24.

References

Aneziri, S. *Die Vereine der dionysischen Techniten im Kontext der hellenistischen Gesellschaft. Untersuchungen zur Geschichte, Organisation und Wirkung der hellenistischen Technitenvereine.* Historia Einzelschriften no. 163. Stuttgart: Steiner Verlag, 2003.

———. 'World Travellers: The Associations of Dionysiac Artists'. In *Poeti Vaganti,* edited by R. Hunter and I. Rutherford. Cambridge: Cambridge University Press, forthcoming.

Ashton, R. J. H. and J. A. W. Warren. 'A Hoard of Western Greek and Pseudo-Rhodian Silver'. *Revue Belge de Numismatique* 143 (1997): 5–16.

Boehringer, C. *Zur Chronologie mittelhellenistischer Münzserien 220–160 v. Chr.* Antike Münzen und Geschnittene Steine V. Berlin: Walter de Gruyter & Co., 1972.

———.,'Beobachtungen und Überlegungen zu den Ären der pamphylischen Alexandreier'. In *Travaux de Numismatique grecque offerts à Georges Le Rider,* edited by M. Amandry and S. Hurter, 65–75, pls. 5–7. London: Spink, 1999.

Bogaert, R. *Banques et banquiers dans les cités grecques.* Leiden: A.W. Sijthoff, 1968.

Bousquet, J. *Études sur les comptes de Delphes.* Athens: École Française d'Athènes, 1988.

Chandezon, C. 'Foires et panégyries dans le monde grec classique et hellénistique'. *Revue des Études Grecques* 113 (2000): 70–100.

Clinton, K. *Eleusis. The Inscriptions on Stone. Documents of the Sanctuary of the Two Goddesses and Public Documents of the Deme.* Athens: The Archaeological Society at Athens, 2005.
Colin, H. J. *Die Münzen von Perge in Pamphylien aus hellenistischer Zeit.* Cologne: Kölner Münzkabinett, Tyll Krohe, 1996.
Czech-Schneider, R. 'Das Apollonheiligtum von Aktion in hellenistischer Zeit: Überlegungen zum wirtschaftlichen Verhältnis zwischen Heiligtum und profanem Inhaber'. *Klio* 84 (2002): 76–100.
Elkins, N. T. 'The Flavian Colosseum Sestertii: Currency or Largess?' *Numismatic Chronicle* 166 (2006): 211–21, pls. 31–32.
Fried, S. 'The Autonomous Silver Coinage of Korkyra from the Earlier Strikings through 229 BC'. Ph.D. diss., Brown University, 1982. Ann Arbor, Mich.: Brown University Microfilms International, 1983.
Fritze, H. von. *Die Münzen von Pergamon.* Berlin: Akademie der Wissenschaften in Kommission bei G. Reimer, 1910.
Gabrielsen, V. 'Brotherhoods of Faith and Provident Planning: The Non-public Associations of the Greek World'. *Mediterranean Historical Review* 22, no. 2 (2007): 183–210.
Gauthier, P. 'Légendes monétaires'. In *Numismatique antique. Problèmes et méthodes. Actes du colloque organisé à Nancy du 27 septembre au 2 octobre 1971 par l'Université de Nancy II et l'Université Catholique de Louvain*, edited by M. Dentzger, P. Gauthier, and T. Hackens, 170–72. Louvain: Peeters, 1975.
Giovannini, A. *Rome et la circulation monétaire en Grèce au IIe s. av. J.-C.* Basle: F. Reinhardt, 1978.
Habicht, C. 'Die Inschriften des Asklepieions'. In *Altertümer von Pergamon* VIII 3. Berlin: Walter de Gruyter, 1969.
Head, B. V. *Historia Nummorum*^{2}. Oxford: Clarendon Press, 1911.
Houghton, A. 'Seleucid Coinage and Monetary Policy of the 2nd Century BCE. Reflections on the Monetization of the Seleucid Economy'. In *Le roi et l'économie. Autonomies locales et structures royals dans l'économie de l'empire seleucide. Actes des rencontres de Lille (23 juin 2003) et d'Orleans (29-30 janvier 2004), Topoi. Orient-Occident*, Suppl. 6. Lyon: Region Centre, Centre Ernest Babelon, Halma, 49–79.
Jones, C. P. 'Diodoros Pasparos and the Nikephoria of Pergamon'. *Chiron* 4 (1974): 183–205.
———.'Diodoros Pasparos Revisited'. *Chiron* 30 (2000): 1–14.
Kinns, P. 'The Amphictyonic Coinage Reconsidered'. *Numismatic Chronicle* 143 (1983): 1–22, pls. I–V.
———.'Asia Minor'. In *The Coinage of the Roman World in the Late Republic. Proceedings of a Colloquium held at the British Museum in September 1985*, edited by A. M. Burnett and M. H. Crawford. *British Archaeological Reports* 326 (1987): 105–19.
Knoepfler, D. 'L'intitulé oublié d'un compte des naopes béotiens'. In *Comptes et inventaires dans la cité grecque. Actes du colloque international d'épigraphie tenu à Neuchâtel du 23 au 26 septembre 1986 en l'honneur de J. Tréheux*, edited by D. Knoepfler, 263–94. Neuchâtel and Genève: Faculté des Lettres/Librairie Droz S.A., 1988.
———.'Tétradrachmes attiques et argent "alexandrin" chez Diogène Laerce'. *Museum Helveticum* 44 (1987): 233–53, and 46 (1989): 193–230.
———.'La chronologie du monnayage de Syracuse sous les Deinoménides'. *Revue Suisse de Numismatique* 71 (1992): 5–40, pls. 1–3.
———.'Alexandreion nomisma. L'apparition et la disparition de l'argent d'Alexandre dans les inscriptions grecques. Quelques réflexions complémentaires'. *Topoi* 7, no. 1 (1997): 33–50.
Kraay, C. M. *Archaic and Classical Greek Coins.* London: Methuen, 1976.
Kroll, J. H. *The Athenian Agora XXVI. The Greek Coins.* Athens and Princeton, NJ: American School of Classical Studies at Athens, 1993.
———.'Observations on Monetary Instruments in Pre-Coinage Greece'. In *From Hacksilber to Coinage: New Insights into the Monetary History of the Near East and Greece. A Collection*

of Eight Papers Presented at the 99th Annual Meeting of the Archaeological Institute of America, edited by M. S. Balmuth, 77–91. Numismatic Studies 24. New York: ANS, 2001.

Lefèvre, F. 'Remarques sur le calendrier des réunions de l'Amphictyonie pyléo-délphique'. *Bulletin de Correspondance Hellénique* 115, no. 2 (1991): 579–94.

———. *Documents Amphictyoniques*. Corpus des Inscriptions de Delphes IV. Athens: Écoles Françaises d'Athènes, 2002.

Le Guen, B. *Les associations des technites dionysiaques à l'époque hellénistique*, vols. 1–2, Études d'archéologie classique 11–12. Nancy: ADRA, 2001.

Le Rider, G. 'Un tétradrachme d'Athéna Niképhoros'. *Revue Numismatique* 128 (1973): 66–79.

———.'La politique monétaire du royaume de Pergame après 188'. *Journal des Savants* (July–December 1989): 163–90.

———.'Sur un aspect du comportement monétaire des villes libres d'Asie Mineure occidentale au IIe siècle: leurs émissions de tétradrachmes de poids attique frappées entre 188 etc. 140'. In *Les cités d'Asie Mineure occidentale au IIe siecle a.C.*, edited by A. Bresson and R. Descat, 37–59, pls. I–III. Paris and Bordeaux: De Boccard, 2001.

———. *La naissance de la monnaie. Politique monétaire de l'Orient ancien*. Paris: Presses Universitaires de France, 2001.

Liampi, K. 'Die Einführung des Kultes des Zeus Eleutherios in Makedonien. Die numismatischen Zeugnisse aus dem Hortfund von Serres'. *Athenische Mitteilungen* 117 (2002): 203–20, pl. 40.

Ligt, L. de. *Fairs and Markets in the Roman Empire. Economic and Social Aspects of Periodic Trade in a Pre-industrial Society*. Amsterdam: J.C. Gieben, 1993.

Ligt, L. de and P. W. de Neeve. 'Ancient Periodic Markets: Festivals and Fairs'. *Athenaeum* 66 (1988): 391–416.

Lorber, C. C. and O. D. Hoover. 'An Unpublished Tetradrachm Issued by the Artists of Dionysos'. *Numismatic Chronicle* 163 (2003): 59–68, pls. 15–17.

Ma, J. *Antiochos III and the Cities of Western Asia Minor*. Oxford and New York: Oxford University Press, 1999.

Mackay, T. S. 'The Major Sanctuaries of Pamphylia and Cilicia'. *Aufstieg und Niedergang der Römischen Welt* II 18, no. 3 (1990): 2048–82.

Marchetti, P. 'La fin des émissions de Larissa et le monnayage Amphictyonique'. *Obolos* 7 (2004): 87–100.

Martin. 'Silver Coins and Public Slaves in the Athenian Law of 375/4 BCE'. In *Mnemata: Papers in Memory of Nancy Waggoner*, edited by W. Metcalf, 21–48. New York: The American Numismatic Society, 1991.

Mattingly, H. B. 'The Ma'aret-en-Nu'man Hoard, 1980'. In *Essays in Honour of Robert Carson and Kenneth Jenkins*, edited by Martin Price, Andrew Burnett, and Roger Bland, 69–87, pl. XVII–XIX. London: Spink, 1993.

Migeotte, L. 'Les pouvoirs des agoranomes dans les cités grecques'. In *Akten der Gesellschaft für griechische und hellenistische Rechtsgeschichte 16. Vorträge zur griechischen und hellenistischen Rechtsgeschichte, Evanston, Illinois, 5–8 September 2001. Symposion 2001*, edited by R. Wallace and M. Gagarin, 287–301. Vienna: Österreichische Akademie der Wissenschaften, 2005.

Mørkholm, O. 'The Era of the Pamphylian Alexanders'. *American Numismatic Society Museum Notes* 23 (1978): 69–75.

———. 'Some Pergamene Coins in Copenhagen'. In *Festschrift für Leo Mildenberg. Numismatik, Kunstgeschichte, Archäologie*, edited by A. Houghton, Silvia Hurter, Patricia Erhat Mottahadeh, and Jane Ayer Scott, 187–92. Wetteren: NR, 1984.

———. *Early Hellenistic Coinage from the Accession of Alexander to the Peace of Apamea (336–188 BCE)*. New York, Port Chester, Melbourne, Sydney, and Cambridge: Cambridge University Press, 1991.

Özhan, T. and M. Tombul. 'A New Hellenistic Decree of τὸ κοινὸν τῶν πόλεων from Ilion'. *Epigraphica Anatolica* 36 (2003): 109–13.

Picard, O. *Chalkis et la confédération eubéenne. Étude de Numismatique et d'Histoire (IVe-Ier siècle).* Athens: École Française d'Athènes, 1979.

———. 'Les Romains et les émissions aux types d'Alexandre'. *Annuario dell'Istituto Italiano di Numismatica* 29 (1982): 245–50.

———. 'Sur deux termes des inscriptions de la trésorerie d'Ai Kanoum'. In *Hommages à Lucien Lerat*, 679–90. Paris: Belles Lettres, 1984.

———. 'Monnaie ὁλοσχερής, monnaie de poids reduit, *apousia* en Eubée, à Délos et ailleurs'. *ΧΑΡΑΚΤΗΡ. Αφιέρωμα στη Μάντω Οικονομίδου*, 243–50. Athens: ΤΑΠΑ, 1996.

Price, M. J. *The Coinage in the Name of Alexander the Great and Philip Arrhidaeus: A British Museum Catalogue.* Zürich: Swiss Numismatic Society in association with the British Museum, 1991.

———. 'The Larissa 1968 Hoard (IGCH 237)'. In *Kraay Mørkholm Essays. Numismatic Studies in Memory of C.M. Kraay and O. Mørkholm. Publications de l'histoire de l'art et d'archéologie de l'Université catholique de Louvain 59. Numismatica lovaniensia* 10, edited by G. le Rider, K. Jenkins, N. Waggoner, and U. Westermark, 233–43, pls. LIV–LV. Louvain la Neuve: Institut supérieur d'archéologie et d'histoire de l'art, séminaire de numismatique Marcel Hoc, Wetteren, 1989.

Psoma, S. 'A propos de drachmes d'argent du décret amphictyonique *CID* IV 127'. *Zeitschrift für Papyrologie und Epigraphik* 160 (2007): 79–88.

———. 'Coinages in the Name of Gods: *Panegyris* Coinages'. Forthcoming.

———. 'Monetary Units in Asia Minor Inscriptions of Hellenistic Date. The Staters of Archippe from Kyme (SEG XXXIII 1035-41)'. Forthcoming.

Raven, E. J. P. 'The Amphictyonic Coinage of Delphi, 336–334 BCE'. *Numismatic Chronicle* 111 (1950): 1–22.

Robert, L. Review of *Samothrace 2.1. The Inscriptions on Stone*, by P.M. Fraser. *Gnomon* 35 (1963): 50–79.

———. *Monnaies antiques en Troade.* Geneva: Droz; Paris: Minard, 1966.

Rostovtzeff, M. 'Some Remarks on the Monetary and Commercial Policy of the Seleucids and Attalids'. In *Anatolian Studies Presented to William Hepburn Buckler*, edited by W. W. Calder and J. Keil, 277–98. Manchester: Manchester University Press, 1939.

Seyrig, H. 'Monnaies hellénistiques. VIII. Pergé'. *Revue Numismatique* 118 (1963): 38–51, pls. IV–V.

———.'Un tétradrachme énigmatique'. *Revue Numismatique* 126 (1971): 24–25.

Sosin, J. 'Agio at Delphi'. *Numismatic Chronicle* 160 (2000): 67–80.

———.'Alexanders and Stephanephoroi at Delphi'. *Classical Philology* 99 (2004): 191–208.

Troxell, H. *Studies in the Macedonian Coinage of Alexander the Great.* American Numismatic Society. Numismatic Studies 21. New York: ANS, 1997.

Trümpy, C. *Untersuchungen zu den altgriechischen Monatsnamen und Monatsfolgen.* Heidelberg: Universitätsverlag C. Winter, 1997.

Tsangari, D. *Νόμισμα ἑλληνικόν. The Alpha Bank Collection.* Athens: Alpha Bank, 2007.

Will, E. *Histoire politique du monde hellénistique* I. Annales de l'Est no. 30. Nancy: Université de Nancy II, 1982.

Wörrle, M. *Stadt und Fest im kaiserzeitlichen Kleinasien.* Studien zu einer agonistischen Stiftung aus Oinoanda. Munich: C.H. Beck, 1988.

On the Road to India with Apollonios of Tyana and Thomas the Apostle

Gary Reger

From Herodotus on, India exercised its 'exotic' fascination on the Greeks. Tales of travels, like the lengthy visit of Megasthenes to the court of Chandragupta, found avid readers. Not every story of a sojourn in India recorded an actual journey, of course; among the many 'fictional' accounts of Roman encounters, the visits of Apollonios of Tyana, the famous sage, and Thomas the Apostle, both written around 220–240 CE, were produced at a time when Roman curiosity about India was especially strong. Neither the *Vita Apollonii* of Philostratos (hereafter *VA*) nor the anonymous *Acts of Thomas* (hereafter *ATh*) has any claim to 'factual' testimony about India.[1] But since 'every text, whatever else it attempts to do, must also seek to procure for itself its recipients' belief in it',[2] I would like to explore the ways in which these two texts exploit readers' experience of networks as facilitators of travel to secure that verisimilitude for the adventures of their protagonists in far-away India. Following Hannah Knox and her collaborators, I take 'network' in the broadest sense, both as social networks and as 'methodological tools, as metaphors for understanding forms of relations and as descriptors of social forms', perhaps less rigorously defined analytically but nevertheless useful—I hope—for conceptualizing ways in which people who otherwise do not know each other interconnect (and a sense treated by Knox and her colleagues as equally legitimate). Networks deployed in this way can be used as 'a device which allows [one] ... to discover latent structure' in social forms.[3]

Discussion about Apollonios in India long focused on positivist questions of the reality of the journey and of his companion Damis, whose *deltoi* Philostratos used to compile his own, more elegant, biography. A turn in the scholarship began as commentators like E. L. Bowie and others began to stress the interplay between 'history' and 'fiction', between 'reality' and 'invention'. This turn has found its latest expression in the new study by Thomas Schirren, whose work saves me the obligation of entering into either the debate about Damis and his testimony, or the equally thorny question of the genre to which the *VA* should be assigned. I will simply take for granted that the trip to India was wholly imagined by Philostratos, but that he sought to make a greater, truer point in his narrative, whose claim to believability was an important tool to convey this greater truth to his readers.[4] This new approach to the *VA* has directed attention particularly at these questions of the creation of verisimilitude: 'the more similar a fiction product is to a real one, the more probable it is.'[5]

Something similar, even convergent, has happened in the scholarship around the *ATh* and its companion *Apocryphal Acts of the Apostles* (hereafter *AAA*). Work placing the *AAA* in their context has emphasized, in understanding that context, links with the Greek novel, particularly in matters of readership. It seems generally agreed that the Greek novel attracted an educated readership familiar with the urban world of Roman Asia Minor, presupposing and reinforcing the cultural and social conventions of the elites of the *poleis* of that world. The *AAA* look more and more like an outgrowth of the novel—a view even more cogent if it is correct, as has been suggested, that women were a conscious target audience for both bodies of work.[6] This view has been questioned, but nevertheless both the novel and the *AAA* often feature strong female characters—like Kallirhoe in Chariton's novel and Thekla in the *Acts of Paul and Thekla*. Women in the *AAA* are often first to recognize the message of Jesus, attracted spiritually—but often in language reminiscent of, or borrowed from, the eroticism of Greek fiction—to his messengers, and brave in facing down the scorn of their menfolk and the contempt of society.[7] In the *ATh* Mygdonia typifies such a character. After conversion she refuses sexual relations with her husband, flees naked from him, and discusses her new relationship with Thomas and Jesus in language that sounds like a woman talking about her adulterous lover—she even bares her head for Thomas when he baptizes her, an act for which women are otherwise condemned several times in the *ATh*. And she remains firm in her new convictions to the end.[8]

This optic that sees the *AAA* as an offshoot of the novel raises then precisely as well the question of how they achieved believability; how the contract of belief was negotiated between book and reader.[9] In the case of the *ATh*, which, unlike the other *AAA*, takes place not in the familiar world of Asia Minor but in exotic India, the issue is more acute. In both the *VA*, then, and the *ATh*, the exploitation of familiar networks may help explain how readers were invited to accept the plausibility of setting and event.

Ethnicity—Jews and Greeks among the Barbarians

The *Acts of Thomas* opens with a gathering of the apostles in Jerusalem. They are dividing up the world by lot to evangelize. When Thomas receives India, he objects on the ground of his ethnicity: 'How can I,' he protests, 'as a Hebrew man, go among the Indians to announce the truth?' But it soon turns out that Thomas' ethnicity provides a key contact early in his travels, when at a wedding feast in Andrapolis he meets a flute-girl who turns out to be Jewish. She serves as his translator at the festival and becomes his first convert in India (*ATh* 8). She disappears after this episode, but Jews evidently remain in Thomas' audience, though only hinted at: for instance, when Thomas takes advantage of a prostitute's description of her visit to hell to proselytize his audience, he lays special stress on the conformity of Jesus' mission and life to the predictions of the prophets (*ATh* 59). This stress on the Hebrew Testament would seem intended to induce in readers a supposition that among Thomas' listeners were Jews to whom the prophets were familiar and meaningful—though of course the passage is aimed also at the readers themselves.

The importance of Jewish converts and Jewish links for early evangelism is a familiar feature of first- and second-century Christianity (e.g. *Acts* 17). Thomas is able to exploit the connection ethnicity affords to demonstrate his powers and gain his first converts. Here, I think, is a very clear instance too of an authorial search for verisimilitude at work. Thomas had tried to beg off from the task Jesus had given him by pointing to his ethnicity, and appropriately his ethnicity seems to be a barrier to this task in his first efforts. But then the network of ethnicity serves an important role in helping him to find his feet in a very foreign land—a nice irony.

A network created by ethnicity recurs also in the Hymn of the Pearl, the most famous episode in the *ATh* and usually regarded as a separate composition, pre-dating the *Acta.*[10] In the Hymn a young Indian prince dispatched to Egypt to snatch a pearl from a dragon loses his way in a foreign land, falls asleep, and neglects his mission. On his journey he meets another young nobleman whom he recognizes as 'a kinsman from the East, a free man, a graceful and beautiful child, son of the greatest men', who becomes his travelling companion and warns him against the Egyptians (*ATh* 109).

For Apollonios, the ethnic connection runs through a complex interconnection of Greek ethnicity and mytho-history. A good example is the episode concerning Eretrians settled in Mesopotamia, descendants of several hundred men captured by Xerxes on Euboia and transported back to Mesopotamia, where they settled in a village. Apollonios notes their tombstones, still composed in Greek fashion, their worship in Greek-style temples, use of an *agora*, remembrance of their Euboian heritage as sailors, and recitation of a Greek elegiac poem over the tomb of some sailors and shipmasters in a kind of commemoration. Apollonios intervenes with the king to better their lot (*VA* 1.25, 35). Apollonios' ethnicity benefits him personally, too. When he arrives in Babylon it is the fact that he is a Greek and a philosopher that gains his entrée to the palace; told that a Greek had arrived, the king Vardanes bethought himself of Themistokles, who had lived with Artaxerxes and brought him great benefit (*VA* 1.29). Ethnicity works here both in favour of the poor Eretrians, whose problems would have remained unsolved without Apollonios, and for the sage himself, who, as a Greek, benefits from the deeds of ancient Greeks in a distant land.

These sketches display the importance of ethnicity as a marker of belonging to a network which both Apollonios and Thomas readily exploit in the unfamiliar environments in which they find themselves. Recent research into Greek ethnicity has revealed the importance of this category for self-definition and identity formation, however difficult the problems of recovering the internal—and probably but rarely expressed—feelings on these matters of citizens of the eastern Roman imperial world. The uses to which ethnicity is put in the *VA* and the *ATh* add another element to our understanding, for clearly the authors of these narratives expected their readers to 'get' the role ethnic networks played in the protagonists' travels.[11]

Trade and Traders

Thomas' journey to India happens because Abbanes, an Indian *emporos*, arrives in Jerusalem looking for an architect to buy and bring back to his king, Gundophoros. While the details here are a little dicey—Jerusalem is not an obvious starting point for a sea journey to India—the presence of Indian traders and merchants in the Near East, in particular in Syria and the Aramaic-speaking provinces of the Roman empire, and Egypt, is well attested. Indeed, the story that Thomas evangelized India 'owes its origin to the lively cultural and commercial relations of Edessa and Osrhoëne with northern India, which particularly at the beginning of the third century were very intensive'.[12]

Several inscriptions from Palmyra shed some light on the trade that passed between this 'caravan city' and the Persian gulf, or India beyond. The normal route for traders from Palmyra was to pass east across the desert via camel caravan, and then to move down the Euphrates to Spasinou Charax (modern Basra) on the Shatt-al-Arab, near the head of the Persian gulf. The route passed near to or through several important centres, including Babylon, Ktesiphon (Seleukeia), and Vologesias. An inscription of 19 CE mentions merchants of Seleukeia (though which one is not certain); another of 24 CE, merchants of Babylon. The route was described in the *Parthian Stations* of Isidoros of Charax, who wrote in the last years BCE as a guide for the mission planned but never executed by Gaius Caesar.[13]

For both Apollonios and Thomas, merchants' networks play another role: access to kings. Apollonios' entry into the region of Babylon begins with his detention at a border post, whose commander—a satrap and 'eye of the king' (1.21)—interrogates him about his intentions, no doubt expecting he has a merchant at hand. Apollonios answers unsatisfactorily until the satrap realizes in astonishment that the sage needed no translator; the conversation proceeds more cordially, with the satrap inviting Apollonios to be his guest and sending him on. Later, after gaining entrée to the king at Babylon, Apollonios receives the camels, guides, and supplies he requires to undertake the difficult overland journey to India itself (*VA* 1.40). At Taxila, Apollonios waits in a temple outside the city for the King Phraotes to be notified of his arrival and to receive him (*VA* 2.22); as we will see in a moment, this is a theme that recurs in accounts of traders. For Thomas, the access to Gundophoros in Andrapolis follows directly from the connection of Abbanes to the king, who had dispatched him to Jerusalem to buy an architect in the first place.

There is explicit evidence from other sources that such links between merchant traders and royalty did indeed exist in the Indian Ocean world. The *Periplous of the Red Sea*,[14] which aims to provide practical information, advises traders travelling via Adoulis on the southwestern Red Sea coast that King Zoskales, who rules this land, though a stingy fellow and tough trader, is otherwise high class and familiar with Greek (5). Under the emperor Claudius a Roman freedman blown off course landed in Taprobane, modern Sri Lanka (Ceylon), where he enjoyed the hospitality of the local king who, so impressed was he by the quality of Roman coinage, that he dispatched an embassy to the emperor (Plin., *Nat. hist.* 6.84–5). Five hundred years later a certain Sopatros landed on Taprobane for business in the company of merchants from Adoulis and a 'Persian elder'. Upon arrival, the local officials and customs officers received them and brought them to the king, who fed them and then asked a series of questions about the character of the kings of his guests' countries, 'Persia' (i.e. the Sasinian kingdom) and Rome. Sopatros' clever response won the king's approval—he told the king to examine the coinage of both countries which the traders had with them. The Roman coins' better quality proved the better character of the Roman kings, and Sopatros was rewarded by being paraded around the city on an elephant to the accompaniment of drums (Kosmas Indikopleustes, *Christian Topography* 11.17–19). Assurances of welcome from Indian kings are not just a *topos*, for a Tamil source recounts an incident of the last quarter of the second century CE, when a king in southern India arrested merchants and confiscated their goods, and the testing of coins was also a standard feature of Indian commercial practice.[15]

The central importance of networks to merchants is abundantly clear—they of all people were most centrally dependent on finding help through contacts in the distant and strange lands they visited. It is therefore no surprise that the author of the *ATh* chose to take Thomas to India in the care of a merchant, whose networks opened to the apostle, or that Apollonios, though himself no merchant (not that trade was an unsuitable occupation for a travelling philosopher—see Plato and Solon!), was at times mistaken for one, and availed himself of the advantages that the networks of traders offered.

Translation and Translators

During his stay in Nineveh, Apollonios met a local named Damis. Damis was so impressed with Apollonios' knowledge and insight that he begged to accompany him on his travels. Damis' self-recommendation lay in his usefulness as a guide and translator: having travelled to Babylon he knew the city and its many villages, and 'the languages of the barbarians, as many as there are, one of the Armenians, another of the Medes and the Persians, another of the Kadousians, I understand them all' (1.19). This recommendation provokes one of Apollonios' oracular comments, that he really does not need a translator because he knows all languages without having learned any—because he understands silence.

Be that as it may,[16] Apollonios rarely needed a translator, but not because he understood silence. For a striking feature of his encounters with Indians and other non-Romans on his travels is the frequency with which he meets people who speak Greek. At the court of Vardanes the king himself addresses Apollonios in Greek, which seems to be the medium of their subsequent conversations. This is perhaps no surprise, given the official use of Greek by the Arsakids for monuments, coin legends, and other texts,[17] but the persistence of Greek farther east may occasion more surprise. Phraotes, king at Taxila in northwestern India, at first uses an interpreter, but after some philosophical conversation dismisses him and confesses his competence in Greek (*VA* 2.27). All the inhabitants of the village at which Apollonios stops before visiting the Indian sages who live on a hill in the Ganges valley speak Greek; it is therefore no surprise to Apollonios when a messenger from the sages sent to guide him to them addresses him in Greek (*VA* 3.13). Likewise the chief sage Iarchas, distinguished by his elevated and decorated stool, speaks Greek (*VA* 3.16, and see Damis' judgement at 3.36). Speaking Greek clearly marks status, and Apollonios' sometime reliance on translators and typically on persons who know Greek reminds us that he and Damis are surrounded by people who speak some other language. The use of Greek, to be sure, was among the conventions of Greek fiction. The ability to speak Greek could distinguish good guys from bad guys, and lamentations on the pain of being surrounded by strangers who cannot speak Greek occur in fiction, especially for characters trapped in the Aramaic-speaking Near East, like Kallirhoe in Chariton's novel, during her sojourn on the banks of the Euphrates (5.1).[18] Speaking Greek may be an outward sign of inward virtue. At the same time, however, Greeks had been visiting India for centuries before Apollonios, and the great infusion of Greek language and culture following Alexander persisted in the region for centuries; Greek documents continue to come to light.[19] The most famous example of Greek in public texts comes of course from the Ashokan Rock Edicts from near Khandahar (Alexandria), Afghanistan, one of which includes versions in both Greek and Aramaic.[20] Greek material culture continued likewise to play a role in the larger cultural construction of this world, as the Begram treasure from Afghanistan, among other finds, reminds us.[21]

Thomas' language situation is more enigmatic. His first language presumably would have been Aramaic. Aramaic had served as a lingua franca of the Achaemenid Empire, which had included the regions where Thomas travelled (in the earlier parts of the *ATh*).[22] Use of Aramaic persisted through the conquest of Alexander and after the concession of the eastern provinces by Seleukos I to the first Mauryan King Chandragupta, and probably left an important trace in the script developed to write Gandhari—the Kharoṣṭhī, which has been argued to have been spun off the Aramaic alphabet.[23]

But Thomas ought not to have worried about language barriers. In a famous scene in the canonical *Acts* (2.5–13), Jews from 'every tribe under the sky' hear the disciples, filled with the Holy Spirit, speaking in their own native languages. The Holy Spirit should thus have provided Thomas with unmediated communication with the locals, in whatever tongue. Indeed, this seems to happen immediately upon his arrival in Andrapolis when, surprised by the evidence of feasting around him, he asks passers-by the occasion and they answer him directly. But Thomas also experiences communication problems. At the banquet, a careless server bumps into the apostle, who curses him in Aramaic (or 'Hebrew'). The curse comes to pass, and Thomas prays a long prayer, again in Hebrew. No one understands him except the female Jewish flute-player, who, when the curse which she had overheard comes true, serves as Thomas' interpreter to the crowd—not all of whom believe her (*ATh* 8). It would be hyper-rational to press too hard on the inconsistency; surely the larger point here is the emphasis on Thomas' power and isolation (clearly brought out in the description of the banquet scene, where Thomas does not understand the rules) and that his first (apparent) convert is Jewish. She disappears after this episode, and language matters are not raised again.

The tensions in language problems are perhaps worth emphasizing. Nothing isolates a traveller in a foreign land more than inability to speak the local language. In Apollonios' case, this problem is finessed by Greek—the language of the educated, of the philosophers: a network of a special kind, to which few are invited. (Damis' linguistic skills are never called on.) Thomas, on the other hand, faces the problem once, but acutely, in a context that strongly emphasizes his isolation. As a follower of Jesus in India, he is a minority of one, not just linguistically, but spiritually too. But ultimately, his goal is to bring that huge majority of unbelieving Indians over to his side; to transform isolation into predominance, into a network to which all can belong. Here we have a first, important hint of divergence between the ways the *VA* and the *ATh* exploit some aspects of the network; we will see more.

Public Banquets and Public Gatherings

Thomas' first opportunity to evangelize came soon after his arrival in Andrapolis. The city was holding a festival to celebrate the marriage of the king's only daughter, and Abbanes and Thomas were invited. As a slave Thomas was seated with the other slaves and some foreigners, while Abbanes as a free man was put in a higher status group. The situation is vivid: Thomas reclines on a couch with a large group of people who are eating and drinking and wearing crowns and perfumes. Thomas dabs the top of his head, his nostrils, his teeth, his ears, and his sternum with perfume; he places a crown of myrtle and other flowers on his head, and holds a reed in his hand. But he refuses food or drink because, he says, he has come for a higher purpose: to fulfil the will of God.

Such public banquets to which were invited not only citizens, but also women, children, visiting foreigners, and even slaves, and might serve as loci for activating network connections, were a standard component of civic life in Graeco-Roman Asia Minor in the imperial period.[24] At the *polis* of Arkesine on the island of Amorgos, an inscription tells us that 'all the Arkesineis and the foreigners who are visiting' were invited to a public banquet (*IG* XII 7 35.4–6); another, from Aigiale, issued an invitation 'to all citizens present in Aigiale and *paroikoi* and *xenoi* and those of the Romans and their wives present' (*IG* XII 7 515.55–57).[25] Inscriptions from Stratonikeia in Karia of the Roman imperial period likewise honour wealthy citizens who had undertaken to subvene the costs of banquets, like Herakleitos and his wife Tatarion, who gave 'three drachmai to each of the citizens three *drachmai* and to the Romans and to the *paroikoi* and to the foreigners (*xenoi*)' in connection with a celebration of the cult of Agrippa.[26]

Wedding banquets and other festivals to which even strangers were invited are a staple also of novels of the Roman period. In Chariton's novel *Kallirhoe*, the whole city of Syracuse, including, Chariton wryly observes, all the women, attend the wedding of Chaireas and Kallirhoe (1.1.11–13).[27] Later in the novel, when Kallirhoe—married again to Dionysios of Miletos after being sold to him as a slave—gives birth to a child, really that of Chaireas, but supposedly his, the city of Miletos throws an 'enormous festival' for Dionysios to celebrate the birth of an heir, and 'embassies arrived from everywhere' to share the joy (3.7.7).

In the end, Philostratos has constructed the series of networks within which Apollonios and Damis operate on Greek models. All the educated people whom Apollonios meets speak Greek; Taxila is depicted as a kind of 'Athens of the east', with its extramural sanctuary and confusing street-plan, ironic inversion of the cultivation of its philosophic inhabitants; the gods are assimilated to Greek deities. The fundamental presupposition upon which this world of fictional India is constructed consists in the view that, as the conquered Greeks conquered their Roman conquerors with their culture, so in the other direction, to the east, all that is good likewise is fundamentally Greek—and so wholly familiar and comfortable. Indeed, from the very start, Philostratos has emphasized Apollonios' Greek heritage: his homeland was Tyana, a Greek city in the *ethnos* of the Kappadokians (*VA* 1.4). The contrast between his Greek homeland and the barbarian region in which Tyana was situated could not be more explicit. Behind this Graecized world lies of course the figure of Alexander. As we have seen, he keeps cropping up on Apollonios' path—whether as conqueror, inscriber of texts, explorer of distant regions, or witness to wonders. The Greek culture Alexander planted grew and by its self-evident superiority conquered, creating a stratum of Greek-infused social institutions and practices Apollonios could recognize and exploit.

Philosophers and God

All the networks, real and imaged, that we have considered in the *VA* and the *ATh* pale in comparison with the network that dominates each work: in Apollonios, the network of philosophers, and in Thomas the network of believers saved by Jesus.

Apollonios' network of philosophers can be traced in part by the letters he writes in the *Life*. For instance, when he rescues the poor Eretrians in Mesopotamia, he writes about it to the sophist Skopelianos (*VA* 1.24). But perhaps the importance of his profession comes most clearly to the fore in his long conversation, occupying the bulk of book III, with the Indian sage Iarchas. The conversation covers most philosophical subjects, including Apollonios' past life as an Egyptian steersman for a merchant ship almost beset by pirates; miracles are also performed, expelling demons and curing the lame; and the marvels of India and Egypt like the phoenix are discussed. The king himself, in an episode like an inverted *Questions of King Milinda*, comes to participate, not always successfully, in the conversation. Apollonios' four-month sojourn with Iarchas ends with the dismissal of the guides and camels that King Phraotes had provided in exchange for camels and guides supplied by Iarchas.

Apollonios moves, then, in a world whose networks are most fundamentally defined by his life as a philosopher. It is not his Pythagoreanism—as opposed to, say, Stoicism or Middle Platonism—that matters, but the magnetism by which his wisdom draws to him people that he needs. Damis—whose worldly knowledge of strange languages is his strongest recommendation to his master—hears Apollonios and offers his services on the spot. At Babylon, Apollonios enjoys the hospitality of the king, who provides all the necessities for his further travels, including camels, money, and guides. When word of his arrival at Taxila reaches the ears of the king there, he is swept up again into the palace, and stays, at his host's most urgent insistence, a day longer than the allotted three for strangers in the city. Iarchas, the Indian sage, recognizes his brother in wisdom immediately, and speaks to him in fluent Greek. In a certain sense, Philostratos' point is that Apollonios can dispense with the sorts of networks ordinary travellers depend on to make trips to countries as far away and strange as India. He moves not, or at least not regularly and obviously, in a world of merchants whose *periploi* recommend kings who may be greedy but show hospitality to strangers, or who set lavish dinners at which guests may be quizzed about the moral qualities of their rulers on the basis of portraits on coins. His letters of recommendation are otiose. Apollonios' disconnection from the social networks on which ordinary people depend forms part of Philostratos' claim for his status, even if in narrating his travels the author cannot help sometimes deploying those networks to make the connections that his hero needs.[28]

The claim of the *ATh* for its most fundamental network, however, is far more radical. For Thomas, belief in Jesus and the theology of salvation that the *ATh* urges create a network of believers that both reveals itself 'magically' in the presence of believers, and that will endure forever, even into life after death. At the same time, the articulation of a Christian network created real connections in the here-and-now that Thomas and his followers could exploit. Word of Thomas' preaching reached a certain captain named Siphor, whose wife and daughter had been afflicted by demons. He sought help from Thomas. Thomas expelled the demons and brought the women back to life (*ATh* 62–80). The captain—who had accepted Thomas' preaching even before the miracles—now became a stalwart supporter of the apostle. But the captain himself had important connections, for he was extremely rich and served the King Midaios as one of his most important counsellors. Siphor provided Thomas a place to teach in his own house (*ATh* 131), no doubt supported the congregation that Thomas created under the deacon Xenophon (*ATh* 65 and 67), and, perhaps most importantly, mediated access to the highest levels of local society and a degree of protection. Thus Siphor attests to Thomas' saving of his wife and daughter before King Midaios, when another powerful noble, Charisios, seeks to destroy the apostle because he has alienated his wife Mygdonia (*ATh* 104). In the end, Siphor cannot prevent Thomas' arrest and imprisonment, but the importance of the network lies not so much in questions of protection as access to those very high levels in Indian society that lead, ultimately, not only to the apostle's martyrdom but also to the conversion of the king himself, when Thomas' martyred dust saves one of his son's from demonic possession.

Indeed, in some sense, ultimately Thomas' mission calls for a complete abandonment of the ties of the human community that create the threads holding networks together, in favour of a new community, with its own, heavenly ties, and promise of a network that never dies, nor ever fails:

> on account of you [Jesus] we have left our homes and our parents, and on account of you we have become strangers, sweetly and willingly ... we have left behind those who matter to us by birth, in order to associate with your kindred ... we who have left our own fathers and mothers and nurses to regard you as our father and to satisfy ourselves with your divine food ... on account of you we have left our wives of the body and our earthly fruits, in order to share in that permanent and true association, and to bear true fruit whose nature lies above, of which no one can rob us, with which we remain and remains with us. (*ATh* 61)

This is the language of social ties and networks—based most fundamentally in the family, with its promises of connection through parents and spouses to relatives, and through children to the future. This call for abandonment of the social ties that mattered most to the majority of people entailed inevitably the disruption of the longer, somewhat more tenuous social strands of connection by virtue of family, class, occupation, ethnicity, language, and social position, that social networks required to operate. And Thomas himself sets a sterling, and rather terrifying, example. Sold as a slave, a deracinated person, unable to control his movements or his duties, he depends, to find his feet at the beginning of his mission, on exactly the kinds of benefits that belonging to social networks confer: his Jewish ethnicity, his skills and reputation as an architect and builder, the entrée to the highest social levels that association with a merchant brings. But his message is a radical call to forsake the security of those ties and the very networks that enabled him to operate in a land where he knew no one and did not speak the language, in favour of a radically different kind of network whose nodes were, in some ultimate sense, not on earth, but in heaven. And even though the community of believers does constitute a new, earthly family, a source of networks itself, in the end it cannot prevent Thomas' final martyrdom. His death is of course necessary to bring his mission to a satisfying end, but certainly in some sense his network has, ultimately, failed. But he also leaves behind Siphor and Midaios' son Vazan as officials in the new Indian church—seed of a new Christian network.

Some Final Reflections

Deploying networks lends verisimilitude to the *VA* and the *ATh*. Apollonios and Thomas, two strangers in a strange land, rely on the entrée granted by belonging to these networks to ease their way into the Indian world, and to accomplish their missions. There is nothing in the networks they exploit that would have been unfamiliar to any of their contemporary readers. Thus our exploration of networks in the two texts opens a window on a technique the authors have used to conjure that claim to 'believability' that is the first duty of fiction—a believability perhaps especially important in narratives whose central features include experiences far removed from the lives of their readership. In important ways, Apollonios and Thomas share a kind of religio-philosophic culture in which networks of belief and achievement are centrally important. At the same time, there are profound differences in at least some of the ends to which these networks are put. Apollonios seeks a transcendence to which but few gain access; we admire and wonder, but few emulate. Thomas, in contrast, is calling on everyone not to admire but *precisely* to emulate. Thomas as a personality is irrelevant, unimportant, vanishing. It is the imitation of Jesus, open to everyone, that matters; and indeed Thomas, the 'twin' who is even sometimes mistaken for Jesus, is the perfect model, precisely because his is the perfect imitation. Networks lend verisimilitude to the travels, but in the end, the most important networks diverge dramatically. The network of philosophers ultimately contains Apollonios, Iarchas, and—really no one else, as even the gymnosophists that Apollonios meets in Ethiopia, despite being exiled Indians, do not make the grade. Thomas' network of believers could—indeed aims to—encompass the entire world.

Notes

[1] For the *VA*, I use the new Loeb edited by C. P. Jones, who sees a composition date in the 220s or 230s as 'likely' (Jones, *Philostratus*, I.3). For the *ATh*, I use the edition of Lipsius and Bonnet, *Acta Apostolorum Apocrypha* II.2, 99–288; English translation with important discussion by H. J. W. Drijvers in Schneemelcher, *New Testament Apocrypha* [2] 2.322–411. For the date, see Bremmer, '*Acts of Thomas*' and 'Apocryphal Acts'; cf. Klijn, *Acts of Thomas*, 15: 'beginning of the third century'. I owe the impetus to this diversion from my more usual interests to the invitation from Katerina Panagopoulou and Christy Constantakopoulou to participate in the Networks conference; I am grateful to the audience, and especially Christina Kokkinia, for very useful questions and critique. An early version of this article was presented to an attentive and very helpful audience in the Department of Classical Studies at the University of Pennsylvania, where Jeremy McInerney, Joseph Farrell, Ann Kuttner, Sheila Murganhan, Cam Roshan, and Robert Kraft offered incisive questions and comments. The version published here has by necessity been considerably shortened; many points deserving detailed discussion have been only touched upon; it has been materially improved by the suggestions of the two anonymous readers. I hope to return to some of these issues in a future publication. I am grateful to Trinity College for appointment to the Charles A. Dana Research Professorship, during the tenure of which this article was written.

[2] Most, *Doubting Thomas*, 7.

[3] Knox *et al.*, 'Social Networks and the Study of Relations', 114. I am very grateful to Hannah Knox for providing me with a pre-publication copy of this article. From a huge bibliography

I point only to Karttunen, *India and the Hellenistic World*; Schneider, *L'Éthiopie et l'Inde*; Schuol, *Charakene*; and Winter, *Bardesanes*.

[4] For the turn, see especially Bowie, 'Apollonius'; Morgan, 'Make-believe and Make Believe'; Bowersock, *Fiction as History*. On Apollonios: Francis, 'Truthful Fiction' (very important) and now Schirren, *Philosophos Bios*, esp. 38–50, who cites all the earlier bibliography; see also Van den Bosch, 'India and the Apostolate of St Thomas' and, for a symbolic reading of Apollonios' travels, Elsner, 'Hagiographic Geography'. For a recent attempt to set the *VA* in a genre, see Burridge, *What Are the Gospels?*, 150–84; Montiglio, *Wandering in Ancient Greek Culture*, 213–20, for Apollonios as a kind of wandering god.

[5] Schirren, *Philosophos Bios*, 30.

[6] Bowie, 'Ancient Readers', 'The Readership of Greek Novels', and 'Lecteurs du roman grec'; Hägg, *Parthenope*, 109–40; Stephens, 'Who Read Ancient Novels?'; Morgan, 'The Greek Novel'; Cueva, *Myths of Fiction*, 91–92. Bremmer, 'The Apocryphal Acts', 160–70; Cooper, *Virgin and the Bride*, 51–56, 115.

[7] Egger, 'Women and Marriage', 264–65. If the novel and the *AAA* derive from oral tales, the link is still more compelling: MacDonald, *Legend and the Apostle*. For the idea that Chariton may have envisioned two readers, male and female, reading *Kallirhoe* differently, see Konstan, *Sexual Symmetry*, 78.

[8] Bremmer, '*Acts of Thomas*', 83–85.

[9] Schirren, *Philosophos Bios*, 23–37, 40, with references to earlier discussions.

[10] Poirier, *L'Hymne de la perle*.

[11] I hope to return in more detail to the role of ethnicity and identity. For a good summary of the state of ethnic studies in Greek antiquity, see Ruby, 'Peuples, fictions?'

[12] Drijvers, 'Introduction', 325, with Drijvers, 'Hatra, Palmyra und Edessa', 893–94.

[13] See Texidor, 'Port romain', 36–39; Gawlikowski, 'Palmyra as a Trading Centre'; Schuol, *Charakene*, 427–45; Schneider, *L'Éthiopie*, 473–87. For another commonplace of merchants, the toll-collector, see *VA* 1.20, where Apollonios' declaration of the virtues he is exporting leads the toll-collector to suppose he has prostitutes. Hartmann, *Das palmyrenische Teilreich*, 52–64, 76–77; Young, *Rome's Eastern Trade*, 136–86.

[14] Casson, *Periplous Maris Erythraei*.

[15] See generally de Romanis, 'Romanukharaṭṭha and Taprobane'; for the king, de Romanis, 'Rome and the *Notia*', 143, n.108 (text) and 103 (date). De Romanis makes a strong case for taking these accounts as historical rather than mere *topoi*.

[16] 'Silence' is fundamental to the Pythagoreanism that lies behind the *VA*: see Schirren, *Philosophos Bios*, 40.

[17] *VA* 1.31; cf. Charpentier, *Indian Travels*, 34, n.3.

[18] Létoublon, *Lieux communs du roman*, 200–01.

[19] Arora, *Greeks on India* for pre-Alexander material; for other examples, see *IK Iscrizioni dello estremo Oriente Greco* 457 and 459; Lerner, 'Correcting the Early History', 391–95. Bernard *et al.*, 'Deux nouvelles inscriptions grecques.' On Alexander: Holt, *Into the Land of Bones*, and Bosworth, *Alexander and the East*.

[20] Fraser, *Cities of Alexander*, 132–40; *IK Iscrizioni dello estremo Oriente Greco*, 291, 292, and 290.

[21] Jarrige and Cambon, *Afghanistan*.

[22] See the survey in Briant, *Histoire de l'Empire Perse*, 764–68.

[23] Salomon, *Indian Epigraphy*, 51–54.

[24] See generally Schmitt Pantel, *La cité au banquet*.

[25] *IG* XII 7 22 for the Itonia, a similar invitation, also 33.13–16.

[26] Çetin Şahin, 'New Inscriptions from Lagina', 91–93 no. 10; see also *IK Stratonikeia* 527.

[27] On the role of weddings in the novels, see Egger, 'Women and Marriage'.

[28] Alexander's long shadow, of course, lay across Apollonios' path. But the sage trumped his precursor's achievements by crossing the Indus and making his way to the Ganges valley. I leave aside further discussion of this very important matter, as it would break the bounds of this study.

References

Arora, Udai Prakash. *Greeks on India: Skylax to Aristoteles*. Bareilly: Isgars, 1996.

Bernard, Paul, George-Jean Pinault, and George Rougemont. 'Deux nouvelles inscriptions grecques de l'Asie centrale'. *Journal des savants* (2004): 227–356.

Bosworth, A. B. *Alexander and the East: The Tragedy of Triumph*. Oxford: Oxford University Press, 1996.

Bowersock, G. W. *Fiction as History. Nero to Julian*. Berkeley: University of California Press, 1994.

Bowie, E. L. 'Apollonius of Tyana. Tradition and Reality'. *ANRW*, II 16, no. 2, 1652–99. Berlin and New York: Walter de Gruyter, 1978.

———. 'Les lecteurs du roman grec'. In *Le monde du roman grec*, edited by M.-F. Baslez, Philippe Hoffmann, and Monique Trédé. Paris: Presses de l'École normale supérieure, 1992.

———. 'The Readership of Greek Novels in the Ancient World'. In *The Search for the Ancient Novel*, edited by J. Tatum. Baltimore and London: Johns Hopkins University Press, 1994.

———. 'The Ancient Readers of Greek Novels'. In *The Novel in the Ancient World*, edited by G. Schmeling. Leiden: Brill, 1996.

Bremmer, Jan N. 'The *Acts of Thomas*: Place, Date and Women'. In *The Apocryphal Acts of Thomas*, edited by Jan N. Bremmer. Leuven: Peeters, 2001.

———. 'The Apocryphal Acts: Authors, Place, Time and Readership'. In *The Apocryphal Acts of Thomas*, edited by Jan N. Bremmer. Leuven: Peeters, 2001.

Briant, Pierre. *L'histoire de l'Empire Perse. De Cyrus à Alexandre*. Paris: Fayard, 1996.

Burridge, Richard A. *What Are the Gospels? A Comparison with Graeco-Roman Biography*. 2nd ed. Grand Rapids, Mich: W. B. Eerdmans, 2004.

Casson, Lionel. *The Periplous Maris Erythraei: Text with Introduction, Translation, and Commentary*. Princeton, NJ: Princeton University Press, 1989.

Çetin Şahin, M. 'New Inscriptions from Lagina'. *Epigraphica Anatolica* 29 (1997): 83–106.

Charpentier, J. *The Indian Travels of Apollonius of Tyana*. Uppsala and Leipzig: Almqvist & Wiksells Boktryckeria, 1934.

Cooper, Kate. *The Virgin and the Bride: Idealized Womanhood in Late Antiquity*. Cambridge, Mass: Harvard University Press, 1996.

Cueva, Edward P. *The Myths of Fiction: Studies in the Canonical Greek Novels*. Ann Arbor: University of Michigan Press, 2004.

De Romanis, Federico. 'Romanukharaṭṭha and Taprobane: Relations between Rome and Sri Lanka in the First Century AD'. In *Crossings: Early Mediterranean Contacts with India*, edited by F. de Romanis and A. Tchernia. New Delhi: Manohar, 1997.

———.'Rome and the *Notia* of India: Relations between Rome and Southern India from 30 BC to the Flavian Period'. In *Crossings: Early Mediterranean Contacts with India*, edited by F. de Romanis and A. Tchernia. New Delhi: Manohar, 1997.

Drijvers, H. J. W. 'Hatra, Palmyra und Edessa. Die Städte der syrisch-mesopotamischen Wüste in politischer, kulturgeschichtlicher und religionsgeschichtlichter Beleuchtung'. In *Aufstieg und Niedergang der römischen Welt* II. *Principat* 8, edited by Hildegard Temporini and Wolfgang Haase. Berlin and New York: Walter de Gruyter, 1977.

———. 'Introduction' to the *Acts of Thomas*'. In *New Testament Apocrypha*, II. *Writings Relating to the Apostles, Apocalypses and Related Subjects*, edited by Wilhelm Schneemelcher, translated by R. MacLachlan Wilson. Cambridge: James Clarke & Co., 1992.

Egger, Brigitte. 'Women and Marriage in the Greek Novels: The Boundaries of Romance'. In *The Search for the Ancient Novel*, edited by James Tatum. Baltimore and London: Johns Hopkins University Press, 1994.

Elsner, John. 'Hagiographic Geography: Travel and Allegory in the *Life of Apollonius of Tyana*'. *Journal of Hellenic Studies* 117 (1997): 22–37.

Francis, James A. 'Truthful Fiction: New Questions to Old Answers on Philostratus's *Life of Apollonius*'. *American Journal of Philology* 119 (1998): 419–41.

Fraser, P. M. *Cities of Alexander*. Oxford: Oxford University Press, 1996.

Gawlikowski, M. 'Palmyra as a Trading Centre'. *Iraq* 56 (1994): 27–33.

Hägg, Tomas. *Parthenope: Selected Studies in Ancient Greek Fiction*, edited by Lars Boje Mortensen and Tormod Eide. Copenhagen: Museum Tusculanum Press, 2004.

Hartmann, Udo. *Das palmyrenische Teilreich*. Stuttgart: Franz Steiner, 2001.

Holt, Frank. *Into the Land of Bones: Alexander the Great in Afghanistan*. Berkeley: University of California Press, 2005.

Jarrige, Jean-François, and Pierre Cambon. *Afghanistan. Les trésors retrouvés*. Paris: Musée des arts asiatiques Guimet, 2006.

Jones, C.P., ed. and trans. *Philostratus: The Life of Apollonius of Tyana*, 2 vols. Cambridge, Mass: Harvard University Press, 2005.

Karttunen, Klaus. *India and the Hellenistic World*. Helsinki: Finnish Oriental Society, 1997.

Klijn, A. F. J. *The Acts of Thomas: Introduction, Text, Commentary*. 2nd (rev.) ed. Leiden: Brill, 2003.

Knox, Hannah, Mike Savage, and Penny Harvey. 'Social Networks and the Study of Relations: Networks as Method, Metaphor and Form'. *Economy and Society* 35 (2006): 113–40.

Konstan, David. *Sexual Symmetry: Love in the Ancient Novel and Related Genres*. Princeton, NJ: Princeton University Press, 1994.

Lerner, Jeffrey D. 'Correcting the Early History of Āy Ḵānom'. *Archäologische Mitteilungen aus Iran und Turan* 35–36 (2003/4): 373–410.

Létoublon, Françoise. *Les lieux communs du roman. Stéréotypes grecs d'aventure et d'amour*. Leiden: Brill, 1993.

Lipsius, R. A., and M. Bonnet. *Acta Apostolorum Apocrypha*, vol. II, no. 2. Leipzig: Herrmann Mendelssohn, 1903.

MacDonald, Dennis R. *The Legend and the Apostle: The Battle for Paul in Story and Canon*. Philadelphia: The Westminster Press, 1983.

Montiglio, S. *Wandering in Ancient Greek Culture*. Chicago: University of Chicago Press, 2005.

Morgan, J. R. 'Make-believe and Make Believe: The Fictionality of the Greek Novels'. In *Lies and Fiction in the Ancient World*, edited by C. Gill and T. P. Wiseman. Austin: University of Texas Press, 1993.

———. 'The Greek Novel: Towards a Sociology of Production and Reception'. In *The Greek World*, edited by A. Powell. London and New York: Routledge, 1995.

Most, Glenn W. *Doubting Thomas*. Cambridge, Mass: Harvard University Press, 2005.

Poirier, P. H. *L'Hymne de la perle des actes de Thomas. Introduction, texte, traduction, commentaire*. Louvain: P. Pierier, 1981.

Ruby, Pascal. 'Peuples, fictions? Ethnicité, identité ethnique et sociétés anciennes'. *Revue des études anciennes* 108 (2006): 25–60.

Salomon, Richard. *Indian Epigraphy. A Guide to the Study of Inscriptions in Sanskrit, Prakrit, and Other Indo-Aryan Languages*. New York and Oxford: Oxford University Press, 1998.

Schirren, Thomas. *Philosophos Bios. Die antike Philosophenbiographie als symbolische Form. Studien zur* Vita Apollonii *des Philostrat*. Heidelberg: Universitätsverlag Winter, 2005.

Schmitt Pantel, Pauline. *La cité au banque. Histoire des repas publics dans les cités grecques*. Rome: École française de Rome, 1992.

Schneemelcher, W., ed. *New Testament Apocrypha*², English ed., edited by R. MacLachlan Wilson. Cambridge: James Clarke, 1992.

Schneider, Pierre. *L'Éthiopie et l'Inde. Interférences et confusions aux extrémités du monde antique (VIII*ᵉ *siècle avant J.-C.—VI*ᵉ *siècle après J.-C.).* Rome: École française de Rome, 2004.

Schuol, M. *Die Charakene. Ein mesopotamisches Königreich in hellenistich-parthischer Zeit.* Stuttgart: Franz Steiner Verlag, 2000.

Stephens, Susan. 'Who Read Ancient Novels?'. In *The Search for the Ancient Novel*, edited by James Tatum. Baltimore and London: Johns Hopkins University Press, 1994.

Texidor, Javier. 'Un port romain du désert'. *Semitica* 24 (1984): 7–127.

Van den Bosch, Lourens P. 'India and the Apostolate of St Thomas'. In *The Apocryphal Acts of Thomas*, edited by Jan N. Brenner. Leuven: Peeters, 2001.

Winter, Frantz. *Bardesanes von Edessa über Indien. Ein früher syrischer Theologe schreibt über ein fremdes Land.* Thaur bei Innsbruck: Druck- und Verlagshaus Thaur, 1999.

Young, G. K. *Rome's Eastern Trade: International Commerce and Imperial Policy.* London and New York: Routledge, 2001.

Via Egnatia after Egnatius: Imperial Policy and Inter-regional Contacts

Yannis Lolos

Routes, whether terrestrial or maritime, are the actual means of communication, and physical expression of networking.[1] The Via Egnatia was the first Roman highway built east of the Adriatic, in a way the continuation of the Via Appia—the *regina viarum* which led from Rome to Brindisi. It was built after a long period of military upheaval, when the Roman Republic confronted major powers of the eastern Mediterranean. The purpose of this article is to trace the impact that the Egnatia had in the long run on the interaction between the Italian peninsula and the Greek world, and among the cities and regions that it crossed. I shall begin with background information on the history of the road as outlined by literary and epigraphical sources, before briefly describing the route of the Egnatia, reconstructed on the basis of physical remains in conjunction with ancient itineraries, milestones, early modern maps and travellers' accounts.[2] The following section is devoted to the emperors' initiatives regarding the road and the areas it crossed, more specifically the imperial *adventus* and the foundation of cities and colonies. I then explore the impact of the Egnatia on a commercial, social and cultural level. In this context, I discuss the presence of Italian businessmen and western imports in cities bordering the road, the Romanization of personal names and the Roman elite network among cities linked by the Egnatia, and the cultural imprint of Rome on the cityscape of the cities affected by the road. In concluding, I attempt a general appraisal of the Egnatia as a road representing a high technical achievement, and of its contribution to the growth of certain cities down to the late Imperial era and beyond.

Historical Background

Our main source on the Via Egnatia is Polybios, who in his 34th book had an extensive account of the road, of which the full text is unfortunately lost. Only parts of Polybios' account are preserved, quoted by Strabo in book 7, frag. 13 (C 322). To repeat the much-quoted passage:[3] 'From Apollonia to Macedonia one travels the Egnatian Road, towards the east: it has been measured by Roman miles and marked with milestones [κατεστηλωμένη] as far as Kypsela and the Hebrus river—a distance of 535 miles'; and later on (7.7.4): 'Now although the road as a whole is called the Egnatian Road, the first part of it is called the Road to Candavia (an Illyrian mountain), and passed through the city of Lychnidos and a place called Pylon, which marks the boundary between the Illyrian country and Macedonia. From Pylon the road runs to Barnus through Heracleia and the country of the Lyncestae, and that of the Eordi into Edessa and Pella, and as far as Thessalonikeia; and the length of this road in miles, according to Polybios, is 267.' And then Strabo goes on to describe the many 'ethne' crossed by the Egnatia. There are two points that I want to stress in this passage: first that the Via Egnatia from Epiros to Macedonia and Thrace crossed the territories of a number of tribes; and second, the special place of Thessalonike within the Egnatian route, so much so that Strabo (following Polybios, no doubt) mentions the distance of this city alone from Apollonia. Thessalonike, besides marking the point of transition where the Via Egnatia changed character from an *over*land route linking Adriatic to Aegean to an *in*land one shadowing the Aegean coast, was also a major Aegean port.[4] On the basis of the advantageous position and importance of this city, some scholars had argued that originally the Via Egnatia stretched from the Adriatic to Thessalonike, and was extended to Kypsela in a later period. However, the discovery of a *miliarium* outside Kavala (170 km to the east of Thessalonike) bearing the name of the road's eponymous builder—the proconsul Cn. Egnatius—proves beyond any doubt that the entire route was designed and built from the start.[5]

To the Kavala *miliarium* we can add one found in Thessalonike also erected by Cn. Egnatius as proconsul of the Roman province of Macedonia, sometime between 146 and 120 (when Polybios died), possibly in the mid-140s.[6] This is to say that the Via Egnatia was built during the third quarter of the second century BC, after a long period of unrest in Macedonia with the three wars against Rome followed by the revolt of Andriskos and the establishment of Macedonia as a Roman province. After the conclusion of the Third Macedonian War and the defeat of Perseus, the Roman senate divided Macedonia into four separate administrative districts (μερίδες). It seems that the first district extended to the east as far as the Hebrus river,[7] which accords with the eastern terminal of the Egnatia—namely, Kypsela. Almost two decades after the defeat of Perseus, a certain Andriskos, who claimed direct descent from the Macedonian royal family, stirred the Macedonians against the Romans, who eventually crushed the revolt in 148. The decision to establish a highway from the Adriatic to the eastern border of Macedonia probably followed this event. The new *via militaris*, to use Cicero's expression (*Prov.* 2.4), would allow the Romans to readily dispatch their army against any future revolt, and to protect the Macedonian borders against barbarian threats.

The road was put to test on a number of occasions during the late second and first century BC, in the context of Roman expeditions against Thracian invaders and during the Roman civil wars.[8] Around 100 BC the Roman province of Macedonia most likely expanded to include eastern Thrace as far as the Bosporus.[9] If so, we could envision the Egnatia extending all the way to Byzantion soon after that time.[10] The upsurge of the Thracian kings during the last decades of the pre-Christian era must have caused the disruption of the Via, and the withdrawal of the Roman province to the Hebrus river.[11] Things began to change during early Imperial times, when the Romans conquered Illyricum, extended their control over the Danube area (AD 9), and made Moesia and Thrace regular provinces (AD 44). At the same time Macedonia was a rather peaceful public province. Roman troops would move north of the Adriatic essentially bypassing the Egnatia, which ceased to be a *via militaris* (and did not assume this role again before the civil wars of the late Empire). From now on the Egnatia served primarily the official business of the Roman Empire and its *cursus publicus*, trade, and travelling generally for touristic or other purposes.[12]

Inscriptions help us to reconstruct the history of Egnatia during the Empire. Thanks to a *miliarium* found to the northwest of Philippi, we know that Trajan undertook the repair of the road *longa intermissione neglectam* from Dyrrachium to a station called Acontisma, which is located near Nea Karvali.[13] Another *miliarium* found outside Thessalonike mentions the repair by the same emperor of the Egnatia from Dyrrachium to Neapolis—modern Kavala.[14] The same emperor had undertaken the radical reconstruction of the Via Appia in Italy, which linked Rome to Brindisi and now became the Via Traiana. If we take the evidence of the *miliaria* at face value, we can assume another repair of the Via Egnatia during the reign of Septimius Severus and then under Caracalla.[15] Using the same, rather inconclusive, evidence we can argue along with Collart that the next repair of the Egnatia occurred during the second tetrarchy, this time instigated by the setting of Thessalonike as Galerius's capital. From the mid-fifth century onwards, with the collapse of the western Roman Empire, the Via Egnatia will suffer from a number of invasions, resulting in the disruption of its route for extended periods of time, and dwindles in importance, for it is no longer the single through-route linking Italy and the Bosporus.[16]

Route of the Egnatia

We reconstruct the route of the Egnatia based on ancient *itineraria*, pictorial as well as descriptive, the find-spots of *miliaria*, physical remains, early modern maps and the topography itself (Figure 1). The Roman *itineraria* that are of use to us are the Peutinger Table, and the Antonine and Bordeaux Itineraries.[17] The *Tabula Peutingeriana*, which goes back to the fourth century, remains the sole pictorial itinerary surviving from the Roman period. Routes throughout are shown as a series of continuous lines, with numerals indicating distances in Roman miles between places. A line, which corresponds to the Egnatia and is drawn in segments, begins at Dyrrachium and ends at Perinthos (west of Byzantion) via Lychnidos, Thessalonike, and Apri (i.e. the colony established by Claudius).[18] Besides cities, the map shows the staging-posts along the route (*mutationes, mansiones*), but given the map's deformation—its total width is about twenty times its height, and as a result north–south distances look very short, while east–west distances seem much too long—we can hardly draw from it any information on the geographical location of these posts. Cities and stations are also listed under each journey in the land section of the Antonine Itinerary (*Itinerarium provinciarum Antonini Augusti*), which seems to be a somewhat disjointed, early-fourth-century AD compilation of earlier, regional itineraries. The starting and finishing points of each journey are indicated affront as well as the road distance in Roman miles. Concerning the Egnatia, the Antonine Itinerary describes two routes (both west to east), one starting from Dyrrachium (*Itin Ant.* 317,3–323,8), and another from Apollonia (329,3–332,9), both ending at Byzantion. Of the two, the Apollonia–Byzantion route does follow what is assumed to be the direct line of the Via Egnatia through Kypsela and Apri.[19] From Dyrrachium to Byzantion the overall length is given as 754 Roman miles (317.7). Finally, the *Itinerarium Hierosolymitanum* or *Burdigalense*, an itinerary from Bordeaux to Jerusalem of AD 333 (the earliest surviving Christian pilgrim account), describes on the return a route that leads (east–west) from Herakleia (Perinthos) to Aulon (*Itin. Burd.* 601,6–609,3), which was south of Apollonia and apparently the most important harbour of Epirus at that time.[20] In listing the places the road went through, the compiler of this *itinerarium* always indicates which is a town (*civitas*) and which is a station (*mansio, mutatio*), and this helps us to assess the importance and size of the settlements crossed by the Egnatia at that period.

Figure 1 Schematic reconstruction of the route of the Egnatia. Source: author.

The *miliaria*, usually cylindrical stones 1.3–1.5 metres high, with inscribed mileage either from Dyrrachium/Apollonia or from the main city (*caput viae*) of a given region, occasionally provide the only information on the route of the Egnatia, simply by virtue of their find-spot. If we are to trust Polybios that the Egnatia was 'κατεστηλωμένη' from Apollonia to Kypsela, then we should think that no less than 535 *miliaria* were set along the original route. Of these, less than forty have been recovered so far.[21]

Physical remains of the Roman road, either of the road surface itself or of the bridges built along the route, are not abundant. Of the western part of the Egnatia, the sections from Dyrrachium or Apollonia to Herakleia Lyncestis (near Bitola, FYROM), have been investigated by Austrian and then Albanian archaeologists, followed by Hammond, and have been the subject of a monograph, as well as the subject of a recent topographical survey.[22] Within the borders of modern Greece, the section from Florina to Thessalonike has been studied by Edson, Hammond, and Hatzopoulos.[23] The eastern part, from Thessalonike to Kypsela, has been studied in sections by various scholars: from Thessalonike to Amphipolis by Moutsopoulos and Hatzopoulos; the section around Philippi by Collart, Samartzidou, and Otatzis; and the part from Kavala to the Hebrus river by Pantos, Dadiras, and Tsatsopoulou-Kaloudi, who most recently produced a guide for this section.[24] In April 2006 I undertook in the company of Leda Costaki the mapping of all known remains of the Egnatia from Florina to the Hebrus river. We shall soon have the first accurate map of the preserved sections of the Egnatia within Greece.

In areas where no traces have been reported so far, the reconstruction of the route is based on early modern maps and accounts of European travellers of the eighteenth and nineteenth centuries, in combination with the physical landscape.[25] From Apollonia and Dyrrachium the Egnatia followed the valley of the river Genusus (modern Shkumbin) as far as Lake Ohrid through a pass in the Candavian mountains. It continued around the northern shore of the lake in order to reach Ohrid (Lychnidos), and thence to Herakleia Lyncestis (south of Monastir). From Herakleia the Egnatia took a southeastern course towards Lake Vegoritis, passing between Mounts Vernon (modern Vitsi) and Varnous (Kaimaktsalan), then proceeded east to Edessa, Pella, Thessalonike, and south of lakes Koroneia and Volvi to Amphipolis. From Amphipolis to Philippi the Egnatia passed north of Mount Pangaion, essentially making a loop in order to avoid the marshy areas, then entered the *pomerium* of the colony of Philippi through a monumental arch and went through the city, most likely serving as its *decumanus maximus*, in order to continue southeast towards Neapolis (Kavala). From Neapolis to Lake Bistonis the route of the Egnatia is still uncertain, but from that point eastwards there are physical traces reported, the best being the section between Sapes and Alexandroupoli over the heights of Zone. More sections are visible between Traianoupolis (east of modern Alexandroupolis) and Pheres (Byzantine Βήρα) in the direction of Kypsela (modern İpsala).

The fact that only a few stretches of the Egnatia are now visible is the result of two factors: a) in many parts its route remained in use up to modern times, and consequently the ancient road bed was either destroyed or covered over; and b) in flat areas where the ancient road deviated from the modern one, it has been ploughed over: this is particularly the case in eastern Macedonia and Thrace, where, after the coming of the refugees from Asia Minor in the 1920s, the lands left by the Turks were distributed among the newcomers, who turned them to more intensive cultivation, gradually removing the stones of the ancient and medieval road.

Via Egnatia and Imperial Policy

Road repairs and occasionally construction of new segments were often instigated by the arrival or the prospect of arrival of the emperor. In the case of the regions crossed by the Egnatia, we have very limited evidence of the coming of an emperor before the second tetrarchy and the establishment of Galerius Maximianus at Thessalonike. Indeed, there is no proof, such as inscriptions mentioning the *adventus* or ἐπιδημία of an emperor, or coins of the *adventus* or *restitutor* type, and we have to rely on circumstantial evidence or the combination of facts to build a case for an emperor using this route. Stronger cases have been made, mostly on the basis of the *miliaria*, for Trajan, who had ordered the repair of the Egnatia as mentioned already, for Hadrian in 131, for Septimius Severus on his way home from Syria in 202, and for Caracalla, who planned to return the same way, but was assassinated prior to departure.[26]

There is no evidence that Augustus ever travelled the Egnatia, but the fact that he founded or refounded three colonies along its route shows the importance he paid to this artery.[27] Dyrrachium—a major harbour on the Adriatic which the *princeps* turned into a colony for veterans of his legions following the battle of Actium—prospered spectacularly in the next three centuries. Augustus also refounded the colony of Pella (originally founded in 40 BC as *colonia Pellensis*), with the name Col(onia) Iul(ia) Aug(usta) Pella, and sent there lower class citizens from Italy whom he had previously removed from their lands in order to distribute those lands among his veterans.[28] The colony struck coins down to the third century, but was never granted the *ius Italicum*, which would have placed the community on the same level as an Italian city. It is possible that Diocletian refounded Pella under the name Diokletianoupolis, which appears in the Antonine Itinerary.[29] The third city, originally founded as a colony under Antony after the battle of 42 BC, was Philippi—the creation of Philip II in 356 BC. The earliest coins of the colony commemorate this first settlement: A(ntoni) I(ussu) C(olonia) P(hilippensium). Some years later, Augustus sent new settlers from Italy and granted the Colonia Augusta Julia Philippensis the *ius Italicum*.[30] The case of Apollonia, a city in Mygdonia between Thessalonike and Amphipolis, presents some similarities to Philippi. Apollonia, which is attested as a city since the reign of Amyntas III, was refounded as a Roman *municipium* at the beginning of the second century.[31] The refoundation of Apollonia is probably linked with the reconstruction of the Egnatia under Trajan.[32]

It was in western Thrace that Trajan took his most important initiatives. He founded Traianoupolis, which evolved into the most important administrative and military centre of Thrace down to the fourteenth century. The city assumed the responsibility of the maintenance of the Via Egnatia, as indicated by two fragmentary inscriptions of AD 202 found in the region and mentioning the repair of sections of this road by villages (*komai*) within the territory of the city.[33] Trajan also refounded Topeiros, an important city along the Via Egnatia, to the west of Traianoupolis, as Ulpia Topeiros.[34] Further east, between Kypsela and Perinthos, Claudius had founded the colony of Apri (*colonia Claudia Aprensis*), soon after the creation of the province of Thrace. Apri, located beside the modern village of Kermeyan to the east of Malkara, was on or very near the route of the Egnatia, and became the only urban centre of the Caenic peninsula.[35]

To the north of Traianoupolis, two more cities were founded in the course of the second century, Plotinopolis and Hadrianoupolis. This changed significantly the land traffic in the Thracian region. The Antonine Itinerary (322,4–323,5) reports an alternative route to Perinthos and Byzantion, which beyond Traianoupolis did not continue east to Kypsela, but followed the right bank of the Hebrus river, to Hadrianoupolis and thence to Perinthos.[36]

During the late Empire, cities founded or refounded along the eastern part of the Egnatia betray the revived imperial interest for this important axis. Anastasioupolis, to the north of Lake Bistonis, apparently so named by Anastasios I, prospered under Justinian I, who gave the city a safe harbour and an impressive aqueduct.[37] Maximianoupolis, to the northeast of Lake Bistonis, was an important settlement already in the fourth century, and had its fortifications rebuilt by Justin I.[38] Both cities became seats of bishoprics, subject to the metropolis of Traianoupolis. Further east, past the Hebrus river and the colony of Apri, Perinthos (now renamed Herakleia) witnessed a spectacular growth. Thanks to the reforms of Diokletian, Herakleia became the capital of the Thracian province of Europe, and from 313 the third most important metropolis of the patriarchate of Constantinople.[39] The increased attention paid to the eastern section of Egnatia at that time was certainly due to the fact that it linked the Tetrarchic imperial courts of Nicomedia and Thessalonike. The route did not really lose its importance with time, since from 441 onwards, when the Praetorian Prefect of Illyricum moved to Thessalonike, it became an important corridor of communication between his court and the imperial court in Constantinople.[40] And several decades later Justinian undertook repairs along sections of the Egnatia in eastern Thrace.[41]

Via Egnatia as an Agent of Commercial, Social and Cultural Interaction

'There is a spot in the gap in the middle,' says Aristotle, 'in which, when a common market is held, Lesbian, Chian and Thassian goods are bought from the merchants who come up from Pontus, and Corcyrean amphoras from whose who come from the Adriatic' (*On Marvellous Things Heard*, 104). Aristotle is describing a terrestrial route between the Pontus and the Adriatic, which served commercial purposes and was already in existence by the fourth century BC, if not earlier.[42] No doubt, Gnaeus

Egnatius incorporated sections of this road in his Via, and the commercial trade was renewed. Already from the second century BC, we witness an increase of commercial operations in Epirus, directed by Roman and Italian tradesmen who settle in major coastal cities and important harbours, Dyrrachium, Apollonia, or Lissos.[43] This explains the increasing volume of Roman coins found in Epirus from the second century BC onwards.[44] Also luxury items, such as glass objects, would arrive in quantities from Italy to Dyrrachium, Apollonia and a few other coastal cities (Lissos, Bouthroton), and from there would make their way inland. Dyrrachium, in particular, was the greatest commercial centre of southern Illyria, and a major distribution centre of Italian imports towards the east.[45] Unfortunately, we are not in a position yet to trace, let alone quantify, the diffusion of Italian imports from the Adriatic to Thrace, that is along the areas crossed by the Egnatia. To the best of my knowledge, no study has so far addressed amphora circulation in these regions during the Roman period. Likewise major pottery publications from the cities on or near the Egnatia are also missing, and we can only rely on excavation reports and scattered articles in order to gain an idea of the circulation of Italian red-wares and other western products in this part of the Empire. Most recently Vaitsa Malamidou examined ceramic assemblages from five sites of northeastern Greece, including Amphipolis and Philippi.[46] Among the finewares, Malamidou noticed particular preference for eastern *sigillata* and Çandarlı ware, whereas Italian *sigillata* does not seem to have been very popular, and no examples of the early Italian *sigillata* were identified.[47] This, according to the author, suggests limited trading relations between northeastern Greece and Italy.[48] I would refrain from such a conclusion for two main reasons: a) The study is based on specific assemblages from these five sites, which might not be representative. In the case of Philippi for example, the assemblage of the theatre contained virtually no finewares, as the author acknowledges.[49] Amphipolis, on the other hand, has yielded a certain amount of Pompeian red-ware, which came from the Campanian region, and the excavations of the agora of Thessalonike produced a large amount of ceramic imports from Italy as well as from Asia Minor;[50] b) Fine pottery was not necessarily the prime item of exchange between eastern Macedonia and Italy, since Asia Minor, which was much closer, had some of the finest ceramic workshops of the Mediterranean. The study of amphoras, which account for ninety-two per cent of the storage and transportation vessels identified by Malamidou, could yield more information in this regard.[51]

We are better off with inscriptions that confirm the presence of Roman and Italian immigrants, mostly from southern Italy, in a number of Macedonian cities.[52] Evidence for the presence of these *negotiatores* in Macedonia is meagre before the middle of the first century BC and the battle of Philippi (42 BC), but increases spectacularly in the next century and a half. The *negotiatores* are mainly attested in the large maritime cities of Macedonia, notably Thessalonike and Amphipolis, but also in cities situated inland and along or near the route of the Egnatia: Herakleia Lyncestis, Beroia, Edessa, Pella, Apollonia in Mygdonia, and Philippi.[53] The activities of the *negotiatores* are not known, but we can assume by analogy with similar communities in the Greek east that they were engaged in commercial and financing businesses. Macedonia was always rich in timber and metals, and we can imagine that these products played a significant role in the commercial transactions.[54] Edessa, the first major city of Macedonia for one travelling the Egnatia from the west, saw an influx of *negotiatores* from the first century BC to the first century AD, who eventually formed a community epigraphically attested as a *conventus civium Romanorum* or ὁι συνπραγματευόμενοι 'Ρωμαῖοι.[55] Thessalonike, which was the seat of the Roman governor, appears to have received the first wave of *negotiatores* in the middle of the first century BC, and a second, more substantial one in the second century AD, when they decided to form a *conventus* attested by three inscriptions.[56] Further to the east, Mygdonian Apollonia seems to have hosted a community of Italian immigrants who probably moved to this city from Chalcidian Akanthos, where they had first settled in the time of Augustus.[57] In Amphipolis, a funerary inscription mentions a *negotiator ab Roma ex horreis Cornific(ianis)*, and another one a procurer of slaves (σωματέμπορος), which shows that Amphipolis too was a trading centre for the region.[58] The second inscription, dated no later than the first century AD, is also revealing of a commercial activity that went on in the region, and would leave hardly any trace in the archaeological record—namely, slave trade.[59] Seen from a wider perspective, we definitely have more evidence for the presence of *negotiatores* in Macedonia than in the whole of the Roman province of Achaia, and the Via Egnatia has certainly to do with this phenomenon.[60]

These groups of Italian immigrants, usually people of low status, including many freedmen, did not stand apart but mingled with the local society.[61] Judging from inscriptions, the majority adopted the Greek language but did not lose their Roman identity. In fact, a prosopographical study for the whole of Macedonia shows the increasing Romanization of personal names from the second century BC to the third century AD.[62] A related phenomenon is the high percentage of Romans belonging to the upper class and holding high offices in the major cities of Macedonia and Thrace.[63] In Herakleia Lyncestis, for example, wealthy Romans served as πολιτάρχαι, ἀγορανόμοι, ταμίαι, and in other distinguished posts already from the late first century BC.[64] In Edessa, to take another example, all persons of high standing in the Imperial era and much of the upper Edessene class in general were Roman citizens.[65] In addition, a good number of the Roman *nomina* found in Edessa have been also traced in other parts of Macedonia, especially in Thessalonike and Beroia, but also in Philippi and elsewhere.[66] Similarly, *gentilicia* of many Roman families of Thessalonike have been attested in Philippi, Beroia, Edessa, Lete, Herakleia Lyncestis, and in a few more cities.[67] It is not coincidental that all of these cities are on or near the Via Egnatia, which is to say that the Roman artery facilitated social intercourse. It also contributed to the creation of a cosmopolitan ambience in the major cities of Macedonia, and above all in Thessalonike, the hub of different ethnic groups, cultures and traditions.[68]

The impact of the Roman arrival on the cultural profile of the cities and regions concerned is certainly the hardest to grasp. Roman-type monuments, associated with

Roman habits, form one kind of evidence. In this respect, it is unfortunate that we know so little about the location, appearance and evolution of the many *mutationes* and *mansiones* that were associated with the route.[69] The best known and best preserved is the complex excavated at Bradashesh, near Elbasan, which has been identified as the *mutatio* Ad Quintum. The complex, in *opus testaceum*, includes a bath and a *nymphaion* and dates from the second century AD (Figure 2).[70] It must have made an impact on travellers on the Egnatia, and may have been imitated elsewhere in the regions that the Egnatia crossed.

Within the cities, characteristic Roman structures, such as amphitheatres, Roman temples, or Romanizing baths, either excavated or epigraphically attested, betray the penetration of Roman ideals and practices, be it gladiatorial shows, the imperial cult, or the bathing culture. Dyrrachium had all types of monuments, as did Thessalonike, plus a 'stadium-theatre' for gladiatorial shows, a temple to the Dea Roma frequently mentioned in inscriptions, and sizeable baths. Colonies may have set the example in terms of architectural prototypes. Philippi is the case in point with its Π-shaped forum going back to the Julio-Claudian period (first century AD). George Velenis has noted the striking similarities between this agora and the agora of Thessalonike, which can be traced back to the second century AD (Figure 3).[71] The *decumanus maximus* of Philippi, that is the intra-mural section of the Via Egnatia, was paved and colonnaded, at least along the excavated stretch that bordered the forum (Figure 4). The same is true about the main, north–south street of Edessa, which was an impressive *via colonnata* (Figure 5). More cultural interactions of this kind could be detected after a thorough comparative examination of the remains and their architectural details.[72]

Figure 2 The complex at Bradashesh, perhaps the *mutatio* ad Quintum. Source: Fasolo, 'Via Egnatia', 146.

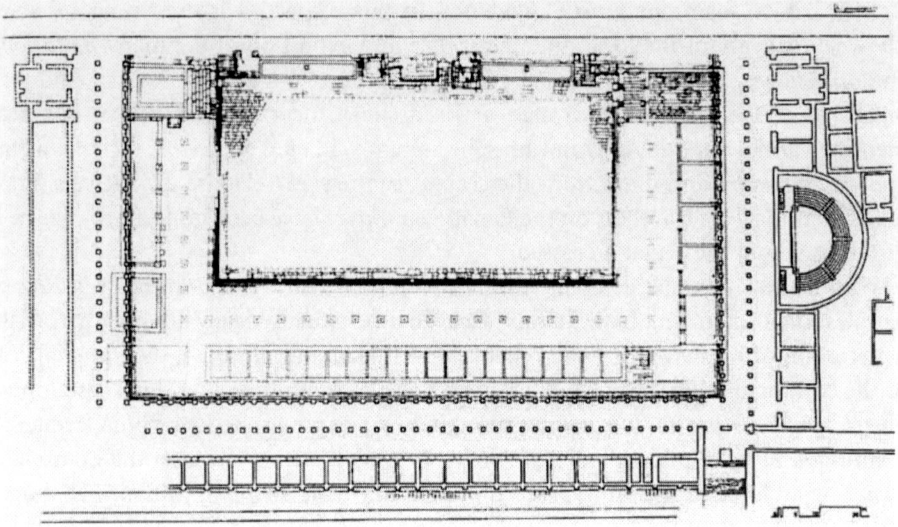

Figure 3 The plan of the forum of Philippi within that of the agora of Thessalonike. Source: Velenis, 'Ἀγορά Θεσσαλονίκης' Εικ. 4.

Via Egnatia and its Impact on Regions and Cities Crossed

Laying the 535 miles (776 km) of the Egnatia was not easy; it certainly imposed great challenges to the engineers, and represented a major investment for the central government. Its western section, past the Genusus river valley, went against the natural passes which have a north–south direction, whereas the Egnatia had to go east.

Figure 4 Philippi: the via Egnatia. Source: author.

Figure 5 Edessa: the monumental *via colonnata*. Source: author.

To do so the road reached an altitude of 720 metres just before Lake Ohrid, then climbed up to 1,190 metres (near Bukovo) to descend to 570 metres at Herakleia. In Macedonia the challenges were of a different kind, as the route followed for the most part areas of plain, some of which were subject to flooding, as at Lake Vegoritis, the area of Philippi, and the Nestos plain.[73] To prevent the flooding of the road in these areas, the Roman builders created a causeway on which the road ran. Praschniker and Schober observed such remains of the Via Egnatia, which they described as a 'Dammstrasse', in their archaeological explorations in Albania during the early twentieth century.[74]

The large rivers that the road had to cross, of the size of the Genusus, Axios, Strymon, Nestos, and Hebrus, were no less of a headache for its designers and builders. In the case of the Genusus (Shkumbin), we are fortunate to have the actual piers of the bridge preserved and, what is more, excavated.[75] The length of the bridge must have reached some 450 metres, and involved at least fifteen piers, of which five represent the original (Roman) phase, with solid foundations and massive platforms. A second bridge, over a tributary of the Genusus, was 113 metres long and built in *opus testaceum*.[76] Bridges over other, narrower rivers were also carefully built, as witnessed by the (now gone) bridge of Kourovo, and that of Kadim Köprü over the Angitis river, both showing monumental faces in *opus quadratum*.[77]

The quality of the road bed itself is harder to evaluate, because it is difficult to differentiate between the various phases of the road from antiquity to the early modern period. The width varies, from 2.5 to over 8 metres, as does the construction, which to some degree depended on the topography.[78] The excavation conducted at Amygdaleonas outside Kavala demonstrated that the original Roman road was paved with small slabs and stones, and its foundation consisted of a layer of small stones and gravel topped by packed earth.[79] Elsewhere, near Komaros in the Zonaian mountains, the foundation was as deep as 1.2 metres, and included successive layers of stones and packed earth.[80] Everywhere, the road kerbs were built with larger stones and projected from the surface of the road. Down the middle of the road ran a single row of larger stones, which probably served as a divider between the opposite lanes, since the road was wide enough to accommodate carts passing each other in opposite directions (Figure 6).

The impact that the Egnatia had in the long run on the associated cities can be traced throughout the regions. Some developed more than others, but almost all of them show a *floruit* in the second and third centuries (Flavian, Antonine, and Severan periods), and again in the fourth to the fifth century, when the large ones (Dyrrachium, Herakleia Lyncestis, Edessa, Mygdonian Apollonia, Amphipolis, Philippi, Traianoupolis) became seats of bishoprics, and built impressive basilicas. Notable exceptions are the formerly important Macedonian cities of Pella and Lete. Despite the foundation of the colony, Pella did not thrive under the Empire, and the same is true for Lete, about which we hear next to nothing in the sources. The reason, I argue, must be sought in their proximity to Thessalonike, in a way the focal point of the Egnatia. The continuing growth of the largest city and important port of Macedonia, which eventually surpassed Beroia, its long-term rival and head of the Macedonian Koinon, worked ultimately against its immediate neighbours. In this case, the Via Egnatia became an agent of centralization of prosperity and power within a certain radius, including Pella and Lete, which fell into insignificance. In the case of Apollonia, the proximity of the Egnatia must be responsible for the transfer of the

Figure 6 Stretch of the Egnatia near Traianoupolis. Source: author.

activities of the Italian merchants and businessmen from Akanthos to this city in the course of the first century AD, and the upgrade of the city to a *municipium* in the reign of Trajan.[81] Further east, the important Macedonian cities of Abdera and Maroneia, situated on the Aegean coast but off the Roman artery, lost importance under the Empire to the advantage of Topeiros, Traianoupolis, and later Maximianoupolis.[82] It would be worth investigating a similar tendency in other regions along the Egnatia, particularly in its western part.

And what of the impact of the road on the rural landscape? The territories of the three colonies are the first to be looked at, since we know that they were subject to centuriation, and settlers were given standardized parcels of land to cultivate. In the case of Philippi, this is epigraphically attested; the colony had a strong agrarian character and its inhabitants had settled in the extended plain.[83] A recent examination of the plain surrounding the city, mostly from aerial photographs, made it possible to recognize the line of the Egnatia.[84] The modern land division (1928–60) is not oriented according to this road,[85] but how certain can we be that the parcels of land of today follow their Roman counterparts? In fact, similar research to the south and southeast of Philippi traced parcels of land oriented according to the Via Egnatia.[86] More field investigation is required here, as in the other colonies and cities founded in the Roman period along the Egnatia, in order to assess the impact of the road on the land division of the surrounding areas.

The Via Egnatia, this important *via militaris* of the Roman Republic—which for the most part had been traced over an earlier Greek trade route—gradually developed into a major artery of communication between the eastern and western parts of the Roman Empire. By linking key harbours of the Adriatic and the Aegean and connecting important cities across the southern Balkans, the Egnatia facilitated and renewed commercial activities, encouraged cultural interactions, became a venue of imperial propaganda, and, ultimately, contributed to the reshaping of the human and natural landscape.

Notes

[1] I would like to thank the organizers of the symposium for inviting me to participate. I am also indebted to Selene Psoma for kindly offering to read a draft of this article, making valuable comments and additions. Finally, the article has greatly benefited from the comments of its reviewers, to whom I am most grateful.

[2] The Via Egnatia has been the subject of numerous articles and a few monographs, on which see my references at the end of the article. I am currently preparing a monograph on the subject to be published by Olkos Editions with text in Greek and English.

[3] I am using the Loeb edition of Strabo, vol. 3, translated by H. L. Jones (London, 1924).

[4] Our archaeological knowledge of this important Mediterranean harbour is very limited: see Vitti, *Πολεοδομική εξέλιξη*, 131–33.

[5] As correctly pointed out by Walbank, 'Via illa nostra militaris', and Walbank, 'The Via Egnatia', 9. The *miliarium* is published by Samartzidou, 'Εγνατία οδός'; see also *SEG* XL (1990): 543.

[6] The *miliarium* of Thessalonike is published by Romiopoulou, 'Un nouveau milliaire'; see also Šašel Kos, *Inscriptiones Latinae*, no. 246 and *AnnEpigr* 1976: 643. It is possible that this

Cn. Egnatius was the same as the homonymous senior senator whose name appears in a letter addressed by the praetor P. Cornelius Blasio to the Corcyreans (*SEG* III 451). Blasio was *praetor* most likely in the late 140s: see Walbank, 'The Via Egnatia', 9–10. On these grounds, Brennan (*The Praetorship*, 225) argues that 'the most plausible date for Egnatius command is a date in the mid-140s'.

[7] As argued by Hatzopoulos, 'Les politarques', 141 and 146; cf. Walbank, 'The Via Egnatia', 14.

[8] Those threatening the Via Egnatia included the Maedi, Caeni, Scordisci, Triballi, Daci, and Bessi, some of which had been harassing the Macedonians for a long time; see Walbank, 'Via illa nostra militaris', and Collart, *Philippes*, 489–90.

[9] On the eastern boundaries of Macedonia, see Loukopoulou, '*Provinciae Macedoniae Finis Orientalis*'; Polychronidou-Loukopoulou, 'Colonia Claudia Aprensis', 701–04; Polychronidou-Loukopoulou, 'Η ρωμαϊκή παρουσία', 181–84.

[10] The eastern stretch of the Egnatia was certainly not a creation *ex nihilo*: there is evidence for a pre-existing road, perhaps initiated by the Persians before 480 and expanded by Philip II, but certainly used by Philip V: see Polychronidou-Loukopoulou, 'Colonia Claudia Aprensis', 701–04; Loukopoulou *et al.*, *Επιγραφές της Θράκης*, 123–24, 137.

[11] Thus we can understand Cicero's lament on the fate of the Egnatia in his time: the eastern part was presumably taken over by the Thracians: see Loukopoulou, '*Provinciae Macedoniae Finis Orientalis*', 96–97.

[12] On the demilitarization of Egnatia, see Walbank, 'The Via Egnatia', 13–17.

[13] See Collart, 'Une réfection', 403–07 (=Collart, 'Les milliaires', 197, no. 3).

[14] See Collart, 'Une réfection', 407–11 (=Collart, 'Les milliaires', 197, no. 2).

[15] See Collart, 'Les milliaires', 194–96; cf. Loukopoulou *et al.*, *Επιγραφές της Θράκης*, E433 (535–38). Gounaropoulou and Hatzopoulos, *Les Milliaires*, on the other hand, take a more cautious attitude. The inscriptions carved on the *miliaria*, they argue, do not necessarily imply reconstruction of the road by the named emperor. In some cases, it may have been a matter of replacing a destroyed milestone, and in other cases they could have been part of the imperial propaganda. Collart associates both repairs with imperial *adventus*, an opinion endorsed by Otatzis, 'Μιλιάρια', 189–90.

[16] The major land route between Constantinople and Italy is now the one going through Hadrianoupolis and Philippopolis mentioned by the *Itineraria*.

[17] For a recent discussion of the *itineraria*, see Salway, 'Travel'.

[18] It is noteworthy that the route shown on the Tabula appears to cross the Hebrus river by Ainos, not Kypsela, and that it does not extend to Constantinople.

[19] The first route (from Dyrrachium to Byzantion) diverts up the Hebrus valley to Hadrianoupolis, and then turns back southeast to rejoin the Egnatia at Perinthos-Herakleia (*Itin. Ant.* 322,4–323,5): see below p. 279.

[20] Incidentally, the *Itin. Burd.* does follow the route of the Via Egnatia from Apri to Kypsela.

[21] There is no complete corpus of the *miliaria* found so far along the route of the Egnatia. The *miliaria* known to the early 1970s are published by Collart, 'Les milliaires', while Gounaropoulou and Hatzopoulos, *Les Milliaires*, publish the *miliaria* found between Herakleia Lyncestis and Thessalonike. In addition, it is not certain that all *miliaria* attributed to the Via Egnatia actually belong to this rather than to a different road.

[22] See Fasolo, *Via Egnatia*, with earlier references (particularly the work of H. Ceka). On the most recent work in Albania, see Amore *et al.*, 'The Via Egnatia Project' and 'Via Egnatia Project of 2002'. The section within FYROM was addressed by Bitrakova-Grozdanova, 'La Via Egnatia'.

[23] Edson, 'The Location of Cellae'; Hammond and Hatzopoulos, 'The Via Egnatia I', 'The Via Egnatia II'.

[24] Moutsopoulos, 'Μυγδονική Απολλωνία'; Hatzopoulos, 'The Via Egnatia'; Collart, *Philippes*; Samartzidou, 'Εγνατία οδός'; Otatzis, 'Ενα μιλιάριο'; Pantos, 'The Present Situation';

Dadiras, 'Η Εγνατία οδός'; Tsatsopoulou-Kaloudi, *Εγνατία οδός*. The list is selective. I have been unable to locate a study of the Kypsela-Byzantion section—that is, across eastern Thrace (and within the borders of modern Turkey).

[25] On the travels of western European dilettanti, see O'Sullivan, *The Egnatian Way*, 221–40.

[26] On Trajan, see Collart, 'Une réfection', especially 413–15; on Hadrian, see Halfmann, *Itinera principum*, 208; On Septimius Severus and Caracalla, see Collart, 'Les milliaires', 192–97; Otatzis, 'Μιλιάρια' (*miliarium* Λ 1275 found at Nea Kerdyllia and bearing dedications of the Amphipolitans to Septimius Severus, Caracalla, and Geta); Koukouli-Chrysanthaki, 'Ακόντισμα' (*miliarium* found at the community of Lefki outside Kavala and dedicated to Caracalla = *AnnEpigr* 1974, 590); Halfmann, *Itinera principum*, 244–45. In addition, Collart (*Philippes*, 511) argues that Tiberius took this road ahead of an army to Syria.

[27] As already pointed out by Collart, *Philippes*, 511.

[28] See Papazoglou, *Villes de Macédoine*, 135–39.

[29] *Itin. Ant.* 330,6. In the Bordeaux Itinerary (606,1), Diokletianoupolis is listed as 'civitas Polli'. We know close to nothing on this city, and scholars have often confused it with the Thessalian Diokletianoupolis: see Papazoglou, *Villes de Macédoine*, 140–41.

[30] See Papazoglou, *Villes de Macédoine*, 405–13.

[31] See Papazoglou, *Villes de Macédoine*, 219; Hatzopoulos, 'The Via Egnatia', 210 (with references).

[32] As argued by Hatzopoulos ('The Via Egnatia', 210), following Souris.

[33] See the discussion in Avezou and Picard, 'Inscriptions', 147–53; Mottas, 'Voies de communication', 101–02; Adams, 'Communications', 141–43.

[34] Topeiros was important enough to strike coins labelled ΟΥΛΠΙΑΣ ΤΟΠΕΙΡΟΥ or ΤΟΠΕΙΡΗΤΩΝ between the reign of Antoninus Pius and Geta: see Parisaki and Loukopoulou, 'Topeiros'. The *civitas* is located near the modern village of Paradeisos: see Loukopoulou *et al.*, *Επιγραφές της Θράκης*, 139.

[35] See Polychronidou-Loukopoulou, 'Colonia Claudia Aprensis'.

[36] As argued by Mottas, 'Voies de communication', 101.

[37] The exact location of Anastasioupolis is not certain, but is often associated with Peritheorion, 26 km west of Komitini: see Kortzi and Siametis, 'Mosynopolis/Maximianoupolis'.

[38] The city kept this name down to the ninth century, when it became Mosynopolis: see Kortzi and Siametis, 'Mosynopolis/Maximianoupolis'.

[39] See Sayar, *Perinthos-Herakleia*, 76–77.

[40] On the movement of the Illyrian prefecture, see *CAH* XIV, 710–11.

[41] These repairs are mentioned by Procopius, *De aed.* 4.8–9.

[42] Corinthian ceramic products of the 6th to 4th century BC found in the excavations of the necropolis of Trebeniste prove that the trade from Apollonia and Dyrrachium followed the Genusus river valley, as the Egnatia did a few centuries later. Likewise, the number of Macedonian coins of the Hellenistic period found in Lleschan show that the road along the Genusus valley was busy during that period as well: see Ceka and Papajani, 'La route', 90–94.

[43] On the major harbours of southern Illyria, see Cabanes, 'Les ports d'Illyrie'.

[44] See Papageorgiadou-Banis, 'La diffusion du monnayage', 115–18.

[45] From the 279 Roman glass objects kept in the Tirana museum, 150 come from Dyrrachium, of which the majority come ultimately from Rome, northern Italy, and Dalmatia. In addition, a number of Italian lamps carry the stamps of their manufacturers who had their workshops in Mutina, Puteoli, Aquileia, etc.: see Tartari, 'Les contacts'.

[46] Malamidou, *Roman Pottery*. The other three sites are Kepia, Abdera, and Thasos.

[47] Malamidou, *Roman Pottery*, 76–87.

[48] Malamidou, *Roman Pottery*, 83.

[49] Malamidou, *Roman Pottery*, 77.

[50] Malamidou, *Roman Pottery*, 79 (on Amphipolis); Adam-Veleni *et al.*, 'Αρχαία αγορά', 516–18, and Grammenos, *Ρωμαϊκή Θεσσαλονίκη*, 141–42 (on Thessalonike).
[51] Malamidou, *Roman Pottery*, 80 (such a study is still wanting).
[52] See the still useful, albeit dated, monograph of Hatzfeld, *Les trafiquants*, and now Rizakis, 'L'émigration romaine' (with earlier references).
[53] See Rizakis, 'L'émigration romaine', 110–11.
[54] As argued by Rizakis, 'L'émigration romaine', 123–24.
[55] See Dimitsas, *Η Μακεδονία*, no. 3: 35; Papazoglou, *Les villes de Macédoine*, 129.
[56] See Rizakis, 'Η κοινότητα των συμπραγματευομένων', and Grammenos, *Ρωμαϊκή Θεσσαλο νίκη*, 92–3.
[57] As argued by Polychronidou-Loukopoulou, 'Η ρωμαϊκή παρουσία', 187.
[58] Roger, 'Inscriptions', no. 6: 53–55 (=*AnnEpigr* 1946, 230) and no. 3: 49–51 (=*SEG* XXXVI [1986] 587), respectively; cf. Papazoglou, *Villes de Macédoine*, 395. No date is given for the first inscription (no. 6), but we do know that Eros (the diseased) was apparently so named by his fellow *negotiatores* ('ab suis connegotiatoribus'), suggesting a community and/or network of businessmen in that town.
[59] On slave trade in Macedonia, which perhaps 'inherited' the Delian tradition, see Rizakis, 'L'émigration romaine', 123–24.
[60] On *negotiatores* in Achaia, see Alcock, *Graecia Capta*, 75–77.
[61] See Rizakis, 'Η κοινότητα των συμπραγματευομένων', 520–23.
[62] The Roman-type names represent 15% of the total in the first century BC, 36% in the first century AD, reaching 60% in the second century, and 80% in the third: see Tataki, 'Προσωπογραφία της Μακεδονίας'. On the high percentage of Romans in Herakleia Lyncestis, see Papazoglou, *Villes de Macédoine*, 262–64. On the strong Roman presence in Edessa, see Tataki, *Macedonian Edessa*, 86–92.
[63] This was of course the case in all three colonies. Their higher offices were held by Romans, and there is little trace of the native elite entering the ruling class: see Wilkes, 'Greek and Latin' (for Dyrrachium).
[64] See Papazoglou, *Villes de Macédoine*, 262–63.
[65] As suggested by two ephebic lists, where of the 45 men listed 27 bear regular Roman names: see Tataki, *Macedonian Edessa*, 88–91.
[66] Antonii, Attii, etc.: see Tataki, *Macedonian Edessa*, 86–87.
[67] See Rizakis, 'Η κοινότητα των συμπραγματευομένων', 520–21.
[68] See Grammenos, *Ρωμαϊκή Θεσσαλονίκη*, 92–97.
[69] The majority of these complexes within Greece have not been excavated, and judging from their visible remains they seem to belong to post-Roman periods.
[70] See Hammond, 'The Western Part'; Fasolo, *Via Egnatia*, 146.
[71] See Velenis, 'Αγορά Θεσσαλονίκης', 132–33; see also Vitti, 'Le città della *provincia*', 486.
[72] Vitti ('Le città della *provincia*', 480–83) attempts to show that the monumental gates of Thessalonike ('Porta d'Oro'), Philippi (Kiemer), Amphipolis (gate Δ), and Pella came into existence during the same, Augustan, era, and shared some common architectural characteristics. Yet, the chronology of these structures is far from established, and the architectural details of the gate of Pella, which is known only from its schematic depictions on coins, escape us. The remaining three gates differ considerably in design and decoration.
[73] Hammond and Hatzopoulos ('The Via Egnatia I') convincingly argue that the Egnatia passed right to the south of Lake Vegoritis, which is now under water. In antiquity the Nestos river flooded the surrounding area: see Adams, 'Topeiros Thraciae', 17–18. Our ancient source on the marshy plain (τενάγη) of Philippi, first drained by Philip II, is Theophrastus *CP* 5.14.6.
[74] Praschniker and Schober, *Forschungen*, 61.
[75] See Amore *et al.*, 'The Via Egnatia Project', 386–88.

[76] This is the so-called Keçi bridge: Amore *et al.*, 'Via Egnatia Project of 2002', 343–44.
[77] See Collart, *Philippes*, 494–97; cf. Tsotsos, 'Γεφύρια της ρωμαϊκής περιόδου', 1273–74.
[78] Ceka and Papajani, 'La route', 96–101. The authors claim to have recognized three different phases of the road traces between Ohrid and Dyrrachium. Furthermore, they argue that in the low-lying areas the road was paved with small stones and gravel, whereas on slopes the road was built using flat stones overlaid with gravel.
[79] See Samartzidou, 'Εγνατία οδός'. The date of the road was firmly established by its connection with the in situ *miliarium* mentioning Egnatius. The *miliarium* was set on the western edge of the road, 0.6 m lower than the road bed.
[80] See Tsatsopoulou-Kaloudi, *Εγνατία οδός*, 38. The excavation was conducted 1.5 km due northwest of Komaros.
[81] As argued by Polychronidou-Loukopoulou, 'Η ρωμαϊκή παρουσία', 186–88. According to the author, the idea of relocation of the Italian population from the coast of Chalcidice to the inlands of Mygdonia is strengthened by the heavy Roman presence in the ephebic lists of Kalindoia, perhaps the most important city of the area in the 1st century AD, situated some 12 km southwest of Apollonia.
[82] See Loukopoulou *et al.*, *Επιγραφές της Θράκης*, 139–40.
[83] As Papazoglou (*Villes de Macédoine*, 411–12) observed, nowhere else have been found so many inscriptions referring to the *vici* of Philippi and their inhabitants.
[84] Kaimaris *et al.*, 'Φωτογραμμετρική έρευνα'.
[85] Kaimaris *et al.*, 'Φωτογραμμετρική έρευνα', 123.
[86] Santoriello and Vitti, 'Il paessagio agrario', 994–95 and fig. 4 (1001).

References

Adams, John P. 'Topeiros Thraciae, the Via Egnatia and the Boundaries of Macedonia'. *Ancient Macedonia* 4 (1986): 17–42.

———. 'Communications in South-eastern Thrace'. In *Actes du 2ᵉ Symposium International des études Thraciennes*', Vol. I, 135–44. Komotini: Μορφωτικός όμιλος κομοτηνής, 1997.

Adam-Veleni, P., P. Georgaki, E. Zografou, V. Kalavria, K. Mpoli, and G. Skiadaresis. 'Αρχαία αγορά Θεσσαλονίκης: η στρωματογραφία και τα κινητά ευρήματα'. *Το αρχαιολογικό έργο στη Μακεδονία καιΘράκη* 10 (1996): 501–31.

Alcock, Susan E. *Graecia Capta: The Landscapes of Roman Greece*. Cambridge: Cambridge University Press, 1993.

Amore, M. G., L. Bejko, Y. Cerova, and I. Gjipali. 'The Via Egnatia Project and the Bridge at Topçias'. *Journal of Roman Archaeology* 14 (2001): 381–89.

———. 'Via Egnatia (Albania) Project: Results of Fieldwork 2002'. *Journal of Roman Archaeology* 18 (2005): 336–60.

Avezou, C., and C. Picard. 'Inscriptions de Macédoine et de Thrace'. *Bulletin de Correspondence Hellenique* 37 (1913): 84–154.

Bitrakova-Grozdanova, Vera. 'La Via Egnatia entre Lychnidos et Pons Servili (nouvelles preuves archéologiques)'. *Ancient Macedonia* VI (1999): 153–65.

Brennan, Corey T. *The Praetorship in the Roman Republic*. Oxford: Oxford University Press, 2000.

Cabanes, Pierre. 'Les ports d'Illyrie méridionale'. In *Strutture portuali e Rotte Marittime nell'Adriatico di età Romana*, edited by C. Zaccaria. Triest–Rome: Centro di Antichità Altoadriatiche, 121–36.

Ceka, H. and L. Papajani. 'La route de la vallée du Shkumbin dans l'antiquité'. *Studia Albanica* 9, no. 1 (1972): 85–106.

Chevallier, Raymond. *Les voies romaines*. 2nd ed. Paris: Picard, 1997.

Collart, Paul. 'Une réfection de la Via Egnatia sous Trajan'. *Bulletin de Correspondence Hellénique* 59 (1935): 395–415.

———. *Philippes, ville de Macédoine*. Paris: de Boccard, 1937.

———. 'Les milliaires de la Via Egnatia'. *Bulletin de Correspondence Hellénique* 100 (1976): 177–200.

Dadiras, Chrysostomos. 'Η Εγνατία οδός στην περιοχή της Θράκης κατά τη Ρωμαϊκή και Βυζαντινή εποχή'. *Περί Θράκης* 3 (2003): 101–30.

Dimitsas, Margaritis, *Η Μακεδονία εν λίθοις φθεγγομένοις και μνημείοις σωζομένοις*. Athens (1896).

Edson, Charles. 'The Location of Cellae and the Route of the Via Egnatia in Western Macedonia'. *Classical Philology* 46 (1951): 1–16.

Fasolo, Michele. *Via Egnatia I. Da Apollonia e Dyrrachium ad Herakleia Lynkestidos*. Roma: Istituto Grafico Editoriale Romano, 2003.

Gounaropoulou, L. and M. B. Hatzopoulos. *Les Milliaires de la Voie Egnatienne entre Heraclée des Lyncestes et Thessalonique*. ΜΕΛΕΤΗΜΑΤΑ, no. 1 Athens: Research Center for Greek and Roman Antiquity, 1985.

Grammenos, D. V., ed. *Ρωμαϊκή Θεσσαλονίκη*. Thessalonike: Archaeological Museum Publications, no. 1, 2003.

Halfmann, Helmut. *Itinera Principum: Geschichte und Typologie der Kaiserreisen im Römischen Reich*. Stuttgart: Steiner, 1986.

Hammond, Nicholas. 'The Western Part of Via Egnatia'. *Journal of Roman Studies* 64 (1974): 185–94.

Hammond, N. and M. Hatzopoulos. 'The Via Egnatia in Western Macedonia, Part I'. *American Journal of Ancient History* 7 (1982): 128–49.

———. 'The Via Egnatia in Western Macedonia, Part II'. *American Journal of Ancient History* 8 (1983): 48–53.

Hatzfeld, Jean. *Les trafiquants Italiens dans l'Orient Grec*. Paris: de Boccard, 1919.

Hatzopoulos, Miltiades. 'Les politarques de Philippopolis'. In *Dritter internationaler thrakologischer Kongress zu Ehren W. Tomascheks*, Vol. 2. Wien, 1984. 137–49.

———. 'The Via Egnatia between Thessalonike and Apollonia'. In *Αφιέρωμα στον N.G.L. Hammond*, 199–213. Thessalonike: Εταιρεία Μακεδονικών Σπουδών, 1997.

Kaimaris, D., O. Georgoula, and G. Karadedos. 'Φωτογραμμετρική–φωτοερμηνευτική έρευνα στην πεδιάδα των Φιλίππων'. *Το αρχαιολογικό έργο στη Μακεδονία καιΘράκη* 16 (2002): 119–29.

Kortzi, M. and V. Siametis. 'Mosynopolis/Maximianoupolis', In http://www.xanthi.ilsp.gr/thraki/; INTERNET.

———. 'Peritheorion'. In http://www.xanthi.ilsp.gr/thraki/; INTERNET.

Koukouli-Chrysanthaki, Chaido. 'Via Egnatia—Ακόντισμα'. *Αρχαιολογικά Ανάλεκτα εξ Αθηνών* 5 (1972): 474–85.

Loukopoulou, Louisa. '*Provinciae Macedoniae Finis Orientalis*: The Establishment of the Eastern Frontier'. In *Two Studies in Ancient Macedonian Topography*, edited by M. B. Hatzopoulos and L. D. Loukopoulou, 61–100. Athens: Research Center for Greek and Roman Antiquity, 1987.

Loukopoulou, L., A. Zournatzi, M. G. Parisaki, and S. Psoma. *Επιγραφές της Θράκης του Αιγαίου, μεταξύ των ποταμών Νέστου και Έβρου*. Athens: Research Center for Greek and Roman Antiquity, 2005.

Malamidou, Vaitsa. *Roman Pottery in Context; Fine and Coarse Wares from Five Sites in North-eastern Greece*. British Archaeological Reports Series, no. 1386. Oxford: John and Erica Hedges, 2005.

Mottas, François. 'Les voies de communication antiques de la Thrace égéenne'. In *Labor omnibus unum. Gerold Walser zum 70 Geburtstag dargerbracht von Freunden, Kollegen und Schülern*. Historia Einzelschriften, no. 60, 82–104. Stuttgart: Steiner, 1989.

Moutsopoulos, Nikolaos. 'Η θέση της Μυγδονικής Απολλωνίας και η παραλίμνια (;) χάραξη της Εγνατίας οδού'. *Ancient Macedonia* 5 (1993): 999–1110.

O'Sullivan, Firmin. *The Egnatian Way*. Newton Abbot: David and Charles, 1972.

Otatzis, Moschos. 'Μιλιάρια της Εγνατίας οδού από τα Κερδύλλια Σερρών'. In *Μνήμη Μαν. Ανδρόνικου*, 187–98. Thessalonike: Εταιρεία Μακεδονικών Σπουδών, 1997.

———.'Ένα μιλιάριο από το μικρό Σούλι Σερρών: συμβολή στη μελέτη τη πορείας της Εγνατίας οδού'. In *Πρακτικά του διεθνούς συνεδρίου 'Οι Σέρρες και η περιοχή τους από την αρχαία στη μεταβυζαντινή κοινωνία'*, 113–26. Thessalonike: Δήμος Σερρών (Municipality of Serres), 1998.

Pantos, Pantos. 'The Present Situation of the Studies in Archaeological Topography of Western Thrace'. *Pulpudeva* 4 (1983): 164–78.

Papageorgiadou-Banis, Charikleia. 'La diffusion du monnayage romain dans l'Épire'. In *L'Illyrie méridionale et l'Épire dans l'Antiquité*, Vol. III, edited by P. Cabanes, 115–18. Paris: de Boccard, 1999.

Papazoglou, Fanoula. *Les villes de Macédoine à l'époque romaine. BCH* supplement series, no. XVI. Paris: de Boccard, 1988.

Parisaki, M. G. and L. Loukopoulou. 'Topeiros'. In http://www.xanthi.ilsp.gr/thraki/; INTERNET.

Polychronidou-Loukopoulou, Louisa. 'Colonia Claudia Aprensis: Μία ρωμαϊκή αποικία στη Θράκη'. In *Μνήμη Δ. Λαζαρίδη: πόλις και χώρα στην αρχαία Μακεδονία και Θράκη*, 701–18. Thessalonike: Greek Ministry of Culture, Archaeological Museum of Kavala and the École Française d'Athenes, 1990.

———. 'Η ρωμαϊκή παρουσία στη νοτιοανατολική Θράκη'. In *Actes du 2ᵉ Symposium International des Études Thraciennes*, Vol. I, 181–92. Komotini: Μορφωτικός όμιλος κομοτηνής, 1997.

Praschniker, C. and A. Schober, *Archäologische Forschungen in Albanien und Montenegro*. Vienna: A. Holder, 1919.

Rizakis, Athanasios. 'Η κοινότητα των «συμπραγματευομένων Ρωμαίων» της Θεσσαλονίκης και η ρωμαϊκή οικονομική διείσδυση στη Μακεδονία'. *Ancient Macedonia* IV (1986): 511–24.

———. 'L'émigration romaine en Macédoine et la communauté marchande de Thessalonique: perspectives économiques et sociales'. In *Les Italiens dans le monde grec, IIᵉ siècle av. J.-C.—Iᵉʳ siècle ap. J.-C.; Circulation, Activités, Intégration. BCH*, supplement series, no. 41, edited by C. Müller and C. Hasenohr, 109–32. Paris: de Boccard, 2002.

Roger, Jacques. 'Inscriptions de la région du Strymon'. *Revue Archeologique* XXIV (1945): 37–55.

Romiopoulou, Caterina. 'Un nouveau milliaire de la Via Egnatia'. *Bulletin de Correspondence Hellénique* 98 (1974): 813–16.

Salway, B. 'Travel, Itineraria and Tabellaria'. In *Travel and Geography in the Roman Empire*, edited by C. Adams and R. Laurence, 22–66. London: Routledge, 2001.

Samartzidou, Stavroula. 'Εγνατία οδός: από τους Φιλίππους στη Νεάπολη'. In *Μνήμη Δ. Λαζαρίδη: πόλις και χώρα στην αρχαία Μακεδονία και Θράκη*, 559–88. Thessalonike: Greek Ministry of Culture, Archaeological Museum of Kavala and the École Française d'Athenes, 1990.

Santoriello, A. and M. Vitti. 'Il paessagio agrario del Territorio della Colonia Victrix Philippensium'. *Ancient Macedonia* VI (1999): 987–1001.

Šašel Kos, Marietta. *Inscriptiones Latinae in Graecia Repertae; Additamenta ad CIL III*. Faenza: Fratelli Lega, 1979.

Sayar, MustafaH. *Perinthos-Herakleia (Marmara Ereğlisi) und Umgebung*. Wien: Österreichischen Akademie der Wissenschaften, 1998.

Tartari, Fatos. 'Les contacts de la civilization de l'Illyrie du sud et de l'Épire avec la civilization romaine à la lumière des productions du verre des Iᵉʳ–IVᵉ siècles de notre ère trouvées sur le territoire d'Albanie'. In *L'Illyrie méridionale et l'Épire dans l'Antiquité*, Vol. III, edited by P. Cabanes, 275–82. Paris: de Boccard, 1999.

Tataki, Argyro. 'Από την προσωπογραφία της αρχαίας Μακεδονίας: παρατηρήσεις στα ρωμαϊκού τύπου ονόματα του IG X 2,1'. *Ancient Macedonia* IV (1986): 581–94.

———. *Macedonian Edessa; Prosopography and Onomastikon*. ΜΕΛΕΤΗΜΑΤΑ, no. 18. Athens: Research Center for Greek and Roman Antiquity, 1986.

Tsatsopoulou-Kaloudi, Polyxeni. *Εγνατία Οδός. Ιστορία και διαδρομή στο χώρο της Θράκης*. Αθήνα: ΤΑΠ, 2005.

Tsotsos, Georgios. 'Γεφύρια της ρωμαϊκής περιόδου στη Μακεδονία'. *Ancient Macedonia* VI (1999): 1261–75.

Velenis, Georgios. 'Αρχαία Αγορά Θεσσαλονίκης'. *Αρχαιολογικά Ανάλεκτα εξ Αθηνών* 23–8 (1990–95): 129–42.

Vitti, Massimo. *Η πολεοδομική εξέλιξη της Θεσσαλονίκης από την ίδρυσή τηέως τον Γαλέριο*. Athens: Η εν Αθήναις Αρχαιολογική Εταιρεία, 1996.

———. 'Adattamento e transformazione delle città della *provincia Macedonia*'. In *Constructions publiques et programmes édilitaires en Grèce entre le IIe siècle av. J.-C. et le I^e siècle ap. J.-C. Bulletin de Correspondence Hellénique* supplement series, no. 39, edited by J.-Y. Marc and J.-C. Moretti, 473–94. Paris: de Boccard, 2001.

Walbank, Frank. 'Via illa nostra militaris . . . : Some Thoughts on the Via Egnatia'. In *Selected Papers: Studies in Greek and Roman History and Historiography*, 193–209. Cambridge: Cambridge University Press, 1985.

———. 'The Via Egnatia: Its Role in Roman Strategy'. *Journal of Ancient Topography* 12 (2002): 7–18.

Wilkes, John. 'Greek and Latin in the City and Territory of Dyrrachium'. In *L'Illyrie méridionale et l'Épire dans l'Antiquité*, Vol. IV, edited by P. Cabanes and J.-L. Lamboley, 383–90. Paris: de Boccard, 2004.

Hadrian's *Panhellenion*: A Network of Cities?[1]

Panagiotis N. Doukellis

Introduction

The subject of networks and how they were perceived and employed in antiquity brings us straight to the thorny issue of reading the past through the conceptual and methodological filters and tools of the present, which is one of the main concerns of history as a discipline. Consequently, my aim here will be not simply to observe, analyse and describe the various elements that construct a network, since such an attempt would be in danger of constructing *a posteriori* a historical reality, but also to see whether the methodological tool of network theory can shed new light on many aspects of various phenomena and events of Antiquity.

The issue of networks in the ancient world involves two levels of analysis: the first deals with the possible uses of network theory for reading and interpreting phenomena and situations of the ancient world; and the second hazards the question of whether ancient societies were aware of the implications of the concept of networks as we understand them today. The bottom line is that the term 'network' today denotes an abundance of phenomena and real situations. While on the one hand this fact provides a practical methodological tool for analysis, on the other hand caution is required when network theory is implemented as a means of re-reading existing data.[2]

To begin with, we should look at the relevant terms *diktyo* and *sagene*, meaning 'mesh' and 'net'. A significant characteristic of these two words is their metaphorical use very early on. Given that they are used literally to denote the familiar fishing net, their understanding as an 'ensemble of interwoven fibres or filaments' has fostered a broad metaphorical use—as a look at some of Galen's texts will readily confirm, in which the functions of the brain, of the system of arteries, and of other parts of the human body, are routinely described as a network.[3] Analogies likewise occur in the respective Latin word *retiolus/retis*, which generated such contemporary words as *réseau* (French), and *rete* (Italian), in the Latin-based languages. It is well known that metaphors are widely used in the study and understanding of the mental representations of a society, in the sense that often, as in the case here, certain very specific and tangible realities can be 'mapped' onto distinctly abstract entities.[4]

This metaphorical use of the network concept, which was common in Greek antiquity and in our own society, does not have a merely static content, limited to the image of 'an ensemble of interwoven fibres or filaments', but includes dynamic aspects also. Therefore, as far as *sagene* is concerned, one of the best-known metaphorical occurrences of the word is when Herodotos describes the practice upheld by the Persians when conquering an area. Literally advancing hand in hand, the Persian soldiers would check the land of, say, Samos from one end to the other, to ensure that none of the local enemies had escaped.[5] In this way, they were able to state that the power of the Persian king had de facto spread across the recently conquered territory like a *sagene*, or mesh. Here as in other texts, Herodotos adopts a metaphorical use of *sagene*, and relates it directly with the expansion of political and military power.

Fifteen centuries after Herodotos, the Byzantine emperor Constantine VII Porphyrogenitos uses *sagene* in a slightly different manner: '*Since Roman power spread its mesh* [sagene] *to the ends of the world everything was remade and transformed into one authority and everyone was forced to heed one rule*'; and '*as the mesh* [sagene] *began to deteriorate and break apart, so the things and the names changed*'.[6] Porphyrogenitos uses the term metaphorically, but he also includes it in a framework of relationships of power. If Herodotos used the word *sagene* to describe a very specific action of the Persians, Constantine Porphyrogenitos uses the same metaphor to describe a general and abstract idea. For him, *sagene* signifies both the means and the results of imposing imperial authority. In both cases, we may note that the metaphorical use of the word *sagene* is based on its dynamic use as a tool for catching fish. In the first case, Persian soldiers function as a mesh that catches enemies, in the second case the Roman power itself is presented as a mesh, which imposes order on the world.

This last example brings us to the topic of this article: networks as power mechanisms, with particular reference to the Greek cities within the Roman Empire. How are we to understand the term network? A network can be a subject of historical research, and, on the other hand, the term network also refers to a specific methodological approach, which was developed by sociologists attempting to describe and analyse our own society as it is transformed by major historical events.[7] Network theory developed many interesting approaches which ancient history could use. One of these, the bipolar relationship between 'network' and 'self', is quite prominent in modern sociological works. Manuel Castels, in his three-volume work on networks, discusses networks as a system of organization based mainly on the technological potentials offered to us by the new information and communications technology.[8] At the same time, according to Castels, the individual is attempting to shape his personal, social and cultural identity. In this way, network society is structured through this basic bipolar relationship between the Network and the Self. On the one hand, we have the individual's submission to and his role in shaping global communication systems, and on the other hand, there is the development of the individual's historically defined identity. Castels's approach is quite problematic as it deals with the issue of self-definition and the definition assigned by others to the individual, and because the application of this interpretative scheme (the bipolar relationship between Self and Network) is not always clear.[9] This, however, is something to which we shall return.

Hadrian's *Panhellenion* as a Cities Network

Looking for an Identity

Let us now turn our attention to Hadrian's Pan-Hellenic programme, promoted by the emperor in the beginnings of the second century AD (AD 131), which, in my opinion, could illustrate the function of a network in the ancient world.[10] By network here, I mean an instituted network functioning as a mechanism of political and social stabilization and governance of the Roman imperial rule over the eastern provinces of the Empire. Hadrian instituted a *synedrion ton Hellenon* (council of the Greeks), which was mostly based on the tradition of the Greek *koina*. The object of the *synedrion* was the organization of worship and competition in games, as well as other rituals related to honouring the ruling emperor. Participation in the *synedrion* was possible for the cities of Achaia, Macedonia, Thrace, Asia, and Cyrenaica, as long as they could demonstrate their being Greek in character and origin. For the cities of metropolitan Greece, participation was axiomatic; to this, the cities of the other provinces responded by forwarding evidence, real or fictitious, proving their Greek descent and identity. Some cities, because of insufficiency of tangible evidence, delve into the historical and mythical past in order to produce Greek *oikistai* (founders), whom they then use in order to be accepted to this new organization, which was promoted apparently from above, by the emperor himself.[11] The desire of many cities to display their Hellenicity, in order to be part of the Panhellenion, led them to make identity claims that were particularly problematic, if one approaches the matter from a positivistic angle. The case of Cibyra, which aimed to attach itself to the most prestigious cities of Greek history, is well known: Cibyra claimed to be a colony of the Lacedaimonians and hence *syngenis* (related) to Athens.[12]

We should note at this point that the cities of metropolitan Greece, and especially Athens, had already functioned since the age of Alexander as a model of political organization and cultural identity for the cities of the Hellenistic east.[13] However, what is worth discussing is the desire of the various Asiatic cities for their Greek identity to be explicitly acknowledged by the superior authoritative institutions, whether this was through the decision of a monarch, or, even better, through the resolution of the Delphic amphictyony.[14] Two supplementary procedures were used for the declaration of the identity of a city. First, we have the explicit declaration from the part of those directly interested—that is, by the magistrates of the candidate city. This is a phase of 'identization',[15] a procedure through which the citizens wish to differentiate themselves from the others, the non-Greek cities, and/or promote their Greek character as the only one, eventually concealing any other, non-Greek, element. Second, there is the procedure of 'identification', by the more powerful and authoritative institutions, which have the *auctoritas* to validate such a procedure, and, consequently, classifying and/or integrating the city into broader groups, whether cultural, ethnic, political, or other.[16]

This tradition continued during the imperial years. The emperor, when asked, could verify or not the identity of a city. The case of Ptolemais Barca, which we shall discuss shortly, is indicative of the whole process of 'identization' on the part of the city's magistrates, and its identification by the emperor.[17]

I do not wish to discuss here the relevant epigraphic, archaeological, or literary evidence on the political use of the past during the Hellenistic and imperial periods.[18] Since the context is the discussion of networks in the ancient world, I would like to examine the processes of the organization of the Panhellenion by the emperor himself, as an example of a network of cities aiming to achieve targets which were widely spread in the eastern provinces of the empire. Did the imperial decisions aim at much more specific targets? On which specific principles and traditions was the Panhellenion founded? Certainly, the attempt to create a network of cities on the basis of accepted practices is worth the attention of the historian. These are questions that are often addressed in the existing scholarship. Yet, it seems to me that they can be answered fruitfully using the principles of network theory.

The Function of Identity

Let us discuss first the conditions of admission for the members. The candidate and the active members must prove their Hellenicity and their good relations with Rome. Certainly, even though this Hellenicity denotes a cultural identity, it is constructed through reference to Greek settlers of the archaic period, or through a Greek, often mythical, founder who organized the local populations and bequeathed to them the Greek way of life, language, worship and so on.[19]

It is well-known that since the fourth century BC, the idea that 'Hellenicity' was a property that could be acquired by an individual, a community, or a city, independently of its own biological origins, had began to be accepted. The concept of Hellenicity was directly connected with education and the acceptance of Greek culture and way of life.[20] Any references during the Hellenistic and imperial years to settlers of Greek origin did not necessarily contest the idea of Hellenicity as a cultural quality; these references did not wish to substitute Hellenicity as cultural quality with the idea of Hellenicity as biological kinship. Very often the various myths used did not present the citizens of a city as Greek. They served to recollect the actions of their mythical Greek settler, who, once upon a time, left a mark on the history of their area.[21] It was from the actions of this particular founder that a new phase of their history began—that is, the phase of Hellenization, the move of their city and its geographical area to 'civilization'. Perhaps one of the best examples of this is Dionysius of Halicarnassus's account of Heracles' passage from Italy, a passage that had as a direct consequence the diffusion of Hellenic culture to all Italy and even Rome itself.[22]

We should consider here these references to Greek settlers. We cannot simply examine them as 'false' arguments aiming at legitimating the Greek identity of a city. Besides, Paul Veyne, discussing whether the Greeks believed their myths, showed that there is a great degree of fluidity between what is true and what is false.[23] Rather, we should look at the uses of calling upon heroes of the past as having more than one function.[24] The hero settler signifies for the city's magistrates a Greek way of life, or, in other words, their Hellenicity. At the same time, such a conceptual link between the settler and Hellenicity is interpretative, as it stresses a specific point of reference in the history of the city, creating, thus, the conditions for a reflexive consideration of the past. Some ancient thinkers had a cyclical perception of time; in such an understanding, the deeds of the past were in direct connection with the present and the future. 'The domestication of the savage (mythical) mind' by the dominant discourse of the Hellenistic and Roman periods transformed the distant past from an abstract and foreign idea into a logical and empirical activity, suitable for the expression of current political and cultural experiences and trends of the civic life.[25] The collective past turned from unfamiliar and alien into something intimate. It was adopted by the city's magistrates as a major collective feature of their community; such a recognition allows us to understand better the *ex silentio* binary and limiting opposition between the self-declared Greeks and those from whom they are dynamically differentiated and whom they renounce. We can take this further: all opposition and dispute that there was on the 'Greekness' of a city or of a *koinon* was directed towards similar arguments on ethnic and cultural identity. We can imagine that such arguments of lack of Greekness could be attributed to other cities and *koina*.[26] Calling the others 'barbarians' or simply Syrians, or anything of the sort, could potentially affect the attempt of a city to be recognized as 'Greek'. We can also imagine which elements of identity the self-declared Greeks rushed to get rid of through official decrees or formal petitions.[27] The past is called upon to selectively cater for the collective memory of the city and to define its identity.[28]

The process of 'identization', that is the expression of the desired identity by those who sought it, is linked through the use of the past to the process of identification by the more powerful authority; in the case of the Panhellenion, identification is completed by the emperor himself, upon whom it falls to accept the arguments promoting Hellenicity, or to reject them. Thus, in a letter of Hadrian to the *archon* of the Panhellenion, Hadrian himself[29] regulates the admission of the city of Ptolemais Barca: ['The people of Ptolemais Barca] *must be admitted* [to the Panhellenion]. *However they make an unjustified request* [in desiring the same (privileges) as the people of Cyrene, whose] *ancestry is Achaean and perfectly Dorian. They themselves, though, are true-born* [Greeks], *but got the additional name of* [Ptolemae]*ans from the fact that* [the Macedonians (?)] *when they were masters* [(of the land?) renamed (resettled, restored?)] *the city. Since therefore the Cyrenaeans are sending two delegates,* [it is my decision that the Barcaeans should send only one].'[30]

This inscription is particularly enlightening as to what is a recognized and accepted character of a candidate city. The emperor acknowledges the Hellenicity of Ptolemais Barca, but it is obvious that this Hellenicity has degrees. The Cyrenaeans are more Greek than the Ptolemaean Barcaeans. The Panhellenion seems to mainly include cities of metropolitan Greece and the overseas colonies, provided that they belonged to one of the *gentes*—Doric, Ionian, and Aeolic. Participation in the Panhellenion seems to be less straightforward for cities founded by Alexander and the Hellenistic kings. In this way, memory is selective, and the founder of the Panhellenion network exercised his right to define the criteria of selecting members of the network, by mainly defining the characteristics of the identity of the members of the network. The bipolar distinction between the Network and the Self, which we discussed above, can be specifically applied to the case of the Panhellenion programme. The emperor, as the supreme moderator of the network, used practices and perceptions of cultural identity that prevailed in the cities of the eastern part of the Empire. He made them central to the network, and he subsequently urged the candidate members to align themselves to them using a limiting and selective reading of their own past.[31] I would even propose, even though evidence is insufficient, that integration in the Panhellenion was possible as long as the historically defined identity of the members was in line with the general guidelines which were determined from higher up. Within this framework, any particularities in the construction of identity had to be drawn from a uniform mythological and historical background.

The Display of Memory

The results of these two processes of 'identization' and identification were exposed to public view and reading. And since these were officially written and displayed documents, which relied on the dominant collective memory of the city, we may adopt Michael Rowlands's classification of the various expressions of collective memory. Rowlands distinguishes 'inscribed' memory practices, characterized by repetition and public access, from 'incorporated' memory practices, characterized by opaque symbolism and secrecy.[32] In addition, 'inscribed' memory is manifested in materially visible commemorative activities, such as the construction of monuments. On the contrary, 'incorporated' memory does not leave archaeological traces. As we have seen so far, the foundational principles of the Panhellenion are related either directly or indirectly to the management of official memory and identity of the city members.

To begin with, it seems that the whole network is structured around sacred places. The temple of Zeus in Athens and especially Eleusis constitute the epicentre of this *synedrion ton Hellenon*. Hadrian aimed to link his programme to the tradition of the *koina* of the Greek world, which, as a rule, functioned around federal sanctuaries. Yet, at the same time, we must not underestimate the role generally played by sanctuaries as museums—that is, as places that function as depositaries of the collective memory.[33] If the incorporated memory is manifested during periodic rituals, the inscribed memory is expressed mainly in the spatial frame of the sacred places. So when the Thyatireans wished to thank the emperor, they chose the Athenian Acropolis as the place to set their decree: '*for the benefactions of the greatest Emperor Caesar* [Trajan Hadrian Augustus...] *which* [benefits he encouraged] *cities and nations of all Hellas* [to share in the most venerable] *Panhellenion* [wishes] *to inscribe on the Acropolis at Athens everything which* [it has received...] *and similarly* [to inscribe] *the gifts of the greatest of the kings of all time, Emperor Caesar Trajan Hadrian Olympios Zeus Panhellenios... to inscribe this decree on a stone stele and to place it on the Acropolis, so that it may be clear to all the Greeks how much Thyatira has received from the greatest of kings.*'[34] The decree continues by calling the city of Athens the most brilliant and benefactress and the one who gives reward of the Mysteries to all people equally. In short, we have a text that mainly expresses the loyalty of the Thyatireans towards the Romans and especially towards the emperor, while, at the same time, proclaiming their admiration of Athens as the birthplace of civilization and the centre of the Greek world.

The Thyatira decree fully illustrates what Jack Goody called the visual function of writing. Indeed, the use of writing leads from the simple auditory level of conveying information to the level of visual perception and reading, which in turn allows multiple approaches to, and evaluations of, this information. By being written, the information acquires an additional dynamic: it contributes to the consolidation of organizational forms, to the creation of patterns.[35] The Thyatira inscription expresses precisely the two poles that form the founding concept of the Panhellenion programme: on one hand, we have the distant past and its relation to the official memory and identity of the member, and on the other, we have the present and the future. In other words, the city defines not only its identity, but also its position within the mesh of relationships of power, as it attempts to offer honours to the emperor and verify its submission to Rome. These attestations of inscribed memory are manifested within the spatial context of a sacred space, the Athenian Acropolis, which, in turn, carries its own specific memories. At the same time, the inclusion of a city within the network of imperial power, which the Panhellenion represents, also promotes the historically defined identity of the city. If our reading of the Panhellenion programme through network theory is valid, then we can see the limitations placed on the members in relation to the expression of elements of their identity. On the surface, it seems that there is absence of any exerting violence; indeed, the emperor is presented as a benefactor of the cities. The cities select particular ideological models of collective identity from the local traditions, and they promote them. We are aware that reality, however, was different. It is possible to trace differentiations and pluralisms underneath this image of uniformity; these differentiations, however, did not necessarily aim at questioning Roman politics and military domination.[36] They questioned prevalent religious, aesthetic, communicative models, while also promoting the existence of other models,[37] or they even adopted narrative strategies that attempted to distance them from official choices. The work of Lucian of Samosata, for example, attests to the existence of various trends in various contexts, religious, ethnic, or other, within the borders of the cities of the eastern Empire.[38]

Let us now return to the uses of the historical and mythical past for the creation of cultural identity of the member cities. We should remember that the second century AD was also the century of the Second Sophistic, during which the search for role models in the classical past reached a peak. Local elites may have attempted to state their cultural identity in the context of a globalizing empire. At the same time, the foremost articulator of this globalizing empire aimed to create a network of cities, based on the expression of particular identity. At this point lies the peculiarity of the Panhellenion as a network of cities; in other words, the peculiarity is the concordance between the aims and programmes of the individual cities and their local elites, on the one hand, and the imperial policy, on the other. We may say that we have a mutual instrumentalism between the parts of this programme. The emperor reinforces the Roman *sagene* under his capacity as benefactor, while the local elites reinforce their position in the local and imperial awareness by participating in the network.

Constantine Porphyrogenitos spoke of the Roman *sagene* in terms of absolute authority and power. Hadrian's *sagene*, however, is projected by the Thyatireans as a welcome benefaction. We should also note that it is in Athens, the centre of the network, that we see more evidence of the emperor in action than we see in any of the member cities. We may assume that the memory of the city members and the immemorial Greek practices were incorporated into the rituals of the Panhellenion, while, on the other hand, the material dimension of the Panhellenion belonged to the realm of the emperor's affairs.

The use of sanctuaries as museums, as I have already argued, was not a second-century innovation. What is significant is that this construction of memory by the elites and its display in the sanctuaries functions as a capital placed in the service of Rome in order to advance Roman policies. In this sense, the Panhellenion network constitutes a multidimensional system of communication, while also it shapes the dominant discourses of cultural identity, not only for the member cities, but also for the central Roman policy, at least as far as the eastern provinces are concerned. However, it is worth noting here that we should not underestimate the impact of adoption of Greek ways of life by imperial Rome, as early as the time of Nero, which of course did not cease to provoke the Senate's reaction.

We have, then, the simultaneous shaping of these two concepts: that of dominant discourse and that of cultural identities. This may explain why the Panhellenion was limited to the cities of mainland Greece, and to those who claimed that their foundation was the result of colonization in the early historical period.

The Panhellenion network as an imperial network of Greek cities apparently intended to mirror cities linked with the glorious classical era, with a principal focus on Athens and Sparta, whose names and history were understood to shape Greek culture, perhaps in contrast with some major Hellenistic cities of Asia, Syria, and Egypt. Hadrian and his successors wished to provide the Panhellenion programme with historical and cultural purity, possibly taking into consideration the political trends in Rome. In other words, the Panhellenion should not be 'read' only from the point of view of the Greek local elites.[39]

From Identities to Actions

In the last part of this article, I would like to explore the activities of the network in light of the limits of Castels's theory on the bipolar Self-Network, as discussed above.

As Antony Spawforth and Susan Walker argued, the Panhellenes had to deal with routine business: regulations of membership, organization of festivities, the imperial cult, and so on. Moreover, they had to deal with the settling of differences between Greek cities and also, as Eusebius mentions in his *Ecclesiastic History*, they attempted to exercise a (perhaps) uniform policy regarding more general issues, such as the Christians.[40] In other words, they had to deal with activities that were common to all previous to, or concurrent with, the Panhellenion *koina*. Several decades ago, Paul Graindor discussed the activities of the Panhellenion in terms of his own time: 'il devait sans doute, s'intéresser à toutes les questions de nature à resserer les liens religieux, intellectuels et moraux unissant les cités helléniques'.[41] The various activities of the Panhellenion thus constitute the particular actions performed by the actors of the Panhellenion network.[42] These actions are not neutral. On the contrary, they produce concepts and patterns, they become emblematic of the elite, but they also have a binding nature, and function as codes of conduct for large groups of people.

The significance of collective memory and its coherent appeal may be better understood now. The references to the past function as a setting for the understanding and acceptance of actions that in turn produce new images, or create new connotations for existing images. What is interesting is the combination of elements which have an intense symbolic character with elements that appear to respond to more rational practices. The texts that we have certainly show how difficult it is to discern between the symbolic and rational element, since they both coexist in every word and meaning used.

Another feature of the ceremonies organized by the Panhellenion is their repetitiveness. The significance of the periodicity of public ceremonies in Rome has already been noted;[43] this periodicity symbolized the constant rebirth of the state, and seems to reverberate in Ovid's poems (*Fasti*, V 275–376). Our characterization of the Panhellenic programme as a network may be reinforced by virtue of the periodicity in the organization of its most important celebratory events, athletic, ritual, or others. Inevitably, the constant repetition of particular ceremonies and actions fuelled the internal continuity of a network.

In our discussion of the activities of the Panhellenion network, we inevitably moved away from the bipolar Self-Network. This relation, between the Self and the Network, is useful insofar as it focuses on the identity of the members. It is obvious that by insisting on this bipolar connection (Self—Network), we are considering what composes a network and take an overview of its very existence, which acts uniformly and produces its own messages. And precisely the Actor–Network Theory,[44] based of course on the high technology networks of our own era, asserts that what is of essence is the understanding of the praxis, of the various actions and acts that take place during the realization of a network's function. If we want to examine the various actions and acts that take place during the realization of a network's function, we will find that the Actor–Network theory is rewarding.[45] We can, then, place the role of the emperor, the role of the magistrates of the Panhellenion, or of its members, at the centre of our investigation. In the Actor–Network theory, even artefacts produced by the network in question become important. Consequently, one could wonder anew about the historical significance of Hadrian's building programme in Athens and Eleusis. The erection and function of such, mainly religious, monuments created a dynamic that may have influenced developments within the framework of the Panhellenic network, but also could create links between this network and other, similar ones. In any case, the Panhellenion functioned as a model of values and practices, which would subsequently guide the political and cultural choices even of cities that were not members.

The official actions that took place within the context of the Panhellenion programme were not so different from those taking place outside it. We are all aware of the significance of the emperor's benefactions towards various cities, the worship honours offered to him by associations, *koina*, cities and so on, the games and other rituals that were held everywhere, even outside the Panhellenion. So what was the true meaning of this programme beyond our assumptions regarding its connection with the Greek policy of the Antonines? I wonder whether in Hadrian's Panhellenion one could see elements of self-reference of actions, activities, political and ideological intentions and perceptions which were widely spread in the Greek world of the Empire. The Panhellenic programme would then constitute an attempt to create a discernible entity, which could mobilize human and material resources for the production and reproduction of messages and actions that were common to all.[46] In this sense (and of course there is no evidence regarding this), the Panhellenion itself could constitute an indication of a conscious decision of creating a network from the point of view of the parties involved.

The practices of self-assessment of the network's active members become important. For instance, how the Thyatireans presented themselves in the decree from the Athenian Acropolis and how they defined their identity through their participation in the Panhellenion can illustrate this interplay between the Self and the Network as a bidirectional relationship. The members do not simply participate in the network displaying their identity; they reinforce their identity because they participate in the network.

To conclude: I believe that the Panhellenion is a useful example of a network in the ancient world, and that network theory is an appropriate methodological tool to study it. Historical inquiry can be reinforced and enhanced by the use of contemporary methodological tools.

Notes

[1] The title 'A network of cities' is a simplification of the more complex reality of Hadrian's Panhellenic programme, where, as is well known, there participated also other types of political organization of the Greek and Greek speaking areas of the Empire, such as *ethne* and *koina*.

[2] Knappet, *Thinking through Material Culture*, 64.

[3] Galen, *De venarum arteriarumque dissectione* 9.13 (ed. C. G. Kühn, vol. II); *De usu partium* 1.10,4 (ed. C. G. Kühn, vol. III) '*to de kaloumenon hypo ton anatomikon diktyoeides plegma megiston thauma ton entauthoi tetaktai*'.

[4] Tilley, *Metaphor and Material Culture*, 8.

[5] Hdt. 6.31 and 3.149.

[6] *De thematibus*, 45. Cf. Koutrakou, '"Sagene"—"Network"', 177.

[7] Such historical events may be the diffusion and deepening of the technological revolution, the collapse of the Soviet Union and the end of the Cold War, the restructuring of capitalism, the crisis of democratic politics, the appraisal of nationalism, and so forth.

[8] Castels, *The Information Age*.

[9] Cf. van Dijk, *The Network Society*, 221.

[10] The bibliography on this topic is rather extended. It is constantly expanded with the discovery of new, mostly epigraphic, evidence. For a synthetic approach to the issue, see Spawforth and Walker, 'The World of Panhellenion I', and 'The World of Panhellenion II'. See also Willers, *Hadrians panhellenisches Programm*; Spawforth, 'The Panhellenion Again', 339; and Romeo, 'The Panhellenion and Ethnic Identity', 21. The earlier bibliography is mentioned in these studies.

[11] Spawforth and Walker, 'Panhellenion I', 79–82; Oliver, *Marcus Aurelius*, 94–96 and 130. More recently, Romeo, 'The Panhellenion and Ethnic Identity', 22; and Buraselis, 'Zum hadrianischer Regierungsstil', 51.

[12] Orientis Graeci Inscriptiones Selectae 497: '*He Kibyraton polis apoikos Lakedaimonion kai syggenis Athenaion.*'

[13] See Billows, 'Les cités', 266.

[14] Cf. Orientis Graeci Inscriptiones Selectae 234.

[15] Cf. Tap, *Identités collectives et changements sociaux*, 12.

[16] As far as we know, there is no evidence attesting that a city's request regarding Hellenicity was rejected.

[17] The framework of this article does not allow me to discuss fully issues of ethnic, cultural, religious, or other identity in the ancient world. For an introduction, see Goldhill, 'Introduction', 18–20.

[18] See Scheer, 'Présence du passé', 290, with earlier bibliography.

[19] See Scheer, 'Présence du passé'; and Hall, *Ethnic Identity*, 42ff.

[20] Hall, *Hellenicity*, 206ff; Romeo, 'The Panhellenion and Ethnic Identity', 32.

[21] Swain, *Hellenism and Empire*, 73, with bibliography. See also Alcock, 'The Heroic Past', 43 ff.

[22] D. H. *Antiqu. Rom.* I.39–44; cf. Jourdain-Annequin, 'Être un grec en Sicile', 165.

[23] Veyne, *Les Grecs*, 133.

[24] I will not discuss here the use of various myths in the tightening of relations between cities. See Erskine, *Troy between Greece and Rome*, 168 ff, and more generally, Jones, *Kinship Diplomacy*, passim.
[25] Goody, *La raison graphique*, 244.
[26] Sugliano, 'Cittadini, pareci, stranieri', 294.
[27] Linguists believe that any text constitutes an answer to a question that has either explicitly or implicitly been posed. Within this context, we can read Strabo 1.4.9, a passage in which the geographer transfers the views of Eratosthenes on mankind, along the lines of good and bad qualities of people, and sets them against the well-known division between Greeks and barbarians.
[28] See Ricoeur, *La mémoire, l'histoire, l'oubli*, 97–105, for a discussion of the relationship between memory and identity.
[29] See Spawforth and Walker, 'The Panhellenion I', 81, for registering new members of the Panhellenion.
[30] Translated by Jones, 'The Panhellenion', 45.
[31] Woolf, 'Becoming Roman, Staying Greek', 134–35.
[32] Rowlands, 'Role of Memory', 142.
[33] Shaya, 'The Greek Temple as Museum', 423.
[34] Jones, 'A Decree of Thyatira', 12; see also Follet and Peppas-Delmousou, 'Le décret de Thyatire', 291.
[35] Goody, *La raison graphique*, 145 and 186.
[36] Hingley, *Globalising Roman Culture*, 92–93.
[37] Goldhill, 'The Erotic Eye', 193.
[38] Elsner, 'Describing Self', 149.
[39] See, for example, Pausanias' description of Greece, which is subject to patterns that are mainly Roman: Lafond, 'Lire Pausanias à l' époque des Antonins', 391.
[40] Spawforth and Walker, 'Panhellenion I', 82.
[41] Graindor, *Athènes sous Hadrien*, 102–04.
[42] The scope of this article does not permit the repetition of the relevant data, which are in any case widely known.
[43] Clavel-Lévêque, *L'empire en jeux*, 24.
[44] Latour, *Reassembling the Social*, 141.
[45] Knappet, *Thinking through Material Culture*, 74. See also www.ensmp.fr/~latour.
[46] van Dijk, *The Network Society*, 234.

References

Alcock, S. E. 'The Heroic Past in a Hellenistic Present'. In *Hellenistic Constructs. Essays in Culture, History, Historiography*, edited by P. Cartledge, P. Garnsey, and E. Gruen. Berkeley: University of California Press, 1997.

Berns, C., H. von Hesberg, L. Vandeput, and M. Waelkens. *Patris und Imperium: kulturelle und politische Identität in den Städten der römischen Provinzen Kleinasiens in der frühen Kaiserzeit.* Leuven: Peeters, 2002.

Billows, R. 'Les cités'. In *Le monde hellénistique. Espaces, sociétés, cultures 323–31 av. JC.* Rennes: Presses Universitaires de Rennes, 2004.

Boffo, L. 'Sentirsi Greco nel mondo romano: espressioni epigraphiche'. *Studi Ellenistici* 13 (2001): 275–98.

Buraselis, K. 'Zum hadrianischer Regierungsstil im griechischen Osten oder vom kaiserlichen Agieren und Reagieren'. In *Staatlichkeit und politishes Handeln in der römischen Kaiserzeit*, edited by H.-U. Wiemer. Berlin and New York: Walter de Gruyter, 2006.

Castels, M. *The Information Age: Economy, Society and Culture*. Vol. 1: *The Rise of the Network Society* (1996); Vol. 2: *The Power of Identity* (1997); Vol. 3: *End of Millennium*. Oxford: Blackwell, 1998.

Clavel-Lévêque, M. *L'empire en jeux. Espace symbolique et pratique sociale dans le monde romain*. Paris: Éditions du CNRS, 1984.

Elsner, J. 'Describing Self in the Language of the Other: Pseudo (?) Lucian at the Temple of Hierapolis'. In *Being Greek under Rome: Cultural Identity, the Second Sophistic and the Development of Empire*, edited by S. Goldhill. Cambridge: Cambridge University Press, 2001.

Erskine, A. *Troy between Greece and Rome*. Oxford: Oxford University Press, 2001.

Follet, S., and D. Peppas-Delmousou. 'Le décret de Thyatire sur les bienfaits d'Hadrien et le "Panthéon" d'Hadrien à Athènes'. *Bulletin de Correspondence Hellénique* 121, no. 1 (1997): 291–309.

Goldhill, S. 'Introduction. Setting an Agenda: "Everything is Greece to the Wise"'. In *Being Greek under Rome. Cultural Identity, the Second Sophistic and the Development of Empire*, edited by S. Goldhill. Cambridge: Cambridge University Press, 2001.

———.'The Erotic Eye: Visual Stimulation and Cultural Conflict'. In *Being Greek under Rome. Cultural Identity, the Second Sophistic and the Development of Empire*, edited by S. Goldhill. Cambridge: Cambridge University Press, 2001.

Goody, J. *La raison graphique. La domestication de la pensée sauvage*. Paris: Minuit, 1979.

Graindor, P. *Athènes sous Hadrien*. Cairo: Imprimérie Nationale, 1934.

Hall, J. M. *Ethnic Identity in Greek Antiquity*. Cambridge: Cambridge University Press, 1997.

———. *Hellenicity, Between Ethnicity and Culture*. Chicago: University of Chicago Press, 2002.

Hingley, R. *Globalizing Roman Culture, Unity, Diversity and Empire*. London and New York: Routledge, 2005.

Jones, C. P. 'The Panhellenion'. *Chiron* 26 (1996): 29–56.

———. 'A Decree of Thyatira in Lydia'. *Chiron* 29 (1999): 1–21.

———. *Kinship Diplomacy in the Ancient World*. Cambridge, Mass: Harvard University Press, 1999.

Jourdain-Annequin, C. 'Être un grec en Sicile: le mythe d'Héracles'. *Kokalos* 34/35 (1988/89): 143–66.

Knappet, C. *Thinking through Material Culture*. Philadelphia: University of Pennsylvania Press, 2005.

Koutrakou, N. '"Sagene"—"Network": A Byzantine Perception of the International Legal Order'. In *The Eastern Roman Empire and the Birth of the Idea of State in Europe*, edited by S. Flogaitis and A. Pantélis. London: Esperia Publications, 2005.

Lafond, Y. 'Lire Pausanias à l'époque des Antonins. Réflexions sur la place de la *Périégèse* dans l'histoire culturelle, religieuse et sociale de la Grèce romaine'. In *Editer, traduire et commenter Pausanias en l'an 2000*, edited by D. Knoepffler and M. Piérart. Neuchatel: Université de Neuchatel/Droz, 2001.

Latour, B. *Reassembling the Social. An Introduction to Actor-Network-Theory*. Oxford: Oxford University Press, 2005.

Oliver, J. H. *Marcus Aurelius. Aspects of Civic and Cultural Policy in the East*. Princeton, NJ: Hesperia Supplement 13, 1970.

Ricoeur, P. *La mémoire, l'histoire, l'oubli*. Paris: Seuil, 2000.

Romeo, I. 'The Panhellenion and Ethnic Identity in Hadrianic Greece'. *Classical Philology* 97 (2002): 21–40.

Rowlands, M. 'The Role of Memory in the Transmission of Culture'. *World Archaeology* 25, no. 2 (1993): 141–51.

Scheer, T. S. 'Présence du passé dans le monde hellénistique: mythe et tradition locale'. In *Le monde hellénistique. Espaces, sociétés, cultures 323–31 av. JC*, edited by A. Erskine. Rennes: Presses Universitaires de Rennes, 2004.

Shaya, J. 'The Greek Temple as Museum: The Case of the Legendary Treasure of Athena from Lindos'. *American Journal of Archaeology* 109 (2005): 423–42.

Siapkas, J. *Heterological Ethnicity: Conceptualising Identities in Ancient Greece*. Uppsala: Universitet, 2003.

Spawforth, A. 'The Panhellenion Again'. *Chiron* 29 (1999): 339–52.

Spawforth, A., and S. Walker. 'The World of Panhellenion I. Athens and Eleusis'. *Journal of Roman Studies* 75 (1985): 78–104.

———. 'The World of Panhellenion II. Three Dorian Cities'. *Journal of Roman Studies* 76 (1986): 88–105.

Stephan, E. *Honoratioren, Griechen, Polisbürger. Kollektive Identitäten innerhalb der Oberschicht des kaiserzeitlichen Kleinasiens*. Göttingen: Vandenhoek und Ruprecht, Hypomnemata Heft 143, 2002.

Sugliano, A. 'Cittadini, pareci, stranieri: Le categorie giuridiche e sociali nelle città greche d'Asia Minore fra III e I secolo a.C'. *Mediterraneo Antico* 4 (2001): 290–312.

Swain, S. *Hellenism and Empire: Language, Classicism and Power in the Greek World AD 50–250*. Oxford: Clarendon Press, 1996.

Tap, P. *Identités collectives et changements sociaux*. Toulouse: Privat, 1994.

Tilley, C. *Metaphor and Material Culture*. Oxford: Blackwell, 2000.

van Dijk, J. *Network Society*. London: Sage Publications, 1999.

Veyne, P. *Les Grecs ont-ils cru à leurs mythes?* Paris: Seuil, 1983.

Willers, D. *Hadrians panhellenisches Programm*. Basle: Antike Kunst 16th Supplement, 1990.

Woolf, G. 'Becoming Roman, Staying Greek: Culture, Identity and the Civilizing Process in the Roman East'. *Proceedings of the Cambridge Philological Society* 40 (1994): 116–43.

Merchant Networks in the Greek World: The Impact of Rome

Dominic Rathbone

Introduction

By merchant networks I mean the institutions, whether public or private, which provided linking mechanisms for individuals involved in maritime commerce. I assume that networks mattered more for maritime than for land-based commerce because merchants usually travelled further from their home society, often into places with different legal traditions, and because of the greater uncertainty, in particular unexpected variations of route caused by adverse weather. In this article I seek to identify and assess developments in merchant networks in the maritime commerce of the eastern Mediterranean following its political unification under Roman rule. The process of unification and its influences have their roots in the second century BC, but my focus is on the outcomes discernible in the first to second centuries AD, when the whole area had become subject to Rome. Hitherto there has been little study of the day-to-day business functioning of maritime commerce in the Hellenistic and Roman worlds, so this article is an exploratory study to suggest issues for further research.

I take as given some broad features of Roman government which pervaded the whole empire despite the persistence of local practices. First, direct and indirect agents of the Roman state: senatorial and equestrian officials, soldiers, imperial freedmen and slaves, state contractors (*publicani*). Second, an overarching scheme of taxation, which included—apart from land taxes and a novel poll tax—Roman customs dues and sales taxes, and the state-organised systems of supply for the city of Rome and Rome's armies (the *annona*). Third, a unitary monetary system based on the denarius, to which the continuing drachma-based coinages were made equivalent. Fourth, the provision of Roman-style justice and the influence of Roman legal principles, overlaying the development of a Hellenistic *koine* in Greek private law. I also take as given that political unification and peace facilitated the potential development of a Mediterranean-cum-European market, and that maritime commerce in the first to second centuries AD was significantly higher than in most pre-industrial periods.[1] I also accept, following previous studies, that this maritime commerce was a cumulation of mostly small-scale transactions, primarily because most ships were relatively small and were owner-operated, even if backed by third-party finance.[2] It is a reasonable hypothesis that networks played a crucial role in enabling the development and functioning of a commerce that was large in aggregate, but composed of small agents and transactions.

There is little direct evidence for this topic. We have nothing like the letters from the Cairo Genizah, which illuminate the working practices of Jewish merchants operating from middle Egypt through north Africa to Sicily in the eleventh century.[3] Our sparse and scattered documents include some inscriptions, the Campanian tablets, and papyri from Roman Egypt. However, we can use institutional and structural considerations to create hypothetical frameworks in which to locate our fragmentary data.

Agents of the Roman State

Contractors (Publicani)

Although the experiment of using *publicani* to collect direct taxes in some eastern provinces in the late Republic had been terminated by the end of the first century BC, the Roman government of the Principate still relied heavily on *publicani*, also called *mancipes* or *conductores*, in two main areas.[4] First, they collected imperial indirect taxes, notably the *portoria* (customs dues) of 2.5% in and out of the internal customs zones and of 25% for imports into the empire. Second, they provided ships for the *annona*, the state-directed supply of grain and other foodstuffs to Rome and, on occasion, other cities, and of food and other supplies to Rome's armed forces. Depending on the nature of the task, contracts could be one-off or renewable for years, made with individuals or with groups. My interest here is in the larger and more long-term arrangements.

The organisation of tax-collecting *publicani* in the Republic is best illustrated by two sources, Cicero's *Verrines* for Sicily in the 70s BC, and the *lex portorii Asiae* which gives, as updated down to AD 62, the 123 BC regulations for the collection of customs in and out of the zone of the province of Asia. In both cases the contractors formed *societates* ('partnerships') based in Rome and fronted by Roman *equites*, which included investing members, executive officers (*magistri*), and local agents permanently stationed in the provinces. In the Principate individual contractors become more prominent, such as the father of the emperor Vespasian in Asia, and the men attested as *arabarches* in Egypt, who collected the 25% import duty and the fees listed in the Koptos tariff. It has been argued that this represents an organisational change to a system of *Einzelpächter*, but I suspect it is partly a question of having different evidence—that is, career inscriptions suitable for prosopography.[5] Just as the Republican *publicani* of the *portorium* of Asia had fixed customs posts and staff—which seem to have been taken over by new *socii*, or a new *societas*, if there was a total change of contactors—so the *arabarchia* in Egypt had its own warehouse at Alexandria, and agents, it seems, at Alexandria, Koptos and the Red Sea ports, which must have passed from one contractor or group of contractors to the next.[6] Shippers under regular contract to the *annona* had a somewhat different organisation, in that they were allowed and encouraged to form *collegia* ('associations') for main routes to which the state granted a corporate legal entity (*corpus*), apparently to aid its own management of them.[7] These *corpora naviculariorum* probably comprised a mixture of private *societates* and individuals, but had some common institutions, such as their second-century offices, labelled by mosaics, in the Piazzale delle Corporazioni at Ostia.

The activities of *publicani*, whether tax collectors or shippers, were not confined to their state function. Shippers could carry other freight on their return voyages and, if there was room, outward on top of their state cargo. Tax collectors could at the same time be involved in private commerce and finance. In the late Republic, for example, *publicani* were used by Cicero to transfer and exchange money for him between Cilicia and Rome, which they presumably were able to do in connection with their transfers of tax revenues.[8] In the *lex portorii Asiae*, paragraphs of 123 and 75 BC specify that *publicani* are themselves exempt from customs dues only on crops they are transporting on behalf of the Roman people, and on items for personal use.[9] The families of two first-century AD *arabarchai*, Tiberius Julius Alexander and Annius Plocamus, were involved in the trade through the Red Sea with Arabia and India, which was taxed by the *arabarchia*; Alexander, a prominent Jew of Alexandria, had business agents or links in Puteoli through whom he could arrange from Alexandria a loan of 170,000 drachmas to be collected in Puteoli by King Agrippa I.[10] Admittedly, the *corpora naviculariorum* seem to have been mainly a phenomenon of the western provinces, and the only securely known eastern *corpus* is, unsurprisingly, the shippers of Alexandria who carried Egyptian wheat to Rome and the west, but they probably played a prominent role in the shipment of goods from Italy and west to the eastern Mediterranean through the commercial hub of Alexandria.

Soldiers and Veterans

Under Augustus the Roman state developed a standing army of some 400,000 men which, although around a half of it was concentrated in frontier zones, permeated the whole empire. Roman soldiers were not well paid and seem, where possible, to have supplemented their salary by extortion and business. Tacitus' denigration of the Syrian legions under Nero as 'sleek businessmen' more at home in the city than the camp cannot simply be dismissed as a topos.[11] A clause in the *lex portorii Asiae* exempts soldiers from customs dues only on official supplies they are delivering or items for their personal use; a ruling by Nero in AD 58 implies that it was a perennial problem that they tried to use this exemption to evade duties on commodities they were trading privately.[12]

Among the Fayyum villages in Egypt which have produced abundant papyri of the Roman period, Karanis and Philadelphia had sizeable military communities—that is, families of veterans and serving soldiers, frequently interrelated. One family archive of the mid-second century AD, gradually being reconstructed through prosopographical links, tells us something of the business affairs and associates of two brothers, both veterans, Longinus Priscus and Longinus Celer, who owned property and spent much time in the Fayyum but were active also in Alexandria. I give two examples. One contract has Celer and a Ti. Claudius Chares making a maritime loan of almost eight talents to a group of four shippers from Ascalon, presumably for a round trading trip from Alexandria, which was in addition to a previous maritime loan by Celer to the same men. One letter reports that a friend called Sempronius had won a lucrative contract to supply bread to the garrison at Taposiris.[13] The archive is full of references to men with Roman names—mostly, it seems, soldiers or veterans. Military service must have been a prolific progenitor of networks—small and close in your unit or home village, weaker but potentially empire-wide through the shared military experience and status—which may have given soldiers and veterans a collective edge over civilian businessmen.

The Imperial Familia

The common claim that the Roman empire had no bureaucracy ignores the pervasive presence of the imperial *familia*. Within a few years of the Roman annexation of Egypt Strabo notes that the equestrian officials of the province had standing staffs of imperial slaves and freedmen.[14] Although this was a novelty, inscriptions show that it rapidly spread to all other provinces, partly in association with imperial procurators in imperial provinces, partly with the ubiquitous spread of the *patrimonium* ('private' property of the emperors). The *patrimonium* itself engaged in maritime commerce, sometimes as part of the *annona*, but also, it seems, selling on the open market. For instance, in a fragment of one day's register of ships arriving in a port of Alexandria, one of the nine ships had a cargo of olive oil and another a half-cargo of pine trunks belonging to our lord Caesar.[15] Like soldiers, imperial slaves and freedmen also engaged in commerce in their own right. I select two examples. Recently published receipts on ostraka from the Red Sea port of Berenike confirm the participation in the later first century AD of individuals from the imperial *familia* in the trade with India.[16] Tablets from the archive of the Sulpicii, bankers at Puteoli, record various transactions involving imperial slaves and freedmen: in AD 48, for example, a Sulpicius purchased 58 pounds of ivory from the imperial freedman Ti. Julius Myrtilus, and in AD 51 Phosphoros Lepidianus, slave of the emperor Claudius, deposited HS 94,000 with the bank for a month for them to lend out—a sum equivalent to the gross annual pay of a hundred legionaries.[17] We know little about the organisation of the imperial *familia* outside Rome; one papyrus, for instance, attests a *collegium* of imperial slaves at Alexandria in 6 BC.[18] However, it is reasonable to hypothesise that they had close local ties and a weaker wider sense of institutional collegiality not unlike those of soldiers.

Private Individuals

Roman Social Institutions

The use by Romans of their slaves and freedmen as 'agents' (*institores*) was common in maritime commerce as in other business. A legal question cited in the *Digest* imagines that 'a provincial has a slave agent for selling goods at Rome'; another imagines a Roman with a slave commercial agent in Beirut.[19] This is the legal principle under which slave agents of *publicani* and the imperial *familia* operated. Insofar as slaves and freedmen had a long-term relationship with their master or ex-master and his heirs, the Roman conception and practice of slavery enabled Romans regularly engaged in commerce in the eastern Mediterranean to develop networks of servile agents. We may, however, wonder how many Romans found this worthwhile outside the *publicani* who ran private interests alongside their public business, and also note that agency was a Roman legal concept, applicable also to the freeborn, which was reinforced, not created, by Roman slavery.

In a recent book, Verboven has argued that Romans of the later Republic did business through established social routes of *amicitia* and *clientela*. That, he says, is how Cicero presents his affairs.[20] In the Principate similar behaviour is attested: Seneca, for example, managed his gift-estates (*ousiai*) in Egypt through 'tenants-in-chief' of some social standing whom he calls 'my people' (*mei*).[21] However, a Roman like Cicero would use *amicus* or *cliens* to denote anyone with whom he happily did business, whether it was a one-off deal or a regular relationship. I do not think that Roman business was done through pre-existing networks of *clientes* and *amici*, but these terms were a polite way to envisage and manage relationships with freeborn men whom one hoped to use or was already using as long-term agents and associates. Business networks, I believe, even when slaves and freedmen were involved, were founded on contractual arrangements, whether a maritime or normal loan, appointment as an *institor* or participation in a *societas*, the standard Roman legal form for a 'partnership' in any enterprise, which by the first century AD had developed very flexible rules—for instance, permitting unequal inputs of labour and resources by *socii*, and unequal shares of profit.[22] On another level, members of the Roman elite doubtless tried to use their contacts with provincial governors and other officials to help if their agents had problems, as Brutus had deceitfully tried to do with Cicero as governor of Cilicia and Cyprus in 51/50 BC, although attested cases in the Principate are unsurprisingly rare.[23]

Roman Associations (Collegia)

Both in the Hellenistic world and in Republican Rome men engaged in a particular occupation might form a private 'association', *sunodos* or *koinon* in Greek, *collegium* in Latin, although some of them were named after a locality or deity, leaving it open whether or not they were restricted to a single occupation or set of related occupations. The social and religious activities of these groups have generated most of the information we have about them, in particular premises for dinners and dedications to their patron deities or emperors. Thus there has been a trend to argue that they were primarily cult associations which could develop a business aspect, whereas other scholars, myself included, would put it the other way round.[24] Roman merchants in the eastern Mediterranean could form their own *collegia*, like those on Delos in the later Republic that were named after deities. Local associations also continued or were created, such as the shippers at Nicomedia, who built their clubhouse as a sanctuary dedicated to Vespasian; 'the shippers and merchants of Ephesus' who through a star Ephesian athlete made a dedication at Rome to Marcus Aurelius; and the third-century *sunodos* at Koptos and Tentyra of 'the Red Sea shippers and merchants of Hadrianic Palmyra'.[25] These associations, where they existed, provided networks that linked businessmen, and seem to have maintained bases at important destinations where visiting members of the association could receive information and support. One problem, in the absence of an up-to-date systematic collation of the inscriptions, is to know how common they were. Another is what effect Roman rule had on them. Probably there was some Roman influence on their institutional structure. It is disputed how real Roman suspicion was of the potential political role of such private associations; although known bans, such as Trajan's order to Pliny in Bithynia, seem to be local and periodic, we have seen that the only known eastern association to which the Roman state allowed the legal status of a *corpus* was that of the annonal shippers of Alexandria.[26]

Roman Business and Banking

Contractual and social links with a powerful man or belonging to an association with members in the principal centres of one's trade could serve a merchant or shipper who stuck to a pattern, like Flavius Zeuxis, who made thirty-six annual return voyages from Hierapolis in Asia Minor to Italy.[27] But what could help the small man of Cicero's imagination who plied irregular routes and did not have people to vouch for him in every port, or was forced off course, as in the imaginary juristic case of the shipper bound for Ostia, whose mast and rigging were burnt by lightning and who had to put into Hippo for emergency repairs?[28] The answer, I think, is banks. In Egypt the Roman period saw a boom in the number of private banks, and it is unlikely to have been peculiar in that respect. Although Roman commercial practice owed much to Greek influence, such as the transfer between accounts (*diagraphe*, Latin *perscriptio*) in banking or the rules of jettison in law, the eastern Mediterranean under Roman rule was in turn influenced by distinct Roman practices, mediated, I will suggest, through banks, which created an increasingly integrated commercial world.[29]

In part this was because Greeks came to Italy and did business there with Roman citizens. I cite three examples from the archive of the Sulpicii, bankers at Puteoli. First, *TPSulp* 78, a contract made at Puteoli on 11 April 38. Menelaos from Keramos (Caria) acknowledges in Greek, but in Roman legal form and format (a diptych of waxed tablets), receipt of a loan of 1,000 denarii from Primus, a slave of P. Attius Severus, a known importer of *garum* from Spain. The loan is described as arising from a sealed freight contract (i.e. recorded as a *testatio* in a diptych) with Primus, and is to be repaid in accordance with it. In Latin an illiterate local, M. Barbatius Celer, acts as surety for Menelaos; a Q. Aelius Romanus signs on Celer's behalf. If the loan here, as I suspect, was to cover the customs dues payable on the cargo, the contract was a means for Primus to try to ensure that Menelaos actually paid the dues; indeed, the loan may have been fictive. Second, in *TPSulp* 60–2 of AD 42/43 Euplia from Melos borrows two sums from the bank and a third from Titinia Anthrax, later repaid, recorded in Roman procedures (*nomina transscripticia* reinforced by *testationes* in diptychs), with Epichares from Athens acting as her *tutor* and *fideiussor*. Third, in *TPSulp* 106 of AD 57 a shipper from Sidon puts his ship up for Roman-style auction through the bank, probably following a default on another contract, or a customs payment.

In part, integration developed because Greeks in the east in some contexts and various ways began to adopt or imitate Roman business practices, whether on their own initiative, deliberate or subconscious, or because of official Roman encouragement. The clearest examples I know concern sales, particularly auctions of slaves. By general Greek legal practice, payment at sales and auctions had to be immediate and direct to effect transfer of ownership (whether or not it always was), whereas Roman sellers often contracted collection of the price due to a *coactor* (collector). By the late Republic most auctions relied on the presence of collectors, who also functioned as bankers and lent money to purchasers and could effect paper payments through their clients' accounts.[30] There is only one attestation of *coactores* in the Roman east Mediterranean outside Egypt: a first-century BC dedication to Hermes at Magnesia (Asia Minor) by the '*komaktores* and auctioneers and attendants'.[31] However, in Syene (Upper Egypt) in 15 BC a slave-girl sold at auction by an auxiliary soldier was paid for through two *komaktores*, one called Solon, and five other texts of the second and third centuries AD from Middle Egypt attest collection of the *komaktoria*—that is, the '*coactor*-tax'—alongside the *enkuklion*, the standard property sales tax of Hellenistic origin.[32] Two of these five texts also concern sales of slaves; in two others the tax is said to be due on property auctioned by the *fiscus* and, unhelpfully, 'other property'. My hypothesis is that *coactores* had followed Roman-Italian businessmen into the eastern Mediterranean from the second century BC, and had probably been prominent in, for example, slave-sales on Delos. In the Principate they virtually disappeared because local Greek banks began fulfilling the same function. In Egypt, and perhaps other provinces, a sales tax called the *komaktoria* was distinguished from the standard *enkuklion* because it related to Roman-style auctions at which the vendor, not the purchaser, was responsible through the collecting bank for paying the sales tax.[33]

Another area of some Roman influence was in the format of contracts. A striking case is two very similar contracts drawn up in the marketplace (*agora*) of Side (Pamphylia) in the mid-second century AD which look to have come from an archive recording purchases by a group of Alexandrians, perhaps slave-dealers.[34] *P. Turner* 22 (AD 142) records the sale of a young slave girl by a Greek, perhaps local, to a citizen of Alexandria. It is written in Greek but in *stipulatio* format translating Roman formulae—for instance, that the slave is 'sound according to the edict (sc. of the aediles)', mechanically written out in the masculine. It is a Roman-style double document, that is an imitation on papyrus of a Roman contract on wax tablets, but it is witnessed only by the purchaser and the *fideiussor* (guarantor) of the transaction, a certain M. Aelius Gavianus, an illiterate for whom a public official signed.[35] *BGU* III 87 (AD 151) is an identical record in Greek of another slave sale by a provincial with Roman citizenship, perhaps a local, to another Alexandrian citizen. This document has more signatories, closer to Roman practice. The *fideiussor* is a Greek, again an illiterate for whom a public official signs. The extraordinary fact that the two contracts, nine years apart, use an identically translated and muddled Greek version of a *stipulatio* contract implies a common source, which I suspect was a bank which financed the two purchases.[36] Also noteworthy is that in both these contracts, as also in the loan to Menelaos at Puteoli in AD 38 (see above), the *fideiussor* was illiterate. I suggest that they were all straw men, provided as part of the bank's service of drafting contracts; in practice the bank was the real guarantor.[37]

A common element in all these cases of the blending of Roman and Greek business formalities appears to be the involvement of banks. The banks, it seems, produced the Roman-style or hybrid contracts which their clients wanted or had been advised by the banks to use. Contracts between Greeks and Roman citizens did not have to use Roman legal forms, let alone contracts between two Greeks such as the parties to *P. Turner* 22. Perhaps using Roman-style contracts was thought advisable for transactions which crossed provincial boundaries, especially sales of slaves for export, which were subject to other Roman regulations (permits, control of status, manumission taxes). Conversely, almost all documented sales of slaves in Roman Egypt are in Greek-style contracts.[38] So are most other contracts, even when one party is a Roman citizen. However, these Greek-style contracts also illustrate the increased role of banks in business. For instance, the maritime loan made by the veteran Longinus Celer and Ti. Claudius Chares to four shippers from Ascalon (see above) was recorded in a Greek-style contract by a bank at Alexandria, which kept 'the copy on land of the contract' and had already handled Celer's previous loan to the shippers; the security was the ship, and there was no personal guarantor. In Egypt, as probably in other provinces, banks had a notarial function too—that is, of making official registration of a change of ownership of property.[39] In Italy there were no state property registers and so no notarial function for banks, but banks were obliged by law to produce their records in legal cases about transactions in which they had been involved, and it is said that often only the bank had any written record of a particular transaction.[40] In practice, if not in law, the involvement of the bank was another tie on the contracting parties. We do not have to believe that the Roman empire had a regular clearing system between banks, although transfers clearly could be effected, to suppose that banks themselves, at least in major commercial centres, formed an information network with bankers elsewhere (the Geniza archive could help us to envisage the mechanics of this), so that if, for instance, the Ascalonites failed to repay Celer and Chares, news of their behaviour would percolate through to banks in Ascalon and elsewhere, making it difficult for them to raise new finance.

Conclusions

This brief survey of possible developments in the organisation of maritime commerce in the eastern Mediterranean under Roman rule is a very preliminary undertaking. Three main points emerge. First, methodological. Granted the paucity of documentary evidence for the day-to-day working of ancient commerce, we need to construct explanatory models. Network theory looks particularly helpful in this case. The picture is complex. A man could simultaneously belong to a tight closed network such as his military unit and the wide weak comrades' network of Roman soldiers in general, and also tap into the equally wide but stronger professional network of bankers, in that it was semi-open to their clients. I have not tried to identify and compare the commercial networks of the Hellenistic east Mediterranean, but I would not be surprised if we found more and further-reaching networks in the Roman period through which maritime commerce was able to function on a grander scale. Second, historical. In assessing the economic impact of the Roman state we tend to forget about the aggregate effect of the petty and not-so-petty economic activities of the state's agents. Network theory highlights the potential advantages conferred on some groups—contractors, soldiers and the imperial *familia*—by their association with the Roman state, although unfortunately it does not allow us to assess the scale of their private business. Third, also historical. Debate about how modern or not banks of the Roman era were has obscured their commercial functions beyond the monetary, which I have tried to bring out here—that is, the provision of information, including inventive legal pseudo-expertise, and the addition of extra security in transactions between third parties who were strangers—in short, the promotion of the *fides* (trust) which was as essential to ancient as it is to modern business. In the Roman period, banks developed as the nodes through which other networks and individuals intersected, connecting the disparate economic corners of the Roman empire into a global village.

Notes

[1] Parker, *Ancient Shipwrecks.*
[2] Rathbone, 'Financing', 201–08.
[3] Goitein, *Mediterranean Society.*
[4] Brunt, 'Publicans'; Malmendier, *Societas publicanorum*; de Laet, *Portorium.*
[5] A selection of views: Brunt, 'Publicans', 366–76; France, *Quadragesima*; Lo Cascio, 'Appaltatori'.
[6] As revealed in *SB* XVIII 13167, the so-called Muziris papyrus; cf. van Nijf, 'Social World'.
[7] De Salvo, *Corpora naviculariorum.* In the Principate, *societates* of tax-collecting and mining contractors were allowed to act as a legal entity (*corpus*), but not, it seems, to form *collegia* (*Dig.* 3.4.1 pr, Gaius).
[8] Jones, *Bankers*, 211–12.
[9] *Lex portorii Asiae* §§25, 31–33 (ll.60–61, 72–78).
[10] Rathbone, 'Financing', 221–22.
[11] Tacitus, *Annales*, 13.35.1–3: 'nitidi et quaestuosi, militia per oppida expleta.'
[12] *Lex portorii Asiae* §27 (ll.64–66); Tacitus, *Annales* 13.51.1; with Rathbone, 'Nero's Reforms'.
[13] *SB* XIV 11850; *P.Mich.* III 206; with Rathbone, 'Financing', 217–20.
[14] Strabo 17.1.12.
[15] *P.Bingen* 77; cf. Statius, *Silvae*, 3.3.86–95, for commodities handled by the *patrimonium* under Nero; the 'Muziris' papyrus shows that the 25% tax on imports into the empire was collected in kind.
[16] *O.Berenike* I p. 27, II pp. 6–7, 10; but note Rathbone, 'Koptos', 188, no. 32.
[17] *TPSulp* 101; 69.
[18] *BGU* IV 1137 = *WChr* 112.
[19] *Dig.* 5.1.91.3 (Labeo); 45.1.122.1 (Scaevola); cf. Aubert, 'Les *institores*'.
[20] Verboven, *Economy of Friends.*
[21] Seneca, *Epistulae*, 77.3; *P.Oxy.* XXXVIII 2873 (AD 62).
[22] Crook, *Law*, 236–43.
[23] Verboven, *Economy of Friends*, 290–315; all the cases he cites are from Cicero's correspondence. We have no comparably unedited correspondence from the Principate, but cf. *P.Gen.* I 1 (AD 213).
[24] Contrast, for example, Rauh, *Sacred Bonds*, with Carrié, 'Associations'; cf. van Nijf, *Civic World*; de Robertis, *Storia delle corporazioni.*
[25] *IGRR* III 4 (70/1); I 147 (184), cf. 151; *I.Portes* 93; 103 = *SEG* XXXIV 1593.
[26] De Ligt, 'Governmental Attitudes', argues for sporadic and diminishing concern.
[27] *Syll.*^{3} 1229; = *IGRR* IV 841.
[28] Cicero, *II In Verrem*, 5.167: Cicero's rhetorical answer was Roman citizenship. *Dig.* 14.2.6 (Julianus); jurists relished debating the potential problems of financial liability if the captain was an agent or partner and not the full owner of the ship or cargo: *Dig.* 14.1.7–11 (various).
[29] On the background see Rathbone, 'Financing'; Rathbone and Temin, 'Financial Intermediation'.
[30] García Morcillo, *Ventas*, 107–36, 157–71.
[31] *I.Magnesia* 217. *Komaktor* was the Greek calque of *coactor.* The *diakonoi* may have been attendants of the *coactores* and 'heralds' or the officers of their local association.
[32] *P.Stras.* I 79; the seller, Didymos, son of Jason, perhaps served in the cohors II(?) Ituraeorum equitata. *P.IFAO* I 3.ii (143); *P.Köln* II 83 (167); *P.Stras.* III 135 = *SB* V 8017 (198–211); *P.Oxy.* XII 1523 (III); *CPR* VII 9 *passim* (III). The *coactores* were not, as some editors claim, the auctioneers (heralds).
[33] Rathbone, 'Nero's Reforms', discussing Tacitus, *Annales*, 13.31.3 (AD 57).

[34] Strabo 14.3.2 notes the slave-market at Side; cf. Nollé, *Side*, I 164–65, who cites these contracts at II 613–22.

[35] For the difference between Roman double documents on papyrus and Hellenistic 'double' documents, see *P.Thomas* 6 commentary (Rathbone).

[36] Possibly even the 'public' bank of Side; the signatory in both texts with a lacunose title might be a 'public [banker]'.

[37] Note the working principle of Roman law known as *receptum argentarii* by which, if a client had instructed a banker to make a payment on his behalf, the banker was legally responsible for making the payment.

[38] Straus, *L'achat.*

[39] Bogaert, 'Liste'; Bogaert, 'Opérations'.

[40] *Dig.* 2.13.9.2 (Pomponius); cf. 2.13.10.1 (Gaius).

References

Aubert, J.-J. 'Les *institores* et le commerce maritime dans l'Empire romain'. *Topoi* 9 (1999): 145–64.

Bogaert, R. 'Liste géographique des banques et des banquiers de l'Égypte romaine, 30A–284'. *Zeitschrift für Papyrologie und Epigraphik* 109 (1995): 133–73.

———. 'Les opérations des banques de l'Égypte romaine'. *Ancient Society* 30 (2000): 135–269.

Brunt, P. A. 'Publicans in the Principate'. In *Roman Imperial Themes*. ch. 17. Oxford: Clarendon Press, 1990.

Carrié, J.-M. 'Les associations professionelles à l'époque tardive: entre *munus* et convivialité'. In *'Humana sapit': études d'antiquité tardive offertes à Lellia Cracco Ruggini*, edited by J.-M. Carrié and R. Lizzi Testa, 309–32. Turnhout: Brepols, 2002.

Crook, J. A. *Law and Life of Rome*. London: Thames and Hudson, 1967.

de Laet, S. J. *Portorium. Étude sur l'organisation douanière chez les Romains, surtout à l'époque du Haut-Empire*. Bruges: De Tempel, 1949.

de Ligt, L. 'Governmental Attitudes towards Markets and Collegia'. In *Mercati permanenti e mercati periodici nel mondo romano*, edited by E. Lo Cascio, 237–52. Bari: Edipuglia, 2000.

de Robertis, F. M. *Storia delle corporazioni e del regime associativo nel mondo romano*. Bari: Adriatica, 1973.

de Salvo, L. *I corpora naviculariorum. Economia privata e pubblici servizi nell'impero romano*. Messina: Samperi, 1992.

France, J. *Quadragesima Galliarum. L'organisation douanière des provinces alpestres, gauloises et germaniques de l'empire romain*. Paris and Rome: Ecole Française de Rome, 2001.

García Morcillo, M. *Las ventas por subasta en el mundo romano: la esfera privada*. Barcelona: Universitat de Barcelona, 2005.

Goitein, S. D. *A Mediterranean Society*. Volume I. *Economic Foundations*. Berkeley and Los Angeles: University of California Press, 1967.

Jones, D. *The Bankers of Puteoli: Finance, Trade and Industry in the Roman World*. Stroud: Tempus, 2006.

Lo Cascio, E. 'Appaltatori delle imposte e amministrazione finanziaria imperiale'. In *Tâches publiques et enterprise privée dans le monde romain*, edited by J.-J. Aubert, 249–65. Geneva: Droz, 2003.

Malmendier, U. *Societas publicanorum*. Cologne: Böhlau, 2002.

Nollé, J. *Side im Altertum*. I–II (*I. Kleinasien* 43–44) Bonn: Habelt, 1993 and 2001.

Parker, A. J. *Ancient Shipwrecks of the Mediterranean and the Roman Provinces*. Oxford: British Archaeological Reports, 1992.

Rathbone, D. W. 'Koptos the *Emporion*: Economy and Society, I–III AD'. In *Autour de Coptos*, edited by M.-F. Boussac, 179–98. Paris: De Boccard, 2002.

———. 'The financing of Maritime Commerce in the Roman Empire, I-II AD', In *Credito e moneta nel mondo romano*, edited by E. Lo Cascio, 197–229. Bari: Edipuglia, 2003.

———. 'Nero's Reforms of *Vectigalia* and the Lex portorii Asiae'. In *The Customs Law of Asia*, edited by M. Cottier, M. H. Clawford, C. V. Clowther, J. L. Ferrary, B. M. Levick, O. Salomies, and M. Wörrle, 251-78. Oxford: Oxford University Press, 2008.

Rathbone, D. W., and P. Temin. 'Financial Intermediation in First-century AD Rome and Eighteenth-century England'. In *Banks, Loans and Financial Archives in the Ancient World*, edited by K. Verboven, K. Vandorpe, and V. Chankowski, 371–419. Leuven: Peeters, 2008.

Rauh, N. K. *The Sacred Bonds of Commerce. Religion, Economy and Trade Society at Hellenistic Roman Delos, 166-87 BC.* Amsterdam: Gieben, 1993.

Straus, J. A. *L'achat et la vente des esclaves en Égypte romaine*. Munich: Saur, 2004.

van Nijf, O. M. *The Civic World of Professional Associations in the Roman East*. Amsterdam: Gieben, 1997.

———. 'The Social World of Tax Collectors and their Personnel'. In *The Customs Law of Asia*, edited by M. Cottier, M. H. Clawford, C. V. Clowther, J. L. Ferrary, B. M. Levick, O. Salomies, and M. Wörrle, 279-311. Oxford: Oxford University Press, 2008

Verboven, K. *The Economy of Friends. Economic Aspects of* Amicitia *and Patronage in the Late Republic*. Brussels: Latomus, 2002.

Index

Greek and Roman Networks in the Mediterranean

Page numbers in **Bold** represent Figures